DESIGN OF AIRCRAFT

DESIGN OF AIRCRAFT

Thomas C. Corke

University of Notre Dame

Pearson Education, Inc.
Upper Saddle River, New Jersey 07458

Library of Congress Cataloging-in-Publication Data

Corke, Thomas, C.
    Design of aircraft / Thomas C. Corke.
        p. cm.
    Includes bibliographical references and index.
    ISBN 0-13-089234-3
    1. Airplanes–Design and construction. 2. Aerospace engineering. I. Title.

TL546.C692 2002
629.134'1–dc21   2002027074

Vice President and Editorial Director, ECS: *Marcia J. Horton*
Acquisitions Editor: *Laura Fischer*
Editorial Assistant: *Erin Katchmar*
Vice President and Director of Production and Manufacturing, ESM: *David W. Riccardi*
Executive Managing Editor: *Vince O'Brien*
Managing Editor: *David A. George*
Production Editor: *Barbara A. Till*
Director of Creative Services: *Paul Belfanti*
Creative Director: *Carole Anson*
Art Director: *Jayne Conte*
Art Editor: *Greg Dulles*
Cover Designer: *Karen Salzbach*
Manufacturing Manager: *Trudy Pisciotti*
Manufacturing Buyer: *Lisa McDowell*
Marketing Manager: *Holly Stark*

About the Cover: An artist's rendition of a twenty-first century "morphing" aircraft that exemplifies abilities for altering its shape and performance *in flight*, in order to meet otherwise conflicting design optimizations.

 © 2003 Pearson Education, Inc.
Pearson Education, Inc.
Upper Saddle River, NJ 07458

The author and publisher of this book have used their best efforts in preparing this book. These efforts include the development, research, and testing of the theories and programs to determine their effectiveness. The author and publisher make no warranty of any kind, expressed or implied, with regard to these programs or the documentation contained in this book. The author and publisher shall not be liable in any event for incidental or consequential damages in connection with, or arising out of, the furnishing, performance, or use of these programs.

Printed in the United States of America
10 9 8 7 6 5 4 3 2 1

ISBN 0-13-089234-3

Pearson Education Ltd., *London*
Pearson Education Australia Pty. Ltd., *Sydney*
Pearson Education Singapore, Pte. Ltd.
Pearson Education North Asia Ltd., *Hong Kong*
Pearson Education Canada, Inc., *Toronto*
Pearson Educación de Mexico, S.A. de C.V.
Pearson Education—Japan, *Tokyo*
Pearson Education Malaysia, Pte. Ltd.
Pearson Education, Inc., *Upper Saddle River, New Jersey*

To Bobbie, Catherine, Laura, and Sarah

Contents

Preface

This book is intended to be a text for a senior-level aerospace engineering course dealing with the conceptual design of aircraft. It is based on my experience in teaching the "capstone" design class in aerospace engineering for the past 15 years. The approach is to demonstrate how the theoretical aspects, drawn from topics on airplane aerodynamics, aircraft structures, stability and control, propulsion, and compressible flows, can be applied to produce a new conceptual aircraft design. The book cites theoretical expressions wherever possible, but also stresses the interplay of different aspects of the design, which often require compromises. As necessary, it draws on historical information to provide needed input parameters, especially at an early stage of the design process. In addition, historical aircraft are used to provide checks on design elements to determine if they deviate too far from historical norms.

The process of the conceptual design of an aircraft is broken down into 14 steps. These are covered in Chapters 2 through 13. The book stresses the use of a spreadsheet approach for iterative and repetitive calculations. Sample spreadsheets, in Microsoft Excel, covering each step of the design are provided for each chapter, except 1 and 13. Each chapter also contains a detailed description of the spreadsheet structure, so that students can easily make modifications. In addition, the input conditions for each spreadsheet correspond to a cohesive conceptual design (supersonic business jet case study) that runs throughout the book. Each part of this case study that relates to the particular chapter topic is discussed at the end of each chapter. Two additional complete case studies that follow the steps outlined in the book are presented in Appendices B and C.

In addition, there are individual problems at the end of each chapter in which the students are asked to utilize the spreadsheet(s) to document different degrees of dependence of the aircraft characteristics on changing input conditions. Some of these problems are "open ended" and require interpretation and discussion.

The book can be used in either of two ways. First, it can be used to develop a *complete* conceptual design of a new aircraft. This is the way that I personally teach this material. Starting at the beginning, the students work in small groups and develop a complete design (similar to the case study) in a step-by-step fashion. This is accomplished over one semester (15 weeks).

The second use of the book is to consider individual aspects of an aircraft without developing a complete design. This approach makes the best use of the problem sets at the end of each chapter. Using the spreadsheets, the effect of different input parameters can be easily investigated, and optimums can be sought. I know of a number of instructors who prefer this approach.

The following is a list of chapters:

1. Introduction
2. Preliminary Estimate of Take-Off Weight
3. Wing Loading Selection
4. Main Wing Design
5. Fuselage Design
6. Horizontal and Vertical Tail Design
7. Engine Selection
8. Take-off and Landing
9. Enhanced Lift Design
10. Structural Design and Material Selection
11. Static Stability and Control
12. Cost Estimate
13. Design Summary and Trade Study

For a complete conceptual design, the chapters are intended to be followed in chronological order. A conscious attempt has been made to include within each chapter, all of the supplementary material that is needed to develop that aspect of the design. This minimizes the need to search for formulas or graphs in other chapters or references.

Two of the chapters have combined topics that are often presented separately. One of these is the chapter on structural design. This chapter includes the load analysis, structure design, *and* material selection. Often material selection is treated separately; however placing it in a chapter on structure design reflects my experience that the two are inevitably tied together. The other chapter is "Static Stability and Control." This includes a section on refined weight estimate, which is also often presented separately. These have been grouped into a single chapter because the magnitude and placement of key weight components inherently affects the stability characteristics.

The last chapter summarizes the case study that runs throughout the text and discusses the role of a Trade Study in a complete design. This is illustrated with the case study design and in the problems at the end of the chapter.

THOMAS C. CORKE
University of Notre Dame

DESIGN OF AIRCRAFT

CHAPTER 1

Introduction

1.1 DEFINING A NEW DESIGN
1.2 DESIGN PROCESS
1.3 CONCEPTUAL DESIGN

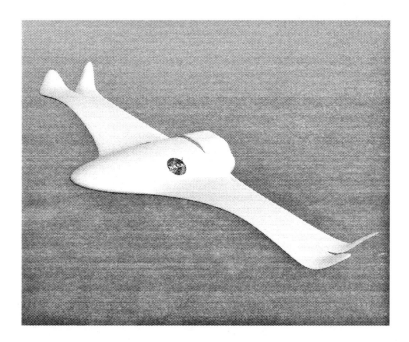

An artist's rendering that illustrates advanced aerodynamic concepts that are envisioned for next generation aircraft designs. (NASA Dryden Research Center Photo Collection.)

When you look at aircraft, it is easy to observe that they have a number of common features: wings, a tail with vertical and horizontal wing sections, engines to propel them through the air, and a fuselage to carry passengers or cargo. If, however, you take a more critical look beyond the gross features, you also can see subtle, and sometimes not so subtle, differences. What are the reasons for these differences? What was on the mind(s) of the designers that caused them to configure the aircraft in this way?

FIGURE 1.1: Photographs of the Fairchild Republic YA-10A, which was developed to be a ground strike and close air support aircraft. (Courtesy of the USAF Museum Photo Archives.)

The photograph at the start of the chapter and Figures 1.1 through 1.5 show aircraft of widely diverse designs. All of them have the same gross features, yet they differ in how these basic elements were incorporated into the complete aircraft.

The YA-10 in Figure 1.1 was designed specifically to be used for close air support of ground forces. Therefore, it needed to be maneuverable at low speeds and low altitudes, and it needed to be able to survive against ground-launched weapons. The YA-10's design and appearance were the direct result of these objectives.

To accomplish this, the designers set the engines high on the fuselage so that they would be less susceptible to ground fire. The engines were also set in front of the horizontal stabilizers so that the hot air jets could be mixed with cooler air in the wake of the horizontal stabilizers and thereby reduce the YA-10's visibility to heat-seeking missiles.

The YA-10 was specifically intended to be able to attack armored (tank) vehicles. To accomplish this, it was designed around the GAU-8/A 30-mm Gatling Gun. This is a large gun that produces a significant recoil (20,000 pound) force when fired. As a result, the gun had to be located on the fuselage longitudinal centerline so that the recoil would not produce an adverse directional (yaw) force. This gun placement required an unusual placement of the nose landing gear. As seen in the bottom photograph in Figure 1.1, the nose landing gear is located to one side of the fuselage centerline. This placement makes the aircraft more difficult to land, but was a necessary compromise to meet the other design objectives.

FIGURE 1.2: Photograph of the Douglas XB-42, which was developed as part of a top security program in World War II. (Courtesy of The Boeing Company.)

Many of the extreme aircraft designs have come from meeting military objectives. In this context, one can look at the XB-42 in Figure 1.2 and wonder what were the driving objectives leading to its appearance. The most unusual feature is the tail design with two pusher propellers located on the fuselage centerline. In addition, the area of the vertical stabilizer was split into two, and placed above and below the fuselage. This arrangement required extra-long struts on the landing gear to prevent damage to the tail section at take-off and landing; a compromise in the overall design.

Why was this design adopted? The pusher-prop configuration left a completely "clean" main wing, optimizing its performance and undoubtedly increasing the XB-42's range and payload capability. Having the engines on the fuselage centerline also improves the roll maneuverability, another possible design objective.

The X-29 in Figure 1.3 is an example of an aircraft designed for extreme maneuverability. It was a fighter-sized aircraft used to explore concepts such as advanced composite materials, variable camber wing surfaces, a forward swept main wing, control canards, and computerized fly-by-wire flight control, which was necessary with this statically unstable geometry. Most of the aircraft flying today use wing sweep to lower the local Mach number (compressibility effect) over the wing. In a vast majority of aircraft, the wings are swept *back*. The advantage of forward sweep is that the ailerons, near the wing tips, remain unstalled, providing control at high angles of attack. High angle-of-attack flight was one of the principle design objectives for this aircraft, and flight tests demonstrated that the aircraft could fly at angles of attack of up to 67 degrees!

A downside of this design was that because of its forward swept wings, the X-29 was statically unstable. This means that it continually required movement of the control

FIGURE 1.3: Photograph of the Grumman X-29, a technology demonstrator for highly maneuverable fighter aircraft. (NASA Dryden Research Center Photo Collection.)

surfaces to fly level. Previously, this would have made this aircraft unflyable, and the design impractical. However, with present computer flight-control systems, the design of unstable aircraft is an acceptable approach to make use of some of the performance advantages that the design can provide.

One of the considerations that is often a part of every aircraft design is take-off and landing distances. In some cases, a short take-off and landing distance may become the principle design objective. A case in point is the Douglas YC-15, pictured in Figure 1.4. This aircraft was designed to take-off and land in less than 1000 feet. The principle features of an aircraft that affect take-off and landing distances are the thrust-to-weight ratio and the maximum lift produced by the wings. The YC-15 was designed to exploit both of these aspects. It uses four engines for thrust. These are mounted high up on the wing so that the exhaust from the engines can be directed over the wing and flaps to augment the lift. This system can produce three to five times the maximum lift compared to a standard wing and gives the YC-15 a very distinctive appearance.

In some cases, aerodynamic performance is not the main design driver for a new aircraft concept. An example of such a design is the TACIT BLUE that is pictured in Figure 1.5. The objective of this design was to minimize its radar signature. Outgrowths of this technology led to the successful Lockheed F-117A stealth fighter. However, because of their shapes, these aircraft require the assistance of fly-by-wire computer control systems to overcome unstable flight characteristics.

FIGURE 1.4: Photograph of the Douglas YC-15, which was developed as a short take-off and landing (STOL) transport for the U.S. Air Force. (Courtesy of the Boeing Company.)

FIGURE 1.5: Photograph of the Northrop TACIT BLUE, used as a demonstrator for stealth technology. (Courtesy of the USAF Museum Photo Archives.)

Aircraft such as these appear radically different, although they still contain the basic elements common to all aircraft. How these designs come together is ultimately a result of the preconceived mission objectives. In the end, accomplishing these unique objectives gives each aircraft design a distinctive element and appearance.

1.1   DEFINING A NEW DESIGN

The design of an aircraft draws on a number of basic areas of aerospace engineering. As shown in the illustration, these include aerodynamics, propulsion, light-weight structures, and control.

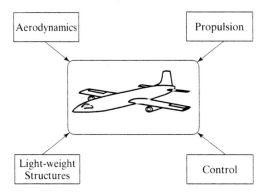

Each of these areas involves parameters that govern the size, shape, weight, and performance of an aircraft. Although we generally try to seek optimums in all these aspects, with an aircraft, this is practically impossible to achieve. The reason is that in many cases, optimizing one characteristic degrades another.

For example, a long-range aircraft should have high aspect ratio (long, narrow) wings, and a high wing loading, in order to minimize lift-induced drag for efficient cruise. Conversely, a highly maneuverable combat aircraft should have low aspect ratio wings and a low wing loading. Thus, the same aircraft cannot be optimized for both of these mission profiles. Table 1.1 demonstrates this on the basis of historical trends for several types of modern aircraft. Figure 1.6 illustrates the effect of changing the aspect ratio on the operating radius of a typical nine-passenger jet commuter aircraft. The parameter structure factor (SF) is the structure factor that corresponds to the ratio of the empty weight to the total take-off weight, which is another aspect of the design. This shows that a significant improvement in the range of an aircraft comes from having larger aspect ratio wings and a lighter structure. An example of an aircraft in which the main design driver was long range was the Voyager (pictured in Figure 1.7), which completed a non-stop flight around the world. This design coupled high aspect ratio wings with an extremely light composite structure (low structure factor).

In most cases, the design objectives are not as focussed as those of the Voyager aircraft. More often, the nature of an aircraft design is **compromise**. That is, the goal is

TABLE 1.1: Comparison of main wing aspect ratio for different aircraft types.

Aircraft Type	Aspect Ratio
Personal	5.0–8.0
Commuter	9.0–12.0
Regional Turboprops	11.0–12.8
Business Jets	5.0–8.8
Jet Transports	7.0–9.5
Military Fighter/Attack	2.4–5.0

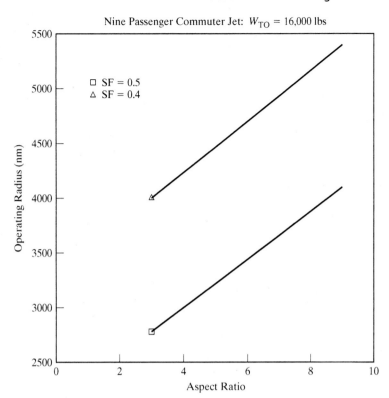

FIGURE 1.6: Effect of wing aspect ratio on operating radius for a typical nine-passenger commuter jet aircraft with different structure factors (SF).

to balance the different aspects of the total performance while trying to optimize a few (or one) based on well-defined mission requirements.

There are many performance aspects that can be specified by the mission requirements. These include

- the aircraft purpose or mission profile;

- the type(s) and amount of payload;

- the cruise and maximum speeds;

- the normal cruise altitude;

- the range or radius with normal payload;

- the endurance;

- the take-off distance at the maximum weight;

- the landing distance with 50 percent of the maximum fuel weight;

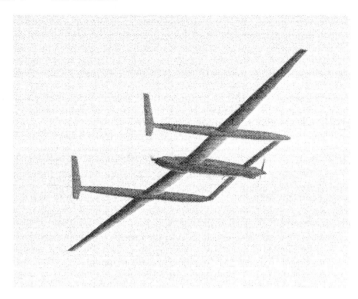

FIGURE 1.7: Photograph of the Voyager aircraft on its return from a non-stop flight around the world. (NASA Dryden Research Center Photo Collection.)

- the purchase cost; and

- other requirements considered important.

1.1.1 Aircraft Purpose

The starting point for any new aircraft is to clearly identify its purpose. With this, it is often possible to place a design into a general category. Such categories include combat aircraft, passenger or cargo transports, and general aviation aircraft. These may also be further refined into subcategories based on particular design objectives such as range (short or long), take-off or landing distances, maximum speed, etc. The process of categorizing is useful in identifying any existing aircraft that might be used in making comparisons to a proposed design.

With modern military aircraft, the purpose for a new aircraft generally comes from a military program office. For example, the mission specifications for the X-29 pictured in Figure 1.3 came from a 1977 request for proposals from the U.S. Air Force Flight Dynamics Laboratory in which they were seeking a research aircraft that would explore the forward swept wing concept and validate studies that indicated such a design could provide better control and lift qualities in extreme maneuvers.

With modern commercial aircraft, a proposal for a new design usually comes as the response to internal studies that aim to project future market needs. For example, the specifications for the most recent Boeing commercial aircraft (B-777) were based on the interest of commercial airlines to have a twin-engine aircraft with a payload and range in between those of the existing B-767 and B-747 aircraft. Table 1.2 summarizes this aspect for the three aircraft.

TABLE 1.2: Range and payload of B-777 aircraft to illustrate how it filled a perceived market gap in large commercial aircraft.

type	B-767-200ER	**B-777-200**	B-747-400
Passengers	181–224	**305–328**	416–524
Range (mi)	6115–6615	**5925–8861**	8400

Since it is not usually possible to optimize all of the performance aspects in an aircraft, defining the purpose leads the way in setting which of these aspects will be the "design drivers." For example, with the B-777, two of the prominent design drivers were range and payload.

1.1.2 Payload

The payload is what is carried on board and delivered as part of the aircraft's mission. Standard payloads are passengers, cargo, or ordnance. The first two are considered non-expendable payload because they are expected to be transported for the complete duration of the flight plan. Military ordnance is expendable payload since at some point in the flight plan it permanently leaves the aircraft. This includes bombs, rockets, missiles, and ammunition for on-board guns.

For personal or small general aviation aircraft, the payload includes the pilot as well as passengers and baggage. For business, commuter, and commercial aircraft, the payload does not include the flight or cabin crew, only the passengers, baggage, and cargo.

1.1.3 Cruise and Maximum Speeds

The mission of an aircraft usually dictates its speed range. Propeller-driven aircraft are usually designed to cruise at speeds between 150 to 300 knots. Jet-powered aircraft have higher cruise speeds that are normally specified in terms of Mach number. The typical cruise Mach number for business and commercial jet aircraft is from 0.8 to 0.85. This range of cruise speeds is close to optimum for maximizing the combination of payload weight, range, and speed. The few supersonic commercial aircraft designs (1) have supersonic cruise speed as their principle design driver and (2) sacrifice range and payload. The cruise Mach number of the Concord is 2.02. It will carry 100 passengers with a range of 3740 miles, which is considerably less than the aircraft listed in Table 1.2, which have high subsonic cruise speeds.

Modern military jet combat and attack aircraft usually have a flight plan that involves efficient cruise at high subsonic Mach numbers. This is usually in the range from Mach 0.85 to 0.90. The maximum speed is usually specified in the context of an intercept portion of the flight plan. This has a Mach number that is typically in the range of 2.0.

1.1.4 Normal Cruise Altitude

The cruise altitude is generally dictated by the cruise speed, propulsion system, and cabin pressurization. An aircraft with an unpressurized cabin would cruise no higher than

10,000 feet. With propeller-driven aircraft, turbo-charged piston engines can maintain a constant horsepower up to an altitude of approximately 20,000 feet. Higher altitudes are possible with turboprop aircraft, such as the Piper Cheyenne, which have a maximum ceiling from 35,000 to 41,000 feet. The decrease in air density with higher altitude lowers the drag, so that for these aircraft, the cruise range increases with altitude.

At higher subsonic Mach numbers, the turbo-jet engine gives the highest efficiency. For subsonic turbo-jet aircraft, there is an optimum altitude where the fuel consumption is a minimum. This occurs at approximately 36,000 feet. Therefore, it is the best altitude for the most efficient, long-range cruise of turbo-jet-powered aircraft.

1.1.5 Range

The range is the furthest distance the aircraft can fly without refueling. In a flight plan, range refers to the distance traveled during the cruise phase.

The choice of the range is one of the most important decisions because it has a large (exponential) effect on the aircraft take-off weight. Table 1.3 lists the average range for different types of aircraft. An aircraft that is intended to fly across the United States (New York to Seattle) should have a minimum range of 2500 nautical miles. A range of 3500 nautical miles would be necessary for transatlantic flights from East coast U.S. cities to coastal cities in Western Europe. All of the aircraft listed in Table 1.2 have the capability of flying non-stop from cities in the interior of the United States to cities in Eastern Europe, or from the West coast of the United States to cities in the Pacific Rim.

Shorter range transports that are designed to fly between major cities in a regional area (e.g., Los Angeles to San Francisco) should have a minimum range of 500 nautical miles. Twice that range would allow an aircraft to fly non-stop between most of the major cities along either coast of the United States.

1.1.6 Endurance

Endurance is the amount of time an aircraft can fly without refueling. With a reconnaissance aircraft, endurance is one of the main design drivers. For a commercial aircraft, a flight plan will include an endurance phase to allow for time that might be spent in a holding pattern prior to landing. For operation within the continental United States commercial aircraft are required to be able to hold for 45 minutes at normal cruise fuel consumption. For international operation, the required hold time is 30 minutes.

TABLE 1.3: Typical range for different types of aircraft.

Aircraft Type	Range (nautical miles)
Personal/Utility	500–1000
Regional Turboprop	800–1200
Business Jets	1500–1800
Smaller Jet Transports	2500–3500
Larger Jet Transports	6500–7200

1.1.7   Take-Off Distance

The total take-off distance consists of the length of a runway needed to accelerate, lift off, and climb to a prescribed obstacle height. The obstacle height is 50 feet for military and small civil aircraft, and 35 feet for commercial aircraft. The take-off distance that is required to accomplish this depends on different factors in the design such as the thrust-to-weight ratio, the maximum lift-to-weight ratio, and the surface of the air field that affects the rolling friction of the landing-gear wheels.

Different designs can fall into standard categories for take-off and landing. A conventional take-off and landing (CTOL) aircraft has distances that are greater than 1000 feet. A short take-off and landing (STOL) aircraft, such as the YC-15 in Figure 1.4, can take off and land in under 1000 feet. Both of these would have a ground roll portion during take-off and landing. A vertical take-off and landing (VTOL) aircraft does not require a ground roll.

Personal and general aviation propeller-driven aircraft, which are intended to operate out of smaller airports, need take-off distances of 1200 to 2000 feet. Larger twin-engine propeller commuter aircraft, which operate out of medium to larger size airports, have take-off distances from 3000 to 5000 feet. Business and smaller commercial jets have take-off distances of 5000 to 7500 feet. Larger commercial jet transport aircraft require take-off distances from 8000 to 11,000 feet. The take-off distance is a function of the altitude of the airport, although the distance at sea level is usually specified. Table 1.4 lists the altitude and runway lengths of some of the major airports in North America. A partial list of smaller airports within the United States is given in Table 1.5.

1.1.8   Landing Distance

The landing distance consists of the length of the runway needed to descend from a specified height of 50 feet, touchdown, and break to a stop. Factors that affect the landing distance are the maximum lift-to-weight and the surface of the air field, which affects the landing-gear wheels' braking friction coefficient. The lift-to-weight ratio directly affects the slowest (stall) speed at which the aircraft can fly. The landing touchdown speed is taken to be a small percentage higher than the stall speed.

For commercial aircraft, in a worst case scenario, the landing distance is determined with half of the fuel weight at take-off remaining and with an additional two-thirds distance to account for pilot variability. Even with these measures, the landing distances are almost always less than the take-off distances. Therefore, with regards to airports with available runway distances, the limiting condition will generally be set by take-off.

1.1.9   Purchase Cost

The purchase cost of an aircraft involves the costs incurred in the research, development, test, and evaluation (RDT&E) phase of the new aircraft design, and the acquisition (A) or production cost of customer-ordered aircraft. The cost of research and development is amortized over an initial fixed number of production aircraft. Therefore, as the number of production aircraft used to distribute this cost increases, the purchase cost per aircraft decreases. The decision on the total number of aircraft to be produced is therefore an

TABLE 1.4: Altitude and runway length of major airports in North America.

City	Airport	Elevation (f)	Runway (f)
Atlanta	Hartsfield	1026	11,889
Boston	Logan	20	10,081
Chicago	O'Hare	667	11,600
Dallas	Dallas–Ft. Worth	596	11,387
Denver	Denver	5431	12,000
Detroit	Detroit Metropolitan	639	12,000
Houston	Houston Intercontinental	98	12,000
Kansas City	Kansas City	1026	10,801
Los Angeles	Los Angeles	126	12,091
Miami	Miami	10	13,002
Minneapolis	Minn–St. Paul	841	10,000
Montreal	Dorval	117	11,000
New Orleans	New Orleans	6	10,080
New York	Kennedy	12	14,574
Oklahoma City	Will Rogers	1295	9,802
Philadelphia	Philadelphia	21	10,499
Phoenix	Sky Harbor	1133	11,001
St. Louis	Lambert	605	11,019
Salt Lake City	Salt Lake City	4227	12,003
San Diego	Lindbergh	15	9,400
San Francisco	San Francisco	11	11,870
Seattle	Seattle–Tacoma	429	11,900
Toronto	Toronto	569	11,000
Vancouver	Vancouver	9	11,000
Washington	Dulles	313	11,500

important factor in establishing the purchase price. In some cases, this price and customer competition may be the final arbiters that determine if a design is to be built.

The cost estimates are based on "cost estimating relationships" or CERs. These are simple model equations that correlate a few important characteristics of a large group of aircraft with their cost. The primary characteristics on which these are based are the weight of the structure of the aircraft, which is a fixed percentage of the take-off weight, the maximum speed at best altitude, and the production rate. From these, we expect that larger, heavier aircraft will cost more than smaller, lighter aircraft. Similarly, aircraft with higher cruise speeds are expected to cost more than slower aircraft. This is illustrated in Figure 1.8 for a sample jet transport. Since the CERs are based on structure weight, there is a cost incentive to use lighter weight materials, such as composite structural elements.

1.1.10   Federal Aviation Regulations

Any aircraft design must consider standards and regulations that are set by government associations. Civil aircraft designed, built, and operated in the United States must satisfy

TABLE 1.5: Altitude and runway length of smaller airports in North America.

City	Airport	Elevation (f)	Runway (f)
Tuscaloosa, AL	Tuscaloosa Municipal	170	6499
Nome, AK	Nome	36	6001
Sedona, AZ	Sedona	4827	5131
Fayetteville, AR	Fayetteville Municipal	1251	6006
Santa Ana, CA	John Wayne	54	5700
Greely, CO	Greely–Weld	4658	6200
New Haven, CT	Tweed–New Haven	14	5600
Wilmington, DE	New Castle County	80	7165
Key West, FL	Key West	4	4800
Macon, GA	Middle Georgia Regional	354	6501
Lihue, HI	Lihue	153	6500
Lewiston, ID	Lewiston–Nez Perce	1438	6512
Decator, IL	Decator	682	8496
Elkhart, IN	Elkhart Municipal	778	6500
Dubuque, IA	Dubuque Regional	1076	6498
Emporia, KS	Emporia Municipal	1206	5000
Paducah, KY	Barkley Regional	410	6499
Lake Charles, LA	Lake Charles Regional	15	6500
Portland, ME	Portland	74	6800
Salisbury, MD	Salisbury–Wicomica	52	5500
Hyannis, MA	Barnstable Municipal	55	5430
Muskegon, MI	Muskegon County	628	6501
Hibbing, MN	Chisholm–Hibbing	1353	6758
Tupelo, MS	Tupelo Municipal	346	5499
Joplin, MO	Joplin Regional	981	6503
Grand Island, NE	Central Nebraska Regional	1846	7188
Ely, NV	Ely	6255	5998
Nashua, NH	Boire Field	200	5550
Trenton, NJ	Mercer County	213	6006
Farmington, NM	Four Corners Regional	5503	6702
Poughkeepsie, NY	Dutchess County	166	5001
Winston-Salem, NC	Smith Reynolds	970	6655
Grand Forks, ND	Grand Forks	844	7349
Akron, OH	Akron Fulton	1068	6338
Stillwater, OK	Stillwater Municipal	986	6002
Medford. OR	Rouge Valley	1331	6700
Allentown, PA	Lehigh Valley	394	7600
Providence, RI	Theodoer F. Green	55	7166
Hilton Head, SC	Hilton Head Island	20	4300
Pierre, SD	Pierre Regional	1742	6891
Chattanooga, TN	Lovell Field	682	7401
College Station, TX	Easterwood Field	320	7000

TABLE 1.5: (*Continued*)

City	Airport	Elevation (f)	Runway (f)
Logan, UT	Logan-Cache	4454	5931
Rutland, VT	Rutland State	787	5000
Roanoke, VA	Roanoke Regional	1176	6802
Wanatchee, WA	Pangborn Memorial	1245	5499
Morgantown, WV	Morgantown Municipal	1248	5199
Racine, WI	John H. Batten	674	6556
Jackson Hole, WY	Jackson Hole	6445	6299

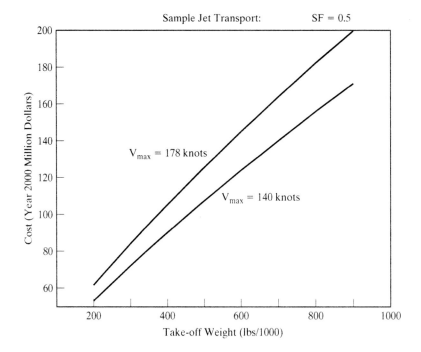

FIGURE 1.8: Example of the effect of take-off weight and cruise speed on the purchase cost of an aircraft.

the provisions of the Federal Aviation Regulations (FARs). The FARs are continually being updated to incorporate additional requirements that come about due to increased time and experience with existing aircraft. Electronic listings of the FARs can be obtained through a World Wide Web link to the Flight Standards Service of the U.S. Federal Aviation Association (FAA). The exact link can be found through a search under the keyword FAA.

Sections of the FARs that are of particular interest to designers are Air Worthiness Standards, General Operating and Flight Rules, and Operations. Air Worthiness Standards Parts 23 and 25 in particular define different categories of aircraft (for example, transport or commuter) based on such characteristics as number of passengers and maximum take-off weight. These categories are important in making comparisons to other aircraft with regard to flight performance, or other design drivers.

1.2 DESIGN PROCESS

The process of designing an aircraft and taking it to the point of a flight test article consists of a sequence of steps, as is illustrated in Figure 1.9. It starts by identifying a need or capability for a new aircraft that is brought about by (1) a perceived market

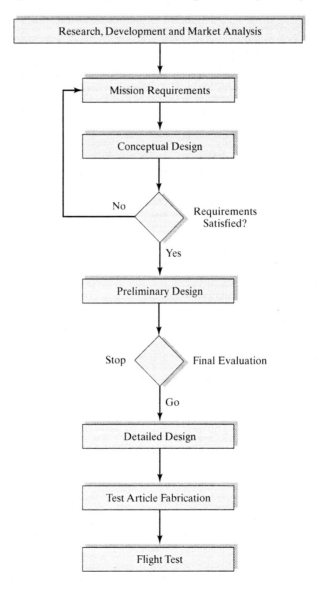

FIGURE 1.9: Design process flow chart.

potential and (2) technological advances made through research and development. The former will include a market-share forecast, which attempts to examine factors that might impact future sales of a new design. These factors include the need for a new design of a specific size and performance, the number of competing designs, and the commonality of features with existing aircraft. As a rule, a new design with competitive performance and cost will have an equal share of new sales with existing competitors.

The needs and capabilities of a new aircraft that are determined in a market survey goes to define the mission requirements for a conceptual aircraft. These are compiled in the form of a design proposal that includes (1) the motivation for initiating a new design and (2) the "technology readiness" of new technology for incorporation into a new design. It is essential that the mission requirements be defined before the design can be started. Based on these, the most important performance aspects or "design drivers" can be identified and optimized above all others. An example of a design proposal follows.

DESIGN PROPOSAL: SUPERSONIC BUSINESS JET (SSBJ)

We propose to design a supersonic mid–long-range business jet. It is intended to have a cruise Mach number of 2.1 and a cruise altitude of 55,000 feet. Its range will be 4000 nm with a full payload. Its nonexpendable payload would consist of passengers and baggage, with a maximum total weight of 4000 lbs. Depending on the internal layout, this will comfortably accommodate from 12 to 15 passengers. The maximum take-off weight is estimated to be 90,000 lbs. Other features of the design include a delta wing planform and the use of control canards. Composite materials will be used extensively to reduce the structure weight. The most critical technology-readiness issue is the propulsion system. An existing engine that has been selected as a reference engine for the design is the GE-F404-100D. Based on the drag estimate at cruise conditions, the aircraft would require four of these engines.

This aircraft would be the only one of its type and therefore would have no other market competitors at this time. Aircraft companies such as Boeing, Lockheed–Martin with Gulfstream, and Dassault have indicated that they are considering designs for a supersonic business jet and therefore could be potential competition. Figure 1.10 shows an artist's rendition of the proposed Dassault design.

The characteristics of the SSBJ are listed in the table that follows. Although there are no existing aircraft to which a direct comparison can be made, some other existing or proposed aircraft with some similar characteristics are cited for reference purposes. One of these is the Russian Sukhoi S-21, which was never built.

The principle design drivers are a supersonic cruise Mach number, and a range and passenger number which are comparable to high-end subsonic business jets. Secondary design considerations include moderate take-off and landing distances which are comparable to existing high-subsonic business jets.

The Dassault Falcon 900B was selected as a representative subsonic business jet. It carries up to 12 passengers and has a range of 3840 nm. To be competitive with aircraft of this class, a capability of 12–15 passengers and a range of 4000 nm is

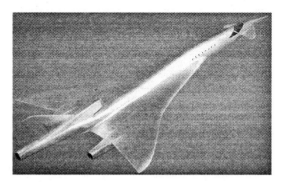

FIGURE 1.10: Artist's rendition of proposed Dassault Supersonic Business Jet (SSBJ). (Courtesy of Dassault Aviation.)

proposed. The Sukhoi S-21 would have been the closest existing aircraft, if it had been built. It was proposed as a 6–10-passenger business jet, with a cruise Mach number of 2.0. Its proposed range was the same as the SSBJ, and its estimated take-off weight was 106,000 lbs. The proposed Dassault SSBJ has a comparable range and slightly lower Mach number. It is also intended to use three engines. The other two comparison aircraft, the Mig-31 and the Tu-22M, are each supersonic bombers. These were used for comparison because of their comparable Mach numbers.

SSBJ and Aircraft with Similar Characteristics

	SSBJ	Sukhoi* S-21	Mig-31	Tu-22M	Dassault Falcon 900B	Dassault[+] SSBJ
W_{TO} (lbs)	90,000	106,000	90,000	273,000	45,500	—
M_{cruise}	2.1	2	2.8	1.9	0.87	1.8
Range (nm)	4000	4000	—	—	3,840	4000
Passengers	12–15	6–10	—	—	12	—

* *Note*: Proposed design never built.
+ *Note*: Proposed design.

Following the design proposal, the next step is to produce a conceptual design. The conceptual design develops the first general size and configuration for a new aircraft. It involves the estimates of the weights and the choice of aerodynamic characteristics that will be best suited to the mission requirements stated in the design proposal. The design will make estimates of the total drag and size the power plant. It will determine the best airframe to accommodate the (1) payload and (2) wing and engine placement. This conceptual design will locate principle weight groups in order to satisfy static stability requirements. It will size control surfaces to achieve a desired degree of maneuverability.

Finally, the conceptual design will estimate the RDT&E and acquisition costs to develop one or more test articles.

The conceptual design is driven by the mission requirements, which are set in the design proposal. In some cases, these may not be attainable so that the requirements may need to be relaxed in one or more areas. This is shown as an iterative loop in the flow chart in Figure 1.9. When the mission requirements are satisfied, the design moves to the next phase, which is the preliminary design.

The preliminary design is a fine tuning of the conceptual design made through parametric wind tunnel tests of scale aircraft models of the design. Some of the more difficult aspects to predict are tested in this phase. This includes the (1) engine inlet interaction with the fuselage and wing and (2) wing interaction on control surfaces.

The preliminary design also involves a more detailed analysis of the aerodynamic loads and component weights. Based on this, the structural design is further refined. Aeroelastic motion, fatigue, and flutter are considered at this stage. Additional confirmation of estimates may require building and testing some of the proposed structural components.

At the completion of this stage, the manufacturing of the aircraft is given serious consideration, and the cost estimates are further refined. At the end of this step, the decision is made whether to build the aircraft. With the decision to build the aircraft, the design is "frozen."

The detailed design involves generating the detailed structural design of the aircraft. This involves *every detail* needed to build the aircraft. Sometimes component mock-ups are built to aid in the interior layout. However, the present use of computer-aided design (CAD) software can substantially minimize the need for mock-ups by providing realistic 3-D views.

1.3 CONCEPTUAL DESIGN

This book deals with the steps involved in the conceptual design of an aircraft. It is broken down into 13 elements, which are followed in order. These consist of

1. preliminary estimate of take-off weight;
2. wing loading selection;
3. main wing design;
4. fuselage design;
5. horizontal and vertical tail design;
6. engine selection;
7. take-off and landing;
8. enhanced lift design;
9. structure design and material selection;
10. refined weight analysis;
11. static stability and control;
12. cost estimate;
13. design summary and trade study.

The development of these elements is illustrated in a case study that consists of the Supersonic Business Jet (SSBJ) defined in the preceding design proposal. The mission requirements of this design are relatively difficult to achieve; therefore, it is a good example in which compromise is needed. Case studies of other types of aircraft are also presented in an appendix of this book.

CHAPTER 2

Preliminary Estimate
of Take-Off Weight

2.1 FUEL FRACTION ESTIMATES
2.2 TOTAL TAKE-OFF WEIGHT
2.3 SPREADSHEET APPROACH FOR TAKE-OFF WEIGHT ESTIMATE
2.4 PROBLEMS

Photograph of Boeing C-17 Globemaster at take-off: Maximum take-off weight equals 585,000 lbs; maximum payload is 169,000 lbs. (Courtesy of the Boeing Company.)

Following the design proposal, the first step in the design of a new aircraft is to obtain an estimate of the take-off weight, W_{TO}. This estimate is one of the most crucial, since it is used in many other parts of the design. Because this is the first step, little is known about the aircraft beyond the objectives for the design. Therefore, at this stage, some of the information used in making an estimate will rely on historic trends of other flying aircraft. The list of comparison aircraft that is cited in the design proposal can also be helpful at this early stage of the design.

The total take-off weight is divided into fuel weight, payload weight, and empty or structure weight:

$$W_{\text{TO}} = W_{\text{fuel}} + W_{\text{payload}} + W_{\text{empty}}. \tag{2.1}$$

The payload is further divided into nonexpendable and expendable types. The nonexpendable payload remains unchanged throughout the flight plan. This includes the crew, passengers, baggage, revenue cargo, etc. Expendable payload is dropped somewhere in the flight plan, before landing. For example, a combat aircraft would include ordnance in this category. The total payload weight is, therefore,

$$W_{\text{payload}} = W_{\text{expendable}} + W_{\text{nonexpendable}}. \tag{2.2}$$

The total weight buildup, which includes all these items, is illustrated in Figure 2.1.

The percentage that each of these weights contributes to the total take-off weight depends on the mission or design objectives. Again we can look to historical trends

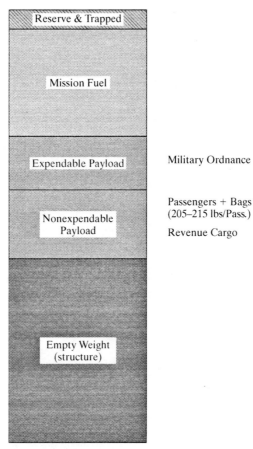

FIGURE 2.1: Typical aircraft weight buildup.

as a guide. However, as a general observation, longer range aircraft devote a greater percentage of their take-off weight to the weight of fuel.

The fuel weight is based on the flight plan. It considers the fuel used in all of the flight phases, including

- engine start-up and take-off;

- acceleration to cruise velocity and altitude;

- cruise out to destination;

- acceleration to high speed;

- combat;

- return cruise;

- loiter;

- landing.

Two representative flight plans are illustrated in Figure 2.2. The top plan is one of the most basic and would generally correspond to a commercial aircraft. It consists of flight phases made up of engine start-up and take-off, climb and accelerate to cruise altitude and speed, cruise out to destination, loiter at destination, and landing. The bottom plan is more specialized and would generally correspond to a fighter aircraft. It consists of the same first four flight phases, but following the cruise out, it includes a high-speed intercept and combat. The combat is generally at a lower altitude and Mach number. Expendable payload may also be dropped during combat. Following combat, the flight plan consists of (1) a climb back to cruise altitude and speed and (2) cruise back. For combat aircraft, the range corresponds to its radius of operation, which implies that it returns to its original point of departure. This is different from that of a commercial aircraft, which is expected to land at a destination other than from where it departed. The final phases of the flight plan for a combat aircraft still include loiter and landing.

All of these phases of the flight plan use a fractional portion of the total weight of fuel that is available at the time of take-off. This is referred to as the mission fuel. In addition to this, a 5 percent allowance for reserve fuel and a 1 percent allowance for fuel trapped in lines is allocated.

In some of the flight phases, historically based empirical relations are used to estimate the fuel fractions. In other phases, analytical approaches are used. For example, the logarithmic fuel fraction for cruise to destination is dependent on the range, lift, drag, velocity, and specific fuel consumption through the Brequet range equation. Since during cruise, the lift equals the weight, we have a coupling with the initial weight estimate. Therefore, an iterative procedure is necessary to determine the total fuel weight required for the complete flight plan that includes cruise.

2.1 FUEL FRACTION ESTIMATES

The approach for determining the total amount of fuel used during a mission is based on considering the individual amounts used within each flight phase. For any of the flight

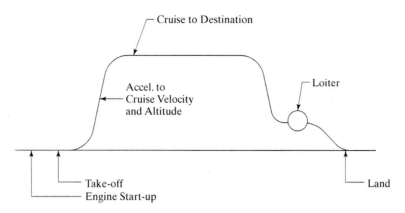

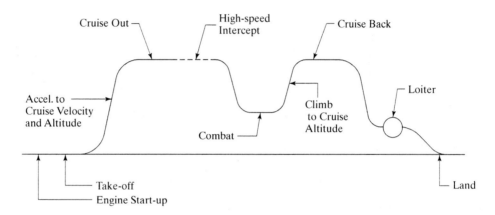

FIGURE 2.2: Representative flight plans for a commercial aircraft (top) and combat aircraft (bottom).

phases listed earlier, the fuel used (by weight) is determined and represented as the ratio of the fuel weight leaving (final) to that entering (initial) that flight phase, namely,

$$\text{Fuel Weight Fraction} = (W_f/W_i)_{\text{fuel}}. \qquad (2.3)$$

The total fuel fraction for the complete flight plan is equal to the products of the individual weight fractions in the respective flight phases, namely,

$$(W_{\text{landing}}/W_{\text{take-off}})_{\text{fuel}} = \frac{W_2}{W_1}\frac{W_3}{W_2}\cdots\frac{W_N}{W_{N-1}}, \qquad (2.4)$$

where $1, 2, \ldots N$ represent the individual flight phases in sequential order in the flight plan, starting with take-off (1) and ending with landing (N). The fuel fractions that correspond to each of the different flight phases are presented in detail in the following sections.

2.1.1   Engine Start-Up and Take-Off

The engine start-up and take-off is the first phase in any flight plan. It consists of starting the engines, taxiing to the take-off position, take-off, and climb out. A good empirical estimate for the weight of fuel used in this phase is from 2.5 to 3.0 percent of the total take-off weight. Therefore,

$$0.97 \leq \frac{W_f}{W_i} \leq 0.975. \tag{2.5}$$

2.1.2   Climb and Accelerate to Cruise Conditions

After take-off, the aircraft will generally climb to cruise altitude and accelerate to cruise speed. The estimate for the weight fraction for this phase of the flight is also found from empirical data. One such set of data from Nicolai (1975) is shown in Figure 2.3. This figure illustrates the weight fraction, W_f/W_i, for a variety of aircraft as a function of the cruise Mach number. The initial Mach number is a small value of the order of 0.1.

Figure 2.3 illustrates that when accelerating from Mach 0.1 up to approximately Mach 1, there is only a small decrease in the weight of fuel, $W_f/W_i \leq 0.95$. Accelerating to higher Mach numbers, however, has a more significant effect on the fuel weight fraction.

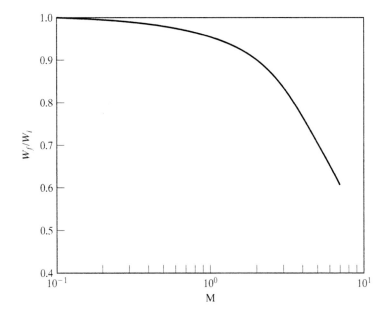

FIGURE 2.3: Weight fractions for different aircraft during a climb and accelerate to cruise condition flight phase.

2.1.3   Cruise Out to Destination

For this flight phase, we do not have to rely as much on historic information. For a cruising aircraft, the fuel weight fraction can be determined quite well from an analytic formulation called the Brequet range equation. Two forms of the equation, for turbo-jet and reciprocating engines, are

$$R = \frac{V}{C}\frac{L}{D}\ln\left[\frac{W_i}{W_f}\right] \tag{2.6a}$$

and

$$R = \frac{\eta}{C}\frac{L}{D}\ln\left[\frac{W_i}{W_f}\right]. \tag{2.6b}$$

Here R is the range specified in nautical miles, V is the cruise velocity, L is the lift, D is the drag, and C is the thrust-specific fuel consumption. For reciprocating engines, the velocity is replaced by the propulsive (or propeller) efficiency, η. Note that the units on C are

1. $lb_{fuel}/hr/lb_{thrust}$ for a turbo-jet engine, and
2. $lb_{fuel}/hr/SHP$ for a reciprocating (propeller) engine.

From these expressions, we see that there is an exponential dependence of the fuel fraction on the range and quantities that are subject to the design, $\frac{V}{C}\frac{L}{D}$. Such extreme sensitivity makes their choice rather critical.

The selection of these quantities might appear to be rather difficult at this point in the design. However, it is important to stress that the conceptual design is an *iterative process*, and initial guesses are likely (and expected) to be improved upon at a later point.

In making choices now, we start with L/D. For efficient cruise, which maximizes range, L/D will be close to L/D_{max}. A reasonable estimate is

$$\frac{L}{D} = 0.94\left[\frac{L}{D}\right]_{max}. \tag{2.7}$$

The problem then reduces to estimating L/D_{max}. For this we turn to additional empirical data compiled by Nicolai (1975), which is shown in Figure 2.4.

This figure shows that below Mach 1, L/D_{max} is dependent on the aspect ratio, A. This is the result of 3-D wing effects. For supersonic wings, 2-D wing theory applies, and there is no dependence on A. For subsonic wings, as the aspect ratio increases, the wings more closely approximate 2-D wings, and we observe an accompanying increase in L/D_{max}.

"Wing-lets" or other modifications to the wing tips can reduce the wing end effects and lower the lift-induced drag. The addition of wing-lets can be incorporated into Figure 2.4 by assuming that they increase the effective aspect ratio by 15 to 20 percent compared to the same wing without wing-lets.

An additional check can be obtained by using the theoretical expression derived for optimum subsonic cruise

$$\left[\frac{L}{D}\right]_{max} = \frac{1}{2\sqrt{C_{D_0}k}}. \tag{2.8}$$

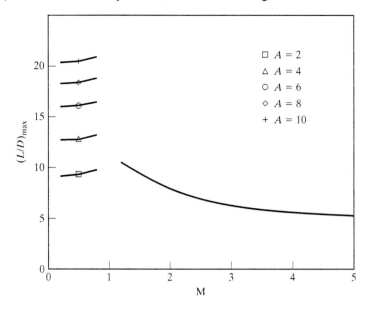

FIGURE 2.4: Variation in $L/D_{\max}$ with Mach number and aspect ratio.

where

$$k = \frac{1}{\pi A e} \qquad (2.9)$$

and

$$e = \text{Oswald's coefficient} \simeq 0.8. \qquad (2.10)$$

C_{D_0} is the minimum 3-D drag coefficient, which is not yet determined in the design, but can be estimated from empirical data for similar aircraft. For a variety of aircraft,

$$0.01 \leq C_{D_0} \leq 0.02. \qquad (2.11)$$

As an additional reference, Table 2.1 lists the values of cruise $L/D_{\max}$ for different aircraft.

To complete the calculation of the weight fraction given by Eq. [2.7], we need to determine V/C. For the cruise phase, it is appropriate to specify the cruise Mach number,

TABLE 2.1: Cruise $L/D_{\max}$ values for various aircraft.

	$L/D_{\max}$ Range	Average $L/D_{\max}$
Propeller Personal/Utility	9.6–14.2	12.1
Propeller Commercial Transport	13.8–18.5	16.3
Business Jet	13.0–15.6	14.3
Commercial Jet Transport	15.0–18.2	14.4
Military Transport/Bomber	17.5–20.5	18.9
Military Fighter (subsonic cruise)	9.2–13.9	11.0

M_c. Therefore, the cruise velocity, V, depends on the local speed of sound, which is a function of the cruise altitude. Values for the speed of sound at different altitudes in a standard atmosphere are listed in Appendix A.

The value of the thrust-specific fuel consumption, C, can only be estimated at this stage by considering comparison aircraft. Values depend on a number of parameters including Mach number, altitude, and bypass ratio. A general range is

$$0.5 \leq C \leq 1.2, \tag{2.12}$$

where the lowest value corresponds to bypass ratios greater than 10 to 12, and the highest corresponds to military aircraft with bypass ratios close to one, with afterburners.

Because of the exponential dependence of the fuel weight fraction on C, if range is an important objective of the design, engine manufacturer data should be examined. Later on, as the design develops, the overall drag and, thus, the required thrust at cruise will be determined. At that point, engines can be selected or scaled to meet the design requirements. If necessary, an improved estimate of C can then be used to update the take-off weight calculations.

For a propeller-driven aircraft, the values for L/D and C are found in the same manner as for the turbo-jet aircraft. The only additional parameter to determine the fuel weight fraction in Eq. [2.7] is the propulsive efficiency,

$$\eta = \frac{TV}{P}, \tag{2.13}$$

where T is the thrust, V is the cruise velocity, and P is the shaft power.

Empirical data for propeller-driven aircraft indicate that the ratio η/C is nearly constant. Therefore, to maximize the range in Eq. [2.7], we want an operating point where η and L/D are maximums. For η, this amounts to finding the velocity where L/D is a maximum. This velocity can be determined analytically as

$$V_{\frac{L}{D}\max} = \left[\frac{2}{\rho}\frac{W}{S}\right]^{0.5}\left[\frac{k}{C_{D_0}}\right]^{0.25}, \tag{2.14}$$

where W/S is the wing loading at cruise, ρ is the air density at cruise altitude, and k and C_{D_0} are given in Eq. [2.9] and Eq. [2.11], respectively. The parameters, W/S and T/P, will be established later in the design. At this step, they can best be estimated by considering comparable aircraft.

As a further guide at this stage in the design, Table 2.2 lists values of η and C for a general set of propeller-driven aircraft. These can be used along with the values of $L/D_{\max}$ in Table 2.1 to estimate the cruise fuel weight fraction.

2.1.4 Acceleration to High Speed (Intercept)

This flight phase involves accelerating from the cruise Mach number to a maximum flight Mach number as part of a high-speed intercept. In order to estimate of the fuel weight fraction required for this, Figure 2.3 is again used. Recall that Figure 2.3 was used to determine the fuel weight fraction corresponding to acceleration from the climb velocity (of the order of Mach 0.1) to the cruise Mach number. We utilize the same approach, but now consider two acceleration phases:

TABLE 2.2: Propulsion parameters for classes of propeller-driven aircraft.

	η	C
Personal/Utility	0.80	0.60
Commuter	0.82	0.55
Regional Turboprop	0.85	0.50

1. acceleration from low speed (Mach 0.1) to cruise Mach number, M_c;
2. acceleration from low speed (Mach 0.1) to the maximum Mach number, $M_{\max}$.

The weight fraction for (1) is

$$\frac{W_f}{W_i} = \frac{W_c}{W_{0.1}}. \tag{2.15}$$

The weight fraction for (2) is

$$\frac{W_f}{W_i} = \frac{W_{\max}}{W_{0.1}}. \tag{2.16}$$

Therefore the weight fraction to accelerate from M_c to $M_{\max}$ is

$$\frac{W_f}{W_i} = \frac{W_{\max}}{W_c} = \frac{W_{\max}}{W_{0.1}} \left[\frac{W_c}{W_{0.1}}\right]^{-1} \tag{2.17}$$

2.1.5   Combat

Combat is defined as a time, t_{combat}, during which the aircraft is flying at maximum thrust, $T_{\max}$, and maximum thrust-specific fuel consumption, $C_{\max}$. The weight of fuel used during combat is

$$W_i - W_f = C_{\max}\, T_{\max}\, t_{\text{combat}}. \tag{2.18}$$

Note that since the left side of the equality has dimensions, care must be taken to use consistent units for the quantities on the right-hand side.

In addition to the fuel used, the combat flight phase could also include the loss of expendable payload, such as ordnance. This change in weight from the start of combat to the end must also be accounted for prior to entering the next flight phase.

2.1.6   Return Cruise

Return cruise refers to a flight plan in which the aircraft returns to its point of origin to land. For a flight plan in which the landing destination is different from where it took off, return cruise can be viewed as the second half of the cruise phase. In either case, return cruise is treated exactly like cruise out with two possible exceptions.

The first comes from the loss of fuel weight, which makes the aircraft lighter. In long-range aircraft, the difference in weight from the start to the end of cruise can be substantial. For the same amount of lift, the aircraft would tend to rise to a higher altitude where the lift again balances the weight. As a result, for the same cruise Mach number, the cruise velocity, V, would be different.

The second comes as a step to counter any increase in altitude due to the decreasing weight. To accomplish this, the pilot would need to adjust the horizontal stabilizer trim, which is a less efficient operating condition and increases the aircraft drag. In this case, L/D would be different from the value used at the beginning of cruise.

Both of these impact the fuel weight fraction for cruise, determined from Eq. [2.6]. The effect will be most significant for long-range aircraft in which the total fuel weight is a larger portion of the take-off weight.

2.1.7 Loiter

The loiter phase consists of cruising for a specified amount of time over a small region. Loiter time is usually built into the flight plan to allow for delays prior to landing. However, reconnaissance aircraft could have loiter endurance as the primary mission.

For this phase, the fuel weight fraction is derived from an analytic expression called the endurance equation. Two forms of the equation, for turbo-jet and reciprocating engines, are

$$ E = \frac{1}{C} \frac{L}{D} \ln \left[\frac{W_i}{W_f} \right] \tag{2.19a} $$

and

$$ E = \frac{\eta}{C} \frac{L}{D} \frac{1}{V} \ln \left[\frac{W_i}{W_f} \right], \tag{2.19b} $$

where E is the endurance (loiter) time.

Eq. [2.20] assumes a fixed altitude and Mach number so that L/D and C are constants with respect to the aircraft weight. Eq. [2.20] is a quite simplified version of the analytic expression for propeller-driven aircraft. This approximate form is better suited to making estimates because some of the parameters that are required are difficult to predict at this early stage of the design.

For Eq. [2.20], it is clear that in order to obtain the maximum endurance for a given fuel-weight ratio, the aircraft should fly at an altitude and Mach number that maximize $L/(DC)$. As an initial approximation, we can take

$$ \left[\frac{L}{DC} \right]_{\text{max}} \simeq \left[\frac{L}{D} \right]_{\text{max}}. \tag{2.20} $$

L/D_{max} is then found as before from Figure 2.4 or Eq. [2.8].

For Eq. [2.20], upon substituting for the propulsive efficiency, η,

$$ E = \frac{T}{P} \frac{L}{DC} \frac{1}{V} \ln \left[\frac{W_i}{W_f} \right]. \tag{2.21} $$

In this case, the maximum endurance for a given fuel-weight ratio occurs when the shaft power, P, is a minimum. This condition occurs at the velocity where L/D is a maximum as in Eq. [2.14].

2.1.8 Landing

The final phase of the flight plan is landing. As an estimate of the fuel weight fraction used at landing, we use the same empirical formula that was used for start-up and take-off, namely,

$$0.97 \leq \frac{W_f}{W_i} \leq 0.975. \tag{2.22}$$

2.2 TOTAL TAKE-OFF WEIGHT

As was given in Eq. [2.4], the total fuel fraction for the complete flight plan is the product of the individual weight fractions for the respective flight phases. The total fuel weight then corresponds to the estimated take-off weight minus the weight after landing minus any expendable (dropped) weight, plus 5 percent reserve and 1 percent trapped fuel.

The available empty weight consists of the initial estimated take-off weight minus all the removable weights including fuel weight and expendable and nonexpendable payload weights. This is then compared to the required empty weight, which is the structure weight we can expect for a particular type of aircraft, based on historical data. The structure weight is determined from the structure coefficient, s, given as

$$s = \frac{W_{\text{empty}}}{W_{\text{TO}}}. \tag{2.23}$$

The historical trend for the structure coefficient as a function of the gross take-off weight, W_{TO}, is shown in Figure 2.5. The scatter in the data indicates the variation

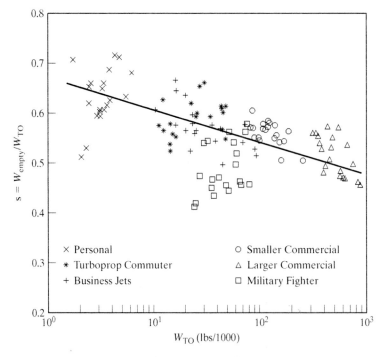

FIGURE 2.5: Structure factor versus gross take-off weight for a variety of aircraft.

TABLE 2.3: Structure factor for selected aircraft as a function of take-off weight.

$s = A W_{\text{TO}}^{C}$	A	C
Sailplane (unpowered)	0.86	−0.05
Sailplane (powered)	0.91	−0.05
Homebuilt (metal/wood)	1.19	−0.09
Homebuilt (composite)	0.99	−0.09
Homebuilt (composite)	0.99	−0.09
General Aviation (single engine)	2.36	−0.18
General Aviation (twin engine)	1.51	−0.10
Twin Turboprop	0.96	−0.05
Jet Trainer	1.59	−0.10
Jet Fighter	2.34	−0.13
Military Cargo/Bomber	0.93	−0.07
Jet Transport	1.02	−0.06

that exists at a given take-off weight. Additional data from Raymer (1992) are given in Table 2.3. These show a general trend of lower structure factors for long-range transport aircraft and higher values for combat aircraft. The former require a larger fuel weight fraction in order to achieve their long range. The latter have a higher structure weight to withstand the high g-loads that occur during combat maneuvers. As a consequence, these fighter aircraft have a lower range.

The final take-off weight is the sum of the fuel weight, required empty weight, and expendable and nonexpendable payload weights. The difference between the available empty weight and the required empty weight gives the surplus empty weight.

The object of the take-off weight estimate is to have a zero surplus empty weight. This requires an iterative approach where an initial take-off weight is guessed. The incremental weights throughout the flight plan are then calculated to give the final weight at landing. This gives the final fuel weight. With this, the surplus empty weight is calculated. Depending on the sign of the surplus empty weight, the initial take-off weight is incremented, and the calculations for the incremental weights are repeated. This process continues until the surplus weight is zero, at which point the take-off weight of the conceptual aircraft is determined.

2.3   SPREADSHEET APPROACH FOR TAKE-OFF WEIGHT ESTIMATE

A spreadsheet approach is useful for performing the calculations used in estimating the take-off weight. Some of the advantages of this approach are that it allows easy entry of parameters and monitoring of intermediate results, which can be useful in exploring the different designs or concepts.

The spreadsheet file that performs the estimate of the take-off weight is called **itertow.xls**. "Iter" refers to the fact that the calculations use iterative steps to reach the solution of the take-off weight. A sample for the case study supersonic business jet described in the design proposal in Chapter 1 is shown in Figure 2.6. The following describes the general spreadsheet structure.

	Mission Requirements				
Max. Mach	2.1				
Cruise Mach	2.1				
Cruise Alt. (ft)	55,000				
Oper. Rad. (nm)	2,000				
Engine: TSFC Min.	0.9				
Engine: TSFC Max.	2.17				
Engine: Thrust (lbs)	108,540				
Aspect Ratio	2				
Combat: Time (min)	0				
Combat: Altitude (ft)	30,000				
Loiter: Time (min)	10				
Loiter: Altitude (ft)	0				
Fuel Reserve (%)	5				
Trapped Fuel (%)	1				
Structure Factor	0.5				
Payload: Exp. (lb)	0				
Payload: Non-exp. (lb)	4000				
		Iteration 1	Iteration 2	Iteration 3	Iteration 4
Weight: T-O (estimated)	40,000	40,000.00	42,232.50	90,523.17	90,523.17
Weight: T-O (final)		42,232.50	44,366.34	90,523.17	90,523.17
Surplus Empty Wt. (lbs)		−2,232.50	−2,133.85	0.00	0.00
1. Start-up & T-O		39,000.00	41,176.68	88,260.09	88,260.09
2. Climb & Accel. to Cruise		36,153.00	38,170.79	81,817.10	81,817.10
3a. L/D		7.59	7.59	7.59	7.59
3b. V (f/s)		1,925.70	1,925.70	1,925.70	1,925.70
3c. Cruise to destination		29,364.58	31,003.49	66,454.38	66,454.38
4. Accel. to high speed		29,364.58	31,003.49	66,454.38	66,454.38
5. Combat		29,364.58	31,003.49	66,454.38	66,454.38
6. Drop Exp. Payload		29,364.58	31,003.49	66,454.38	66,454.38
7. Cruise back		23,850.82	25,182.00	53,976.30	53,976.30
7. Loiter		23,384.13	24,689.26	52,920.15	52,920.15
8. Land		22,799.53	24,072.03	51,597.15	51,597.15
Total Fuel Wt. (lbs)		18,232.50	19,250.10	41,261.58	41,261.58
Available Empty Wt. (lbs)		17,767.50	18,982.40	45,261.58	45,261.58
Required Empty Wt. (lbs)		20,000.00	21,116.25	45,261.58	45,261.58

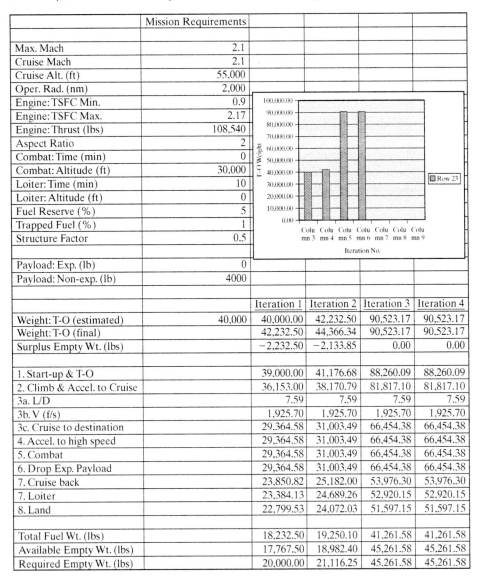

FIGURE 2.6: ITERTOW spreadsheet for SSBJ case study.

2.3.1 Spreadsheet Structure

The input parameters for the spreadsheet include all of the information that is needed to describe the aircraft mission. These include some or all of the following:

1. operating radius;
2. cruise speed and altitude;
3. maximum speed;

4. combat time;

5. loiter time;

6. thrust-specific fuel consumption (cruise and maximum);

7. wing aspect ratio.

The names of the input items are listed in the first column (A) of the spreadsheet. The numeric values or each respective item are placed in the same row in the second column (B). Note that it is important to use the proper units when entering these. For example, operation radius has units of nautical miles.

The input items do not change from one iteration to the next. Therefore, these have also been defined in the spreadsheet with "absolute" names rather than by their row/column index. For example, the maximum Mach number is assigned the name of MMX, cruise altitude is H, and range is RA. This was not done because of convenience, but rather to make it easier to do the iterative calculations.

Different calculated values are located in successive columns starting in Column C. These columns have been labeled Iteration 1, Iteration 2, etc. The flight phases are denoted by name and labeled 1–8 in Column A. The aircraft weight at the end of a flight phase appears in the respective rows.

The flight phases begin with "start-up and take-off." The formula for the weight is based on Eq. [2.5], where $W_f/W_i = 0.025$.

The weight following "climb and accelerate to cruise conditions" is determined next. This is based on Figure 2.3. As a simple approximation to this figure, the curve is represented by two piecewise linear segments with

$$\frac{W_f}{W_i} = 1 - 0.04M_f; \quad M_f < 1 \tag{2.24a}$$

and

$$\frac{W_f}{W_i} = 0.96 - 0.03(M_f - 1); \quad M_f \geq 1. \tag{2.24b}$$

Obtaining the weight following the "cruise out" phase requires some intermediate calculations. The first is the lift-to-drag ratio. This depends on the cruise Mach number and the wing aspect ratio. The value for L/D is found from Figure 2.4. For automated calculations, these curves have been represented by the following simple formulas:

$$\frac{L}{D} = A + 10; \quad M_c < 1 \tag{2.25a}$$

and

$$\frac{L}{D} = 11M_c^{-0.5}; \quad M_c \geq 1. \tag{2.25b}$$

The cruise velocity is also needed for the range equation. This is found by multiplying the cruise Mach number by the speed of sound. The speed of sound depends on the cruise altitude. For the spreadsheet, the standard atmosphere data has been modeled in

$$V = [1036 - 0.0034(H_c - 20,000)] M_c, \tag{2.26}$$

where H_c is the dimensional cruise altitude with units of feet and V has units of feet per second.

Using these values for L/D and V, we can calculate the weight following cruise out based on Eq. [2.7]. Note that this form of the range equation is for a turbo-jet engine. If the aircraft design is propeller driven, the formula needs to be changed to Eq. [2.7]. In that case, a value for η needs to be added to the list of input parameters. An example for a propeller driven aircraft is given in the case study in Appendix B.

The weight following an "acceleration to high speed" is found from Figure 2.3, as represented by Eq. [2.24], in the manner given in Eq. [2.17]. Note that if the maximum Mach number is the same as the cruise Mach number, there will be no acceleration to high speed, and the weight will remain the same.

The weight following "combat" is based on Eq. [2.18]. It is a function of (1) the combat time (minutes) and (2) the maximum thrust (pounds) and thrust-specific fuel consumption. If the combat time is zero, the weight remains unchanged.

Any expendable payload weight that is specified in the input parameters is subtracted from the weight following the combat phase.

For the "cruise back" phase, the same values apply for L/D and V as were used in "cruise out." As pointed out in Section 2.1.6, this is not completely accurate, but at this stage of the design, it is sufficient for estimating the take-off weight.

The weight following the "loiter" is based on Eq. [2.19]. The formula in the original spreadsheet is for a turbo-jet engine (Eq. [2.20]). If the aircraft design is propeller driven, the formula needs to be changed to Eq. [2.20]. Again an example is given in Appendix B.

The final phase of the flight plan is "landing." The weight following landing is based on Eq. [2.22], where $W_f/W_i = 0.025$.

The total fuel weight is taken as the estimated take-off weight, minus the weight after landing, minus any expendable payload weight, plus 6 percent of the initial fuel weight, which corresponds to reserve and trapped fuel.

The "available empty weight" consists of the initial estimated take-off weight, minus all the removable weights including fuel weight and expendable and nonexpendable payload weights.

The "required empty weight" corresponds to the structure weight of the aircraft. This is a fixed percentage of the take-off weight defined as the structure factor, s. The structure factor depends on the type of aircraft and the take-off weight. Estimates can be obtained from Figure 2.5 and Table 2.3.

The final take-off weight is the sum of the fuel weight, required empty weight, and expendable and nonexpendable payload weights. The difference between the *available* empty weight and the *required* empty weight gives the "surplus empty weight."

The objective of the calculation is to achieve a zero surplus empty weight. Therefore, if a surplus weight exists in one iteration, the estimated take-off weight in the next iteration is changed in the direction dictated by the sign (+ or −) of the surplus.

In order to speed up the convergence to the correct solution, a method based on the local slope of the solution from previous iterations is used. Because this method uses the slope, it can only be implemented after the second iteration.

Eventually, an initial estimate of the take-off weight leads to a converged solution, where the surplus weight is zero. With this approach, the final solution has been found to be independent of the initial guessed weight. However, it is possible in some cases, to reach a **nonphysical** solution. This is evident when the converged value for take-off weight is *negative*! When this occurs, it generally requires reducing one or more of the input parameters such as range, payload, TSFC, etc.

2.3.2   Using the Spreadsheet

Using the spreadsheet is relatively easy. The arrow keys or mouse can be used to scan through the rows and columns to see the format of input variables and formulas. There should be enough columns (iterations) for the solution to converge for any type of suitably designed aircraft. The default flight plan was made quite general so that few, if any, changes may need to be made.

Note that the "operating radius" assumes that the point of landing is the same as the point of departure. If this is not the case, then the value of the operating radius should correspond to one-half of the desired range.

The spreadsheet displays a bar graph that shows the estimated take-off weight at each iteration. Convergence occurs when it stops changing from one iteration to the next. If the converged value is negative, the solution is nonphysical and adjustments to the input values need to be made.

2.3.3   Case Study: Take-Off Weight Estimate

This case study corresponds to the conceptual supersonic business jet (SSBJ) described in the design proposal at the end of Chapter 1. In summary, this design is intended to have a cruise Mach number of 2.1 and a cruise altitude of 55,000 feet, with a range of 4000 nm. Its payload would consist of passengers and baggage, with a maximum total weight of 4000 lbs. It is expected to be propelled by four GE-F404-100D engines, or their equivalents. Specifics on these engines were used to obtain an estimate of the thrust-specific fuel consumption. It was intended that composite materials would be used as much as possible in order to reduce the structure factor. As such, a relatively low structure factor of $s = 0.5$ was used in the weight estimate. Other parameters such as aspect ratio were based on comparison aircraft cited in the design proposal.

The spreadsheet output is shown in Figure 2.6. Because there was no intercept flight phase, the cruise Mach number was the same as the maximum Mach number. Since this was a passenger aircraft, the combat time was set to zero. Also there was no expendable payload. It is important to note that the operating radius is set to be one-half of the desired range, because the spread sheet is configured to have cruise out and cruise back flight phases.

An initial guess for the take-off weight of 40,000 lbs was used. This converged to final take-off weight of 90,523 lbs in the third iteration. This is reflected in the iteration history, which is plotted in the figure inset.

Scanning down the column at Iteration 3, one can see how the aircraft weight decrements at the end of each flight phase. Because this is a long-range aircraft, the largest weight change occurs during the cruise phase.

Figure 2.7 documents how the take-off weight would change with the cruise range and Mach number. Sensitivity studies like this are important in determining the impact of different parameters. This study demonstrates a rapid increase in the take-off weight and an increasing sensitivity to cruise Mach number, as the design cruise range increases.

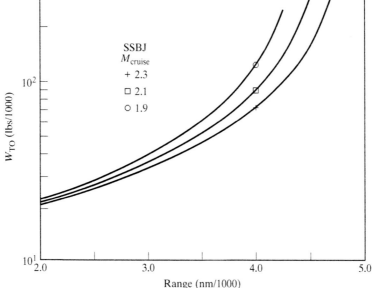

FIGURE 2.7: Effect of take-off weight on range and cruise Mach number for conceptual SSBJ.

The proposed 4000-nm range appears to be near an upper limit before the take-off weight increases more rapidly with range.

2.3.4 Closing Remarks

The approach for determining design parameters using a spreadsheet is quite useful because it allows easy entry of design conditions and provides immediate feedback. This is intentional so that you can investigate the influence of different parameters on the design. You should try to push your designs to maximize the performance in areas such as range, payload, maximum speed, maximum combat time, etc. The value you get for the take-off weight will be used throughout the design. If you find later that you need to revise it, the spreadsheet makes it easy to change some of the input conditions as needed.

2.4 PROBLEMS

2.1. In the spreadsheet, keeping everything else fixed, make a plot of how the take-off weight changes with the following parameters:
   1. range;
   2. aspect ratio;
   3. payload;
   4. endurance;
   5. cruise altitude;
   6. cruise Mach number.
   Discuss the trends, citing the underlying theory.

2.2. In the spreadsheet, for the given case study, what is the largest structure factor possible?

2.3. Consider a jet-powered combat aircraft with the following characteristics:
   1. cruise Mach number $= 2.1$;
   2. max. Mach number $= 1.9$;
   3. cruise alt. $= 60,000$ ft;
   4. oper. rad. $= 300$ nm;
   5. engine TSFC (min/max) $= 0.8/1.8$;
   6. thrust $= 22,000$ lbs;
   7. aspect ratio $= 2.4$;
   8. combat time $= 8$ min;
   9. combat alt. $= 20,000$ ft;
   10. loiter time $= 20$ min;
   11. loiter altitude $= 10,000$ ft;
   12. structure factor $= 0.5$;
   13. exp. payload $= 0$ lbs;
   14. nonexp. payload $= 600$ lbs.
   Determine the final take-off weight. How is this changed if the expendable payload is 500 lbs?

2.4. For the conditions of Problem 2.3, make a plot of how the take-off weight changes with the following parameters:
   1. range;
   2. combat time;
   3. combat altitude;
   4. thrust;
   5. expendable payload;
   6. structure factor;
   7. maximum Mach number.
   Discuss the trends, citing the underlying theory.

2.5. Input the conditions for a Boeing 747 aircraft into the spreadsheet. Most of these can be found in "Jane's All the World Aircraft." Others can be estimated from figures and tables in the textbook. How does the maximum range for this aircraft compare to that of the combat aircraft in Problem 2.3 or the supersonic business jet in the case study? What are the controlling characteristics that determine the maximum range?

C H A P T E R   3

Wing Loading Selection

Photograph of the Gossamer Albatross, which was the first human powered aircraft to fly across the English Channel. The wing loading on this aircraft was only 0.44 lbs/f². (NASA Dryden Research Center Photo Collection.)

TABLE 3.1: Wing loading for aircraft
with different mission requirements.

Mission Requirement	$(W/S)_{TO}$ lbs/f²
Long Range	125 ± 15
Short/Medium Range	95 ± 15
Short TO & L	65 ± 25
Light Civil	20 ± 10
Combat Fighter	55 ± 15
Combat Intercept	135 ± 15
High Altitude	45 ± 15

Once the weight estimate for the conceptual aircraft is completed for each phase of the
flight plan, the next step in the design is the selection of the wing loading. The wing
loading is defined as the ratio of the gross weight of the aircraft to the planform area of
the primary lifting surface, W/S. In most designs, the primary lifting surface is the main
wing, and S is the wing planform area.

The wing loading is selected by considering the principle mission objectives of the
aircraft. All of the following parts of a flight plan are affected by the wing loading:

1. take-off and landing;
2. climb and acceleration;
3. range;
4. combat;
5. flight ceiling;
6. glide rate.

In some cases, the wing loading that optimizes one of these parts has a detrimental effect
on another. For example long-range commercial aircraft traditionally have a higher wing
loading to maximize their range, whereas combat aircraft tend to have lower wing loading to provide better maneuverability. Multipurpose aircraft with more than one principle
mission sometimes require compromises in regards to the wing loading selection. Alter-
natives may lead to variable wing geometries such as flap extensions used at take-off
and landing. Table 3.1 gives values for wing loading at take-off for a variety of aircraft
types. Even the lowest of these is 25 times higher than that of the Gossamer Albatross
pictured at the start of this chapter.

The effect of wing loading on the six flight phases listed earlier will be examined
in detail in the following sections.

3.1 WING LOADING EFFECT ON TAKE-OFF

The wing loading affects take-off through the stall speed, which is defined as

$$V_s = \left[\frac{W}{S} \frac{2}{\rho C_{L_{max}}} \right]^{0.5},$$

(3.1)

where $C_{L_{max}}$ is the maximum 3-D lift coefficient for the aircraft. Methods for augmenting the lift and estimating the 3-D lift coefficient will be covered in Chapter 9.

The velocity required for take-off is defined as

$$V_{TO} = 1.2 V_s = 1.2 \left[\left(\frac{W}{S} \right)_{TO} \frac{2}{\rho C_{L_{max}}} \right]^{0.5}. \tag{3.2}$$

A more refined estimate of the take-off distance will be performed later in the design in Chapter 8 when more of the relevant parameters have been determined. At this point, we wish to demonstrate the influence of wing loading on take-off distance and get the *first estimate* of take-off distance. For this, we use historical data and a take-off parameter, TOP, which has been found to correlate the take-off distances for a wide range of aircraft. The TOP is defined as

$$\text{TOP} = \left(\frac{W}{S} \right)_{TO} \frac{1}{C_{L_{max}}} \left(\frac{W}{T} \right)_{TO} \frac{1}{\sigma}, \tag{3.3}$$

where σ is the ratio of the air density at the take-off site to that at sea level,

$$\sigma = \frac{\rho_{TO}}{\rho_{SL}}. \tag{3.4}$$

Note that the thrust-to-weight ratio, T/W, is a function of altitude as well.

With this correlating factor, the empirical estimate of the take-off distance, s_{TO} is

$$s_{TO} = 20.9(\text{TOP}) + 87\sqrt{(\text{TOP})(T/W)}, \tag{3.5}$$

where the first and second coefficients have units of $f/(lb/f^2)$ and $f/(lb/f^2)^{0.5}$, respectively.

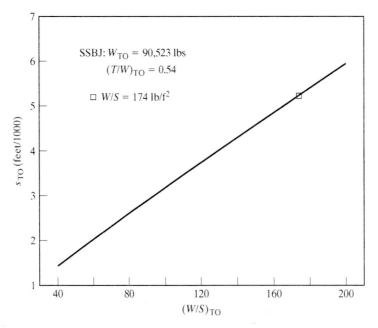

FIGURE 3.1: Effect of wing loading on take-off distance for conceptual SSBJ.

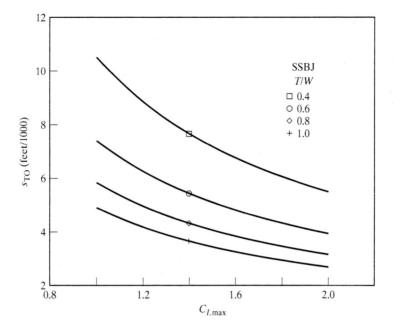

FIGURE 3.2: Effect of maximum lift coefficient and thrust-to-weight on take-off distance for a fixed wing loading of 174 lb/f^2 on conceptual SSBJ.

It is clear from Eq. [3.5] that having an excessively large wing loading at take-off can lead to larger take-off distances. This is illustrated in Figure 3.1, which shows the take-off distance versus wing loading for the case study SSBJ. This is based on a given take-off weight of 90,523 lbs, a thrust-to-weight ratio of 0.54, and $C_{L_{\max}} = 1.6$. For this T/W, the take-off distance is nearly linear with wing loading. The symbol represents an expected wing loading for this aircraft of 174 lb/f^2, which then gives a take-off distance of approximately 5200 feet.

Two parameters that can be used to control the take-off distance are thrust-to-weight ratio and the maximum lift coefficient. This is illustrated in Figure 3.2 for the conceptual SSBJ with the wing loading of 174 lb/f^2. The curves correspond to the effect

TABLE 3.2: Maximum lift coefficient and thrust-to-weight ratio for different aircraft types.

Mission Requirement	$\left(C_{L_{\max}}\right)_{\mathrm{TO}}$	$(T/W)_{\mathrm{TO}}$
Long Range	1.6–2.2	0.20–0.35
Short/Medium Range	1.6–2.2	0.30–0.45
Short TO & L	3.0–7.0	0.40–0.60
Light Civil	1.2–1.8	0.25–0.34
Combat Fighter	1.4–2.0	0.60–1.30

of C_{L_max} on take-off distance for a fixed T/W. The effect of T/W is larger for smaller lift coefficients and does provide an effective means of reducing the take-off distance. However, too large of a thrust-to-weight ratio can lead to poor fuel economy at cruise.

At this point in the design, estimates for T/W and C_{L_max} can be obtained from comparison aircraft. Values for different aircraft types are listed in Table 3.2.

3.2   WING LOADING EFFECT ON LANDING

As with take-off, a more refined estimate of the landing distance will be performed later. At this point, we again utilize historical data that has lead to a correlating factor called the landing parameter that relates the wing loading to the landing distance:

$$\text{LP} = \left(\frac{W}{S}\right)_L \frac{1}{\sigma C_{L_\text{max}}}. \tag{3.6}$$

With this correlating factor, the empirical estimate for the landing distance, s_L is

$$s_L = 118(\text{LP}) + 400, \tag{3.7}$$

where the first and second coefficients have units of $\text{f}/(\text{lb}/\text{f}^2)$ and feet, respectively.

Eq. [3.7] indicates that shorter landing distances can be accomplished by a combination of lower wing loading at landing and a higher C_{L_max}. Since the wing loading affects other parts of the flight plan, obtaining a higher lift coefficient is generally the approach used to minimize the landing distance.

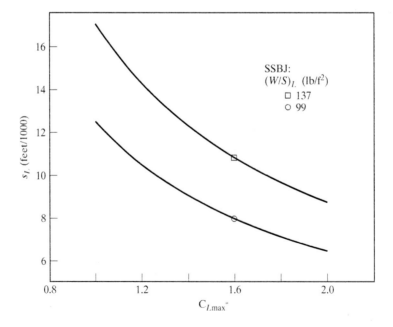

FIGURE 3.3: Effect of maximum lift coefficient and wing loading on landing distance for conceptual SSBJ.

The effect of the maximum lift coefficient on the landing distance is illustrated in Figure 3.3 for the conditions of the conceptual SSBJ. This is shown for two different wing loadings. The lower is representative of the aircraft after flying its maximum range. The latter corresponds to the wing loading if it were to land with half of its take-off fuel weight left.

3.3   WING LOADING EFFECT ON CLIMB

The rate of climb of an aircraft is the vertical velocity given as

$$\frac{dH}{dt} = V \sin \gamma = \frac{P_s}{1 + \frac{V}{g}\frac{dV}{dH}}, \tag{3.8}$$

where P_s is the excess power given as

$$P_s = V\frac{(T-D)}{W}. \tag{3.9}$$

This is schematically represented in Figure 3.4.

If the aircraft climbs at a constant speed so that $dV/dH = 0$, Eqs. [3.8 & 3.9] simplify and combine to give

$$G = \sin \gamma = \frac{(T-D)}{W}, \tag{3.10}$$

where G is called the climb gradient.

The climb gradient represents the ratio between the vertical and horizontal distance traveled by the aircraft. Eq. [3.10] can be rearranged to solve for D/W, namely,

$$\frac{D}{W} = \frac{T}{W} - G. \tag{3.11}$$

For a subsonic climb, the total drag is the sum of the base drag, with drag coefficient C_{D_0}, and the lift-induced drag. Therefore,

$$\frac{D}{W} = \frac{1}{W}\left[qSC_{D_0} + qS\left(C_L^2/\pi Ae\right)\right], \tag{3.12}$$

where q is the dynamic pressure, $\rho V^2/2$; A is the aspect ratio; and e is Oswald's coefficient, as before.

Substituting for the lift coefficient, C_L, Eq. [3.12] becomes

$$\frac{D}{W} = \frac{qC_{D_0}}{W/S} + \frac{W}{S}\frac{1}{q\pi Ae}, \tag{3.13}$$

where we can now easily identify the wing loading, W/S.

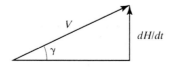

FIGURE 3.4: Coordinate frame for climb.

TABLE 3.3: Take-off climb specifications.

	MIL-C5011A Military	FAR Part 23 Civil	FAR Part 25 Commercial
Gear Up, AEO	500 fpm at SL	300 fpm at SL	—
Gear Up, OEI	100 fpm at SL	—	3% at V_{CL}
Gear Down, OEI	—	—	0.5% at V_{CL}

AEO = all engines operating
OEI = one engine inoperative

Equating Eq. [3.11] and Eq. [3.13] to eliminate D/W and solving for wing loading, we obtain

$$\frac{W}{S} = \frac{[(T/W) - G] \pm \left[[(T/W) - G]^2 - [4C_{D_0}/\pi Ae]\right]^{0.5}}{2/q\pi Ae} \tag{3.14a}$$

with the condition that

$$\frac{T}{W} \geq G + 2\sqrt{\frac{C_{D_0}}{\pi Ae}}. \tag{3.14b}$$

FAR or military requirements specify the rate of climb for different aircraft types, under different conditions, such as one engine out, or landing gear up or down. An example of these for take-off climb are given in Table 3.3.

Eq. [3.14] can be used in the following way for the selection of the wing loading. The first step is to choose the appropriate climb rate, dH/dt, based on FAR or military specifications. Next the minimum thrust-to-weight ratio that satisfies Eq. [3.14b] is calculated. In this, $G = (dH/dt)/V$, where V is the velocity that is appropriate to the climb conditions. For example, $V = V_{TO}$ is the representative velocity for the take-off climb specifications in Table 3.3. Lastly, Eq. [3.14a] is used to determine the wing loading for these conditions.

An example of these calculations is shown in Figure 3.5 for the conceptual SSBJ. This shows the minimum thrust-to-weight ratio for a range of climb rates. The two lines correspond to two different climb Mach numbers, which define the climb velocity, V. The lower Mach number corresponds to the expected take-off velocity of the aircraft. The required wing loading in this case is 62 lbs/f^2. The other line corresponds to a higher Mach number. In this case, for the same climb rate, the minimum T/W is lower, but the required wing loading is higher. This sets a range of conditions that need to be satisfied in the proposed design.

3.4 WING LOADING EFFECT ON ACCELERATION

The approach to maximize the acceleration is to maximize the excess power, P_s. The excess power is defined as the difference between the available power delivered by the engines and the required power needed to overcome drag, namely,

$$P_s = P_a - P_r. \tag{3.15a}$$

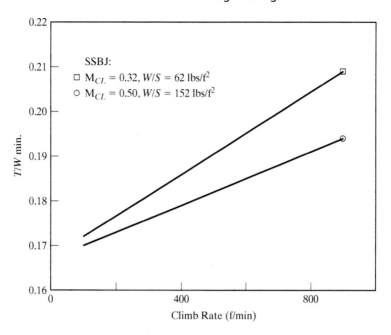

FIGURE 3.5: Effect of aircraft climb rate on minimum thrust-to-weight ratio and wing loading for conceptual SSBJ at take-off conditions.

where

$$P_a = \frac{VT}{W} \qquad (3.15b)$$

and

$$P_r = \frac{VD}{W}. \qquad (3.15c)$$

When the definitions for P_a and P_r are substituted into Eq. [3.15a], the result is identical to Eq. [3.9].

Since the available power is generally limited, an approach that maximizes the excess power is to minimize the required power. Therefore, it is useful to determine the conditions that lead to a minimum D/W.

For a subsonic aircraft, the drag is given by Eq. [3.12]. Introducing k from Eq. [2.9] this becomes

$$D = qS\left[C_{D_0} + kC_L^2\right]. \qquad (3.16)$$

By introducing the load factor, n, defined as

$$n = \frac{L}{W} = \frac{C_L qS}{W}, \qquad (3.17)$$

Eq. [3.17] gives a form for C_L, which involves the wing loading. Substituting this into Eq. [3.16], we obtain

$$\frac{D}{W} = q\frac{S}{W}\left[C_{D_0} + k\left(\frac{n}{q}\frac{W}{S}\right)^2\right]. \qquad (3.18)$$

TABLE 3.4: Wing loading for maximum acceleration for different load factors.

Mission	H (f)	M	q (lb/f^2)	C_{D_0}	k	n	W/S (lb/f^2)
Intercept	25,000	0.8	352	0.025	0.17	1	135
Combat	25,000	0.8	352	0.025	0.17	7	19

In Eq. [3.13], the dependent variable is D/W, and the independent variable with which we are concerned is the wing loading, W/S. Our objective is to minimize the D/W with respect to W/S. This leads to the following relation:

$$\frac{\partial (D/W)}{\partial (W/S)} = 0 = -\left[C_{D_0} + k \left(\frac{n}{q} \frac{W}{S} \right)^2 \right] \left(\frac{W}{S} \right)^{-1} + 2k \left(\frac{W}{S} \right) \left(\frac{n}{q} \right)^2 . \quad (3.19)$$

The condition for wing loading that satisfies Eq. [3.19] is

$$\frac{W}{S} = \frac{q}{n} \sqrt{\frac{C_{D_0}}{k}} = \frac{q}{n} C_{L_{\min D}} . \quad (3.20)$$

Recall that this is the wing loading that minimizes D/W and thereby maximizes excess power.

Eq. [3.20] is used by first deciding on the design load factor, n. At a given altitude and Mach number, the dynamic pressure is then calculated. The parameters k and C_{D_0} are specific to the design and can be estimated at this stage by comparing the design to similar aircraft. Table 3.4 demonstrates such calculations for the same aircraft with different principle mission objectives. The point of this is that the selection of the design load factor dictates the wing loading that is needed for maximum acceleration. Unless a variable wing geometry is used, it is unlikely that an aircraft can have have a primary mission that involves both high-speed intercept and high-load-factor maneuvering. The F14A, such as shown in Figure 3.6, is an example of a variable wing aircraft that was designed to accomplish both.

FIGURE 3.6: Photograph of F14A showing wings in swept out position during landing. (U.S. Navy photograph.)

3.5   WING LOADING EFFECT ON RANGE

The principle formula that defines the range is the Brequet range equation, which was given in Eq. [2.6]. Based on this, for maximum range, $(V/C)(L/D)$ should be a maximum. As pointed out in Chapter 2, at cruise, $L/D \simeq (L/D)_{\max}$. However, we can realize a further optimum by recognizing that the thrust-specific fuel consumption, C, is a function of altitude. Turbo-jet engines generally have an optimum altitude where fuel consumption is a minimum. The wing loading enters into this optimization because it determines the altitude where lift equals weight.

When range is a principle mission objective, the wing loading should be selected so that the altitude where weight equals lift corresponds to where the thrust-specific fuel consumption is also a minimum.

To illustrate this, we consider the conditions of the conceptual supersonic business jet (SSBJ). The top plot in Figure 3.7 shows the variation in the weight of the aircraft with range up to the maximum range. For a fixed wing area, the change in weight results in a reduction in the wing loading during cruise. The corresponding wing loading is read on the right axis.

At cruise, weight equals lift so that the wing loading is

$$\left(\frac{W}{S}\right)_{\text{cruise}} = qC_L. \tag{3.21}$$

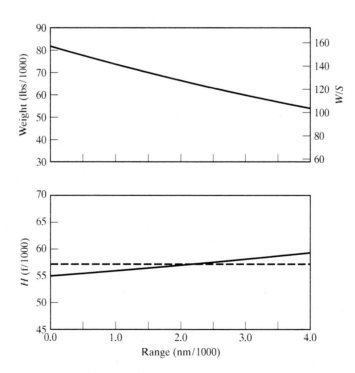

FIGURE 3.7: Variation in wing loading with cruise range and its effect on cruise altitude for conceptual SSBJ.

The lift coefficient can be estimated at this stage using

$$C_L = \sqrt{\frac{C_{D_0}}{2k}}. \tag{3.22}$$

Therefore, substituting for C_L and q, the wing loading at cruise is

$$\left(\frac{W}{S}\right)_{\text{cruise}} = \frac{\rho V^2}{2} \sqrt{\frac{C_{D_0}}{2k}}. \tag{3.23}$$

In Eq. [3.23], assuming that the speed of the aircraft is constant, a change in the wing loading is only balanced by a change in the density, ρ, which is a function of altitude. Therefore, as the wing loading decreases, the aircraft will naturally rise to an altitude where the lift again balances the weight. The lower plot in Figure 3.7 shows how the altitude would change during cruise for the SSBJ. In this case, for most efficient cruise, the engines should be designed to have the minimum TSFC at an average altitude of approximately 57,000 feet.

3.6 WING LOADING EFFECT ON COMBAT

Wing loading enters the combat capability of an aircraft through the instantaneous and sustained turn rates. The instantaneous turn rate is the highest turn rate possible while ignoring loss of altitude or speed. The sustained turn rate is the turn rate for some flight condition at which the thrust is just sufficient to maintain velocity and altitude in a turn.

The turn rate is defined as $\dot{\psi} = d\psi/dt$. A turn rate of 2 degrees per second is considered significant.

3.6.1 Instantaneous Turn Rate

The instantaneous turn rate is limited only by the amount of usable maximum lift, given that $L = W$. The turn rate is dependent on the load factor, n, as

$$\dot{\psi} = \frac{g\sqrt{n^2 - 1}}{V}, \tag{3.24}$$

where g is the gravitational constant. Solving for the wing loading, gives

$$n = \sqrt{\left(\frac{\dot{\psi} V}{g}\right)^2 + 1}. \tag{3.25}$$

The wing loading is introduced into Eq. [3.25] by substituting

$$n = \frac{q C_{L_{\max}}}{W/S}. \tag{3.26}$$

This gives

$$\frac{W}{S} = \frac{q C_{L_{\max}}}{\sqrt{\left(\frac{\dot{\psi} V}{g}\right)^2 + 1}}. \tag{3.27}$$

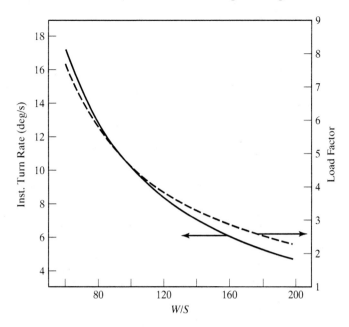

FIGURE 3.8: Variation in instantaneous turn rate and load factor with wing loading for the combat aircraft conditions given in Table 3.4.

The load factor in Eq. [3.27] is usually specified by the structure design or by the limits set on the pilot. Maximum values of $5 \leq n \leq 7$ are fairly typical for piloted combat aircraft. $C_{L_{max}}$ for combat is not the same as that used for take-off and landing. More typically $1.0 \leq C_{L_{max}} \leq 1.5$ for combat maneuvers. Using the appropriate values of n and $C_{L_{max}}$ for the design, the desired instantaneous turn rate can be input into Eq. [3.27] to determine the necessary wing loading.

An example is shown in Figure 3.8. This corresponds to the conditions of the combat aircraft listed in Table 3.4, along with $C_{L_{max}} = 1.2$. This shows the effect of wing loading on the instantaneous turn rate (solid curve) and load factor (dashed curve). It demonstrates that greater maneuverability comes with lower wing loading. Under such conditions, the pilot and aircraft structure have to withstand higher loads.

For an aircraft with design objectives that do not include combat, wing loading that was obtained to optimize other flight phases can be input into Eq. [3.27] in order to estimate the instantaneous turn rate.

3.6.2 Sustained Turn Rate

In a sustained turn, speed is maintained so that $T = D$, and altitude is maintained, so that $n = L/W$ is constant. The maximum sustained turn rate occurs at maximum available thrust.

Substituting $T = D$ into Eq. [3.12] gives

$$\frac{T}{W} = \frac{1}{W}\left[qSC_{D_0} + qS\left(C_L^2/\pi Ae\right)\right]. \tag{3.28}$$

This can be put in terms of the load factor n by substituting Eq. [3.17] into Eq. [3.28] to obtain

$$\frac{T}{W} = \frac{qC_{D_0}}{W/S} + \frac{W}{S}\left(\frac{n^2}{q\pi Ae}\right). \tag{3.29}$$

One objective could be to maximize the T/W with respect to wing loading, W/S, keeping n constant. This is equivalent to minimizing D/W in Eq. [3.18] and leads to the same condition for wing loading given by Eq. [3.20].

The alternate approach is to consider the wing loading needed for a sustained load factor using all the available thrust. This can be found by solving for the wing loading in Eq. [3.29]. This gives

$$\frac{W}{S} = \frac{(T/W) \pm \left[(T/W)^2 - \left(4n^2 C_{D_0}/\pi Ae\right)\right]^{0.5}}{2n^2/q\pi Ae}. \tag{3.30}$$

Note that this is the same as Eq. [3.14] for $G = 0$ and $n = 1$. In the present case, T/W would be the maximum available for combat. At its minimum,

$$\frac{T}{W} \geq 2n\sqrt{\frac{C_{D_0}}{\pi Ae}} \tag{3.31}$$

for W/S to be real in Eq. [3.30].

Now Eq. [3.29] can be rewritten to express the maximum sustained load factor, $n_{\max-s}$, where

$$n_{\max-s}^2 = \frac{q\pi Ae}{W/S}\left[\left(\frac{T}{W}\right)_{\max} - \frac{qC_{D_0}}{W/S}\right]. \tag{3.32}$$

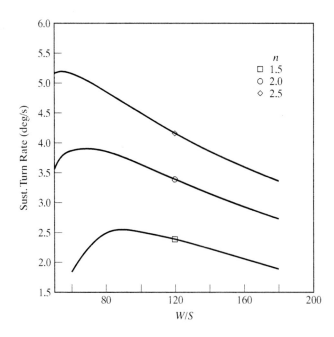

FIGURE 3.9: Variation in sustained turn rate with wing loading and load factor for the combat aircraft conditions given in Table 3.4.

This can then be substituted into Eq. [3.23] to obtain the maximum possible sustained turn rate:

$$\dot{\psi}_{\max -s} = \frac{g\sqrt{\frac{q\pi Ae}{W/S}\left[\left(\frac{T}{W}\right)_{\max} - \frac{qC_{D_0}}{W/S}\right] - 1}}{V} \tag{3.33}$$

An example of the dependence of the sustained turn rate on wing loading and load factor is shown in Figure 3.9. This again corresponds to the conditions of the combat aircraft listed in Table 3.4. Each load factor corresponds to a minimum T/W, which satisfies Eq. [3.31]. For each load factor there is a wing loading that gives the maximum sustained turn rate. The wing loading where this occurs is smaller as the load factor and thrust-to-weight ratio increase. In general, at any wing loading, higher load factors allow higher sustained turn rates.

3.7 WING LOADING EFFECT ON FLIGHT CEILING

The wing loading enters into the determination of the maximum ceiling through the equation for the load factor, such as Eq. [3.17]. For level flight, $n = 1$ so that

$$\frac{W}{S} = C_L q. \tag{3.34}$$

If high-altitude flight is a principle design objective, then Eq. [3.33] is used to determine the wing loading required to achieve a given altitude for a specified Mach number and wing lift coefficient. The lift coefficient should be in the range of $1.0 \leq C_L \leq 1.5$.

For a given wing loading, Mach number, and lift coefficient, the values of air density with altitude determine the maximum flight ceiling. This is illustrated in Figure 3.10 for conditions relevant to the conceptual SSBJ. Plotted is the maximum altitude as a function of wing loading for a fixed $C_L = 1.0$ at a cruise Mach number of 2.1. In general, this demonstrates that high-altitude flight requires aircraft with lower wing loading.

3.8 WING LOADING EFFECT ON GLIDE RATE

For a maximum range during a gliding descent, a minimum glide angle is required. This is achieved when L/D is a maximum. A relation for the drag is given by Eq. [3.12]. To find the condition that maximizes L/D, we find the C_L that minimizes the drag. In performing this operation, we note that q is a function of C_L, namely,

$$q = \frac{1}{C_L}\frac{L}{S} = \frac{1}{C_L}\frac{W}{S} \quad ; \text{ for } n = 1. \tag{3.35}$$

Upon substitution, this gives

$$D = W\frac{C_{D_0}}{C_L} + WkC_L. \tag{3.36}$$

We seek the condition where

$$\frac{\partial D}{\partial C_L} = 0 = \frac{-C_{D_0}}{C_L^2} + k. \tag{3.37}$$

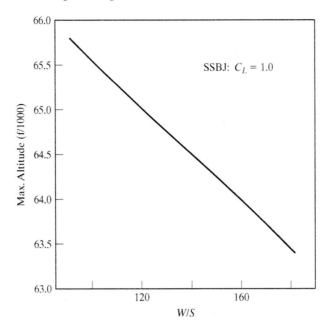

FIGURE 3.10: Variation in maximum altitude with wing loading for $C_L = 1.0$ and a cruise Mach number of 2.1 for conceptual SSBJ.

This gives the condition for C_L that minimizes the drag, namely,

$$C_{L_{\min D}} = \sqrt{\frac{C_{D_0}}{k}}. \tag{3.38}$$

The wing loading enters into the minimum glide condition through the velocity for $C_{L_{\min D}}$. This is given as

$$V_{\min D} = \sqrt{\frac{2}{\rho C_{L_{\min D}}} \frac{W}{S}} = \sqrt{\frac{2}{\rho} \frac{W}{S}} \left[\frac{k}{C_{D_0}} \right]^{0.25}. \tag{3.39}$$

Therefore, the velocity for the most efficient glide is a function of the wing loading given by Eq. [3.39].

We next consider the rate of descent of the aircraft, which following Section 3.3, is the vertical velocity given as

$$\frac{dH}{dt} = V \sin \gamma. \tag{3.40}$$

A schematic for descent is shown in Figure 3.11

Substituting for V in terms of wing loading for $n = 1$, the descent velocity is

$$\frac{dH}{dt} = \sin \gamma \sqrt{\frac{W}{S} \frac{2}{\rho C_L / \cos \gamma}}. \tag{3.41}$$

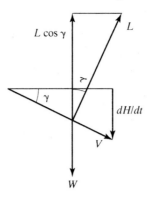

FIGURE 3.11: Coordinate frame for descent.

From a free-body diagram of the forces on the aircraft,

$$\sin \gamma = \frac{D}{L} \cos \gamma. \tag{3.42}$$

Substituting this, using C_L and C_D, into Eq. [3.41] gives

$$\frac{dH}{dt} = \sqrt{\frac{W}{S} \frac{2 \cos^3 \gamma}{\rho C_L^3 / C_D^2}}, \tag{3.43}$$

which for small angles of descent is

$$\frac{dH}{dt} = \sqrt{\frac{W}{S} \frac{2}{\rho C_L^3 / C_D^2}}. \tag{3.44}$$

As with climb, for subsonic descent, the total drag is the sum of the base drag and lift-induced drag. Therefore,

$$C_D = C_{D_0} + k C_L^2. \tag{3.45}$$

Substituting this into Eq. [3.44] gives

$$\frac{dH}{dt} = \sqrt{\frac{W}{S} \frac{2}{\rho} \frac{(C_{D_0} + k C_L^2)^2}{C_L^3}}. \tag{3.46}$$

The condition for the minimum descent velocity is found by taking the partial derivative of Eq. [3.46] with respect to C_L. This gives

$$C_{L_{\text{min descent}}} = \sqrt{\frac{3 C_{D_0}}{k}}. \tag{3.47}$$

Substituting Eq. [3.47] into Eq. [3.46] gives the minimum descent velocity in terms of the wing loading, namely,

$$\left(\frac{dH}{dt} \right)_{\text{min}} = 4 \left[\frac{2}{\rho} \frac{W}{S} \right]^{\frac{1}{2}} \left[\frac{k}{3} \right]^{\frac{3}{4}} \left[C_{D_0} \right]^{\frac{1}{4}}. \tag{3.48}$$

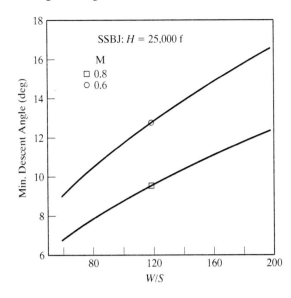

FIGURE 3.12: The effect of wing loading on the minimum glide descent angle for conceptual SSBJ at an altitude of 25,000 feet for two Mach numbers.

The effect of wing loading on the minimum glide angle is demonstrated in Figure 3.12 for the conceptual SSBJ at an altitude of 25,000 feet, for two different subsonic Mach numbers. In general, the results illustrate the need for low wing loading in gliding aircraft.

3.9 SPREADSHEET APPROACH FOR WING LOADING ANALYSIS

A spreadsheet has been constructed to implement the calculations used to estimate the effect of wing loading on the different parts of the aircraft performance. This incorporates the different equations that were presented in this chapter. It includes analysis of take-off and landing, the start and end of cruise, climb and acceleration, instantaneous and sustained turns, and maximum ceiling and glide descent rate. The spreadsheet is called **wingld.xls**. The following describes the spreadsheet structure.

3.9.1 Spreadsheet Structure

The spreadsheet is organized according to the different flight phases. These are denoted by the bold labels. For each of these, the input variables and computed output values are separated by a blank row. The input variables are placed in the top portion. For example, for take-off, the input variables are altitude, H; maximum lift coefficient, $C_{L_{max}}$; maximum thrust, T_{max}; and the take-off weight, W_{TO}. The calculated values are the wing area, S; density ratio, σ; thrust-to-weight ratio, T/W; take-off parameter, TOP; and take-off distance, S_{TO}. All of the dimensions are listed next to the variable.

The determination of velocity as a function of altitude and Mach number follows the same approach used in the spreadsheet in Chapter 2. In addition, a piecewise linear approximation of the change in density, ρ, with altitude was used throughout the spreadsheet. This can be viewed by clicking on any cell containing the numerical value of ρ. The formula has three different functions relating altitude and density based on the conditions: $H < 10,000$ f; $10,000$ f $\leq H \leq 40,000$ f; and $H > 40,000$ f. For the density ratio, σ, the standard density at sea level was $\rho_{SL} = 0.076474$ lbm/f^3. It is also important to note that because of the units of density, whenever it is involved in a formula, $g_c = 32.2$ f-lbm/lbf-s^2 was used to convert units. An example of this is in the calculation of dynamic pressure, q. This calculation can be seen by viewing the formula by clicking on any cell containing the numerical value of q.

In the take-off and landing analysis, the altitude, H, refers to the altitude of the airport. It was assumed that these altitudes would always be below 10,000 feet. Therefore the formula used in relating altitude to density only includes the lower range, $H < 10,000$ f.

There are a few variables that are global to all the flight phases. These are the base drag coefficient, C_{D_0}; the wing area, S; aspect ratio, A; and $k = 1/\pi Ae$, with $e = 0.8$. The wing area is calculated based on the wing loading that gives the most efficient cruise, using Eqs. [3.22] and [3.23], and the weight at the start of cruise. The wing area is kept fixed throughout the spreadsheet, although it can be overwritten if desired.

3.9.2   Case Study: Wing Loading Analysis

A spreadsheet was applied to the analysis of the conceptual supersonic business jet described in the design proposal at the end of Chapter 1. The spreadsheet output is shown in Figures 3.13 and 3.14. The previous spreadsheet, which determined the weight at the different flight phases, was used to provide the input weights for the present spreadsheet.

For this design, the range is the primary design driver. As a result, the wing loading was selected to give optimum cruise conditions. This is based on Eq. [3.23]. For the conditions listed under "Cruise Start," the optimum wing loading is 157 lb/f^2. With this wing loading and weight at the start of cruise, the wing area is 519 f^2. Because cruise is considered the most important part of this design, this wing area is now fixed.

For the full cruise range, the weight of the aircraft decreases by approximately 30 percent. This weight is listed under "Cruise End." At the end of cruise, the actual wing loading, based on the actual weight and wing area, denoted as W/S_{actual}, decreases to approximately 104 lb/f^2.

At the lower wing loading, the tendency of the aircraft will be to increase altitude to where the lift again balances the weight. To determine where this occurs, the altitude input under "Cruise End" can be changed. The object is to find the altitude where W/S_{ideal} equals W/S_{actual}. For this design, that altitude is 59,300 feet.

We now consider take-off and landing. The take-off weight determined in the previous spreadsheet was 90,523 lbs. For the wing area determined for cruise, the wing loading at take-off is 174 lb/f^2.

A relatively conservative $C_{L_{max}} = 1.6$ was selected. Based on Table 3.2, this is on the low end of comparable aircraft. However, because the wings will be highly swept for supersonic flight, enhanced lift devices such as flaps, will not be as effective.

Take-off			Landing		
H (f)	1000		H (f)		1000
CL(max)	1.6		CL(max)		1.6
T(max) (lb)	49000		T(max) (lb)		49000
W_TO (lb)	90523		W_L (lb)		51597
S (f^2)	519.44		W/S (lb/f^2)		99.33
W/S (lb/f^2)	174.27		S (f^2)		519.44
SIGMA	0.97		SIGMA		0.97
T/W	0.54		T/W		0.95
TOP	206.62		LP		63.75
S_TO (f)	5238.44		S_TO (f)		7922.42
Cruise Start			**Cruise End**		
CD_0	0.04		CD_0		0.04
A	2		A		2
H (f)	55,000		H (f)		59,300
Cruise Mach	2.1		Cruise Mach		2.1
W (lb)	81817		W (lb)		53,976
k	0.2		k		0.2
V (f/s)	1925.7		V (f/s)		1895
rho (lbm/f^3)	0.01		rho (lbm/f^3)		0.01
q (lbf/f^2)	531.07		q (lbf/f^2)		347.41
W/S (lb/f^2)	157.51		W/S_optimum		103.04
S (f^2)	519.44		W/S_actual		103.91
Climb			**Acceleration**		
H (f)	0		H (f)		10,000
Climb Mach	0.32		Cruise Mach		0.8
dH/dt (f/min)	900		n		1.7
V (f/s)	353.28				
G (rad)	0.04		V (f/s)		856
Gamma (deg)	2.43		rho (lbm/f^3)		0.06
rho (lbm/f^3)	0.08		q (lbf/f^2)		642.58
q (lbf/f^2)	148.21		W/S (lb/f^2)		158.54
T/W min	0.209				
T/W min - G	0.17				
W/S_ + (lb/f^2)	62.16				
W/S_ − (lb/f^2)	62.16				

FIGURE 3.13: WINGLD spreadsheet for SSBJ case study (Part 1).

Turn-Inst.			Turn-Sustained	
H (f)	25,000		H (f)	25,000
Cruise Mach	0.8		Cruise Mach	0.8
CL_max	0.5		n	1.22
W/S (lb/f^2)	157		W/S (lb/f^2)	157
V (f/s)	815.2		T/W_max	0.2
rho (lbm/f^3)	0.04		V (f/s)	815.2
q (lbf/f^2)	383.8		rho (lbm/f^3)	0.04
psi_dot (rad/s)	0.03		q (lbf/f^2)	383.8
psi_dot (deg/s)	1.59		psi_dot (rad/s)	0.03
n	1.22		psi_dot (deg/s)	1.52
Ceiling			Decent	
W(lb)	81817		W (lb)	53976
S (f^2)	519.44		S (f^2)	519.44
H (f)	64,050		H (f)	25,000
Cruise Mach	2.1		Cruise Mach	0.8
W/S (lb/f^2)	157.51		W/S (lb/f^2)	103.91
V (f/s)	1861.08		V (f/s)	815.2
rho (lbm/f^3)	0.00292		rho (lbm/f^3)	0.04
q (lbf/f^2)	157.3		q (lbf/f^2)	383.8
CL_required	1		dH/dt-min (f/min)	7572.84
			Gamma (deg)	8.91

FIGURE 3.14: WINGLD spreadsheet for SSBJ case study (Part 2).

The thrust was based on the maximum available thrust for the proposed engines. This gives a $T/W = 0.54$, which is on the high end of medium range aircraft listed in Table 3.2.

The take-off conditions went into the calculation of the take-off parameter (Eq. [3.3]), and take-off distance (Eq. [3.5]). In this, the altitude of the airport was considered to be 1000 feet. This is higher than most of the major airports in the United States. (See Table 1.4.) Based on this, the take-off distance was found to be 5238 feet. This is approximately half of the runway distance of all of the major airports in the United States (Table 1.4).

The landing analysis is based on the landing parameter given in Eq. [3.6]. The input includes the weight at landing and the wing area. The other conditions are taken to be the same as for take-off. Based on these, the landing distance is estimated to be 7922 feet. This again is well within the runway distances of all of the major airports in the United States.

A more refined take-off and landing analysis will be completed in Chapter 8. That analysis generally gives longer take-off distances and shorter landing distances than are estimated by the take-off and landing parameters used here.

The analysis for climb used conditions that would be related to the climb phase that follows take-off. The FAR specifications in Table 3.3 are based on Sea Level so that a zero altitude was used. The climb Mach number was set to 0.32, which corresponds to the take-off velocity ($1.2V_s$) at that altitude.

The climb velocity was set based on FAR specification. The spreadsheet calculates the G based on Eqs [3.8] and [3.10]. From this, it determines the minimum T/W based on Eq. [3.14b]. This uses the global parameters, C_{D_0} and k, which are specified in the "Cruise Start" portion of the spreadsheet. The wing loading for the climb angle is determined from Eq. [3.14a].

The results from the climb analysis indicate that the specified climb rate can be achieved with a minimum $T/W = 0.18$ and a $W/S = 62$ lb/f^2. The actual wing loading during the climb phase is approximately 150 lbs/f^2. As indicated in Figure 3.5 (generated using the spreadsheet), the minimum required T/W needed for the specified climb rate is still well below the $T/W = 0.54$, which the engines are expected to provide. Therefore, these values are acceptable.

The analysis of acceleration is relevant to the acceleration to cruise Mach number part of the flight plan. This is based on Eq. [3.20]. For this, an altitude of 10,000 feet and a Mach number of 0.8 were used as input conditions. The load factor, n, was varied in order to give a wing loading that is close to the value between take-off and the start of cruise. With this, $n = 1.7$ was obtained. This load factor is fairly typical of long-range aircraft. It will be recalled later in Chapter 10 in the design of the structure.

The instantaneous and sustained turn analysis are based on the wing loading at the start of cruise. The altitude was chosen to be 25,000 feet. The Mach number was taken as 0.8. At these conditions, a moderate $C_L = 0.5$ gave an instantaneous turn rate of 1.59°/s and a load factor of 1.22. Both of these are very acceptable for an aircraft of this type.

For comparison, the determination of the sustained turn rate considered the same conditions as those for the instantaneous turn rate. This included the same load factor. The required T/W is based on Eq. [3.31]. This gave a value of $T/W = 0.20$, which is well within the capability of the proposed engines. The sustained turn rate was found to be 1.52°/s.

The maximum flight ceiling is based on the weight at the start of cruise, which defines the wing loading. The lift coefficient was taken to be 1, which is estimated to be the largest value without flaps. These conditions give a maximum ceiling of 64,050 feet. As the wing loading decreases during cruise, the flight ceiling increases as shown in Figure 3.10.

The analysis of the minimum gliding descent considers conditions at the end of cruise that might be representative at the start of a landing approach. The altitude is taken as 25,000 feet, and the Mach number as 0.8. These conditions give a minimum descent angle of 8.91 degrees.

3.10 PROBLEMS

3.1. Using the spreadsheet, determine how the optimum wing loading for cruise changes as a function of
1. aspect ratio;
2. cruise altitude;
3. cruise Mach number.
Discuss the trends.
3.2. Using the spreadsheet, determine how the take-off and landing distances vary with
1. wing loading;
2. altitude of the airport;

3. thrust-to-weight ratio;
4. maximum lift coefficient.
Which of these has the most impact on minimizing the take-off and landing distances?

3.3. For the case study conditions in the spreadsheet, plot the wing loading as a function of climb angle. Explain your result.

3.4. For the case study conditions in the spreadsheet, plot the wing loading as a function of load factor based on acceleration. Based on this, describe an optimum intercept aircraft.

3.5. Using the spreadsheet, determine how the instantaneous turn rate depends on
1. Mach number;
2. altitude;
3. wing loading;
4. maximum lift coefficient.
At a fixed altitude and Mach number, which of the others has the most impact? Based on these, describe an optimum maneuverable aircraft.

3.6. Using the spreadsheet, determine how the sustained turn rate depends on thrust-to-weight ratio and wing loading. Which is a more important characteristic?

3.7. Using the spreadsheet, determine what are the most important characteristics to optimize for a reconnaissance aircraft flying at high altitude?

3.8. Determine how the optimum glide velocity or Mach number changes with altitude. Based on this, describe a flight profile that would maximize the gliding range of an aircraft.

3.9. Consider a jet-powered combat aircraft with the following characteristics:
1. cruise Mach number = 2.1;
2. max. Mach number = 1.9;
3. cruise alt. = 60,000 f;
4. oper. rad. = 300 nm;
5. engine TSFC (min./max.) = 0.8/1.8;
6. thrust = 22,000 lbs;
7. aspect ratio = 2.4;
8. combat time = 8 min;
9. combat altitude = 20,000 f;
10. loiter time = 20 min;
11. loiter altitude = 10,000 f;
12. structure factor = 0.5;
13. exp. payload = 0 lbs;
14. nonexp. payload = 600 lbs.
Determine the wing loading based on the factors that you think are most important. Given this, list the following:
1. take-off and landing distances;
2. instantaneous and sustained turn rates;
3. maximum flight ceiling;
4. load factor.

3.10. Input the conditions for a Boeing 747 aircraft into the spreadsheet. Most of these can be found in "Jane's All the World Aircraft." Others can be estimated from figures and tables in the textbook. What values of do you obtain for the following:
1. instantaneous and sustained turn rates;
2. take-off and landing distances;
3. maximum flight ceiling;
4. load factor.

CHAPTER 4

Main Wing Design

4.1 AIRFOIL CROSS-SECTION SHAPE
4.2 TAPER RATIO SELECTION
4.3 SWEEP ANGLE SELECTION
4.4 3-D LIFT COEFFICIENT
4.5 WING DRAG ESTIMATION
4.6 PLANFORM GEOMETRIC RELATIONS
4.7 SPREADSHEET FOR WING DESIGN
4.8 PROBLEMS

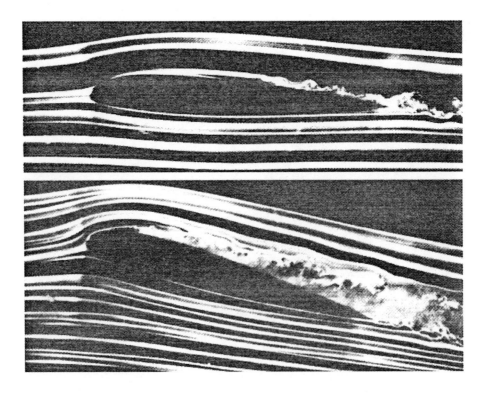

Smoke visualization of the flow over a NACA 23012 airfoil. The upper photograph is for a 3.5° angle of attack, where the flow is attached. The lower photograph is for a 14° angle of attack, where the flow separates (stalls). (From F. N. M. Brown, 1971.)

This chapter deals with the design of the main lifting surface of the aircraft. This is the logical next step in the design of a conceptual aircraft since the weight and wing loading that match the mission requirements have now been determined (Chapters 2 and 3).

In most cases, the main lifting surface is a single wing. The design of the wing consists of selecting the airfoil cross-section shape; the average chord length, $\bar{c}$; the maximum thickness-to-chord, $(t/c)_{max}$; the aspect ratio, $A = b^2/S$; the taper ratio, $\lambda = c_t/c_r$; and the sweep angle, Γ, which is defined for the leading edge as well as the maximum thickness line. A schematic of the wing cross-section and planform views that illustrate these parameters is shown in Figure 4.1. The effect and selection of each of these parameters is presented in the following sections.

Another part of the wing design involves enhanced lift devices such as leading and trailing edge flaps. This will be left to Chapter 9 following the detailed take-off and landing analysis.

In this chapter, a majority of the information used in the selection of the airfoil cross-section shape comes from experimental results. One of the key references is the "Theory of Wing Sections" by Abbott and von Doenhoff. This information comes from

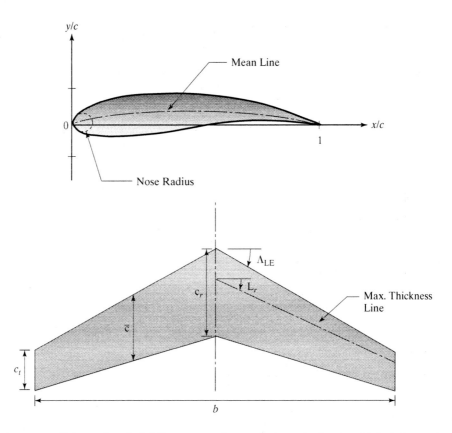

FIGURE 4.1: Schematic of airfoil cross-section and wing planform, which defines wing design parameters.

these authors' 1945 NACA Report 824, which is also available in electronic form on the NASA Internet site.

The ultimate "goals" for the wing design are based on the mission requirements. However, these generally include having a high 3-D lift coefficient versus angle of attack, $dC_L/d\alpha$; a high maximum lift coefficient, $C_{L_{max}}$; a low base drag coefficient, C_{D_0}; a low lift-induced drag; a high wing volume; and a low wing weight. In some cases, these goals are in conflict and will require some compromises.

4.1   AIRFOIL CROSS-SECTION SHAPE

The shape of the wing cross section determines the pressure distribution on the upper and lower surfaces of the wing. The pressure distribution integrated around the wing is the lift force. The lift force normalized by the wing area and dynamic pressure is the lift coefficient, C_l. Note that the lowercase l is used here to denote that this refers to the 2-D section lift coefficient. An uppercase L will be used to signify the lift corresponding to a full 3-D wing.

The lift coefficient increases with angle of attack for an airfoil. A generic representation is shown in Figure 4.2.

Linear airfoil theory indicates that a 2-D (infinite aspect ratio) airfoil section has a lift coefficient, C_l, which is linearly proportional to its angle of attack. The slope of this relation is equal to

$$\frac{dC_l}{d\alpha} = 2\pi \ \text{(radians)}^{-1}. \tag{4.1}$$

This theory suggests that the lift generated by the wing will increase with increasing angle of attack, without limit. In reality, the maximum lift is limited by a maximum angle of attack beyond which the flow can no longer follow the curvature of the airfoil. In this case, the flow is said to be separated, such as in the lower picture at the start of the

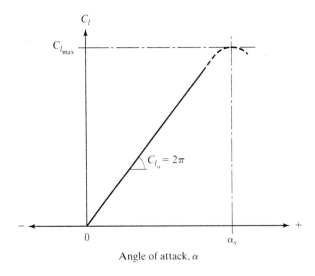

FIGURE 4.2: Lift coefficient versus angle of attack for a symmetric 2-D wing section.

chapter. As a result, the lift coefficient reaches a maximum and then decreases as the angle of attack increases further. The maximum 2-D lift coefficient is denoted as $C_{l_{max}}$. The angle of attack where $C_l = C_{l_{max}}$ is denoted as α_s, where s refers to stall. Stall is represented by a simultaneous loss of lift and increase in drag of an airfoil.

Airfoils with a small leading-edge radius, r_{LE}, can have a tendency for "leading-edge separation" at small angles of attack. This then lowers $C_{l_{max}}$. The selection of the leading-edge radius can therefore be important.

The maximum lift coefficient is also affected by the maximum thickness-to-chord ratio, $(t/c)_{max}$. This is demonstrated in Figure 4.3, which shows $C_{l_{max}}$ as a function of $(t/c)_{max}$ for a variety of common airfoil sections. These data indicate that the largest maximum lift coefficient occurs for $(t/c)_{max} \simeq 14\%$.

Beyond the optimum $(t/c)_{max}$, the decrease in $C_{l_{max}}$ is caused by a flow separation that occurs downstream of the maximum thickness point on the airfoil. This is referred to as "trailing-edge separation." Such flow separations are more likely to occur at lower chord Reynolds numbers. A laminar-separated flow can be induced to reattach through the use of surface roughness, or vortex generators that cause the flow over the airfoil surface to become turbulent or otherwise well mixed.

From a structural point of view, there is a benefit to having thicker wing cross sections. Considering the wing to be a uniform loaded cantilever beam, the maximum deflection is proportional to the thickness to the third power. Therefore, a small increase in the thickness in the wing has a large benefit toward increasing the bending stiffness. This allows the load carrying elements in the wing to be made lighter. As a result, for the same amount of bending, the structure weight of the wing can be reduced by

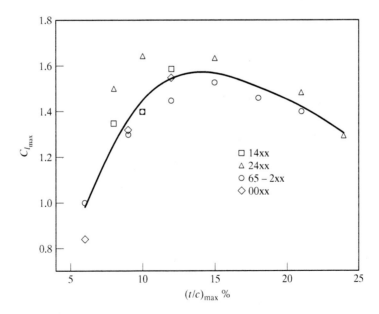

FIGURE 4.3: Effect of the maximum thickness to chord on the maximum lift coefficient for a variety of 2-D airfoil sections. (From NACA Report 824, 1945.)

increasing $(t/c)_{max}$. Historical data indicate that the structure weight of the main wing varies roughly as

$$W_W \propto (t/c)_{max}^{0.5}. \tag{4.2}$$

This depends somewhat on other factors, for example, on the taper ratio and the amount of wing sweep. The design of the wing structure will be discussed in Chapter 10. Also a more explicit formula to predict the weight of the main wing will be presented in Chapter 11 when a refined weight estimate is made.

Increasing the $(t/c)_{max}$ obviously also has the benefit of increasing the internal volume of the wing. In many aircraft, the main wing is used to carry fuel. It also encloses the leading and trailing flaps, and often the landing gear. In some cases, this is a more important design driver than having optimum aerodynamics. Historical data show a predictable outcome that the weight of fuel carried in the main wing, increases as

$$W_f \propto (t/c)_{max}^{1}. \tag{4.3}$$

An aircraft with a subsonic cruise Mach number may have conditions over the wing where the Mach number is transonic or supersonic. This would result in an increase of the base drag, C_{D_0}, of the wing. The cruise Mach number at which the local Mach number over the wing is supersonic is called the critical Mach number, $M_{critical}$. As shown in Figure 4.4, for the same airfoil sections in the previous figure, the critical Mach number for a wing section decreases with $(t/c)_{max}$. Although this problem can be partially offset by using wing sweep, the $(t/c)_{max}$ of wing sections of higher speed aircraft is generally less than the optimum which gives the largest $C_{l_{max}}$. This is illustrated in the historical data plotted in Figure 4.5.

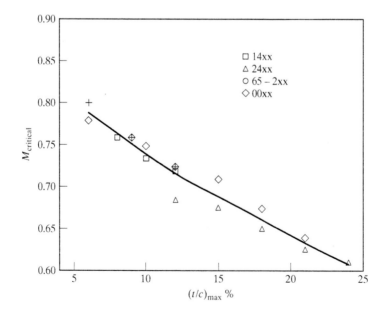

FIGURE 4.4: Effect of the maximum thickness to chord on the critical Mach number for a variety of 2-D airfoil sections.

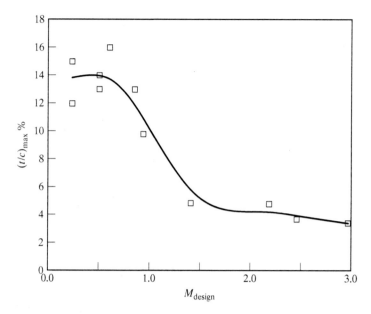

FIGURE 4.5: Historic values of the maximum thickness to chord on the main wing of aircraft as a function of their design cruise Mach numbers. (Data from Raymer, 1992.)

The airfoil sections on supersonic aircraft tend to be thinner and have a smaller nose radius. The purpose is to minimize the base and wave drag components. In addition, the critical Mach number for the wing is designed to be different from that of the other components (horizontal and vertical tail sections and fuselage) so that drag buildup that occurs near Mach 1 does not simultaneously occur for all the components at the same free-stream Mach number.

Camber is a parameter corresponding to a curving of the airfoil section. The mean camber line, illustrated in the airfoil section in Figure 4.1, corresponds to a line that is equidistant from the upper and lower surfaces. The total camber is defined as the maximum distance of the mean camber line from the chord line, which is a straight line drawn from the leading edge to trailing edge. The camber is generally expressed as a percentage of the chord length.

The top part of Figure 4.6 shows the effect of camber on the lift versus angle of attack for a 2-D airfoil. Without camber (solid curve), a wing section produces zero lift at zero angle of attack. In addition, the base drag is a minimum when the angle of attack is zero. This is illustrated in the bottom part of Figure 4.6.

Many airfoil types have a noticeable depression near the minimum drag coefficient. This depression is referred to as the "drag bucket."

The effect of positive camber is to shift the angle of attack for zero lift to negative values. The zero lift angle of attack is referred to as α_{0_L}. Note that an uppercase L is used here because the value will not change when 3-D effects are included.

A more important effect of camber is illustrated in the bottom part of Figure 4.6. This shows that positive camber shifts the value of the lift coefficient for minimum drag

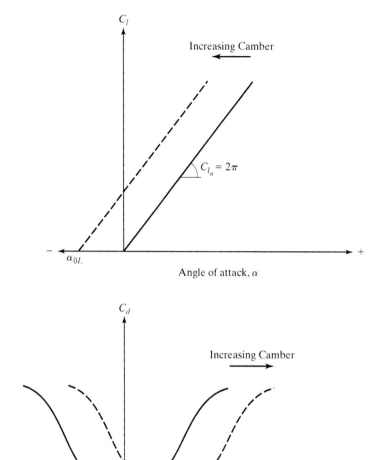

FIGURE 4.6: Effect of camber on the lift versus angle of attack (top), and drag coefficient versus lift coefficient (bottom) for a 2-D airfoil.

(drag bucket) from zero to positive values. This is important in the design of aircraft for efficient cruise.

4.1.1 Airfoil Shape Selection

A good starting point in the selection of the airfoil cross-section shape is to consider the conditions at cruise. There the required lift equals the weight so that

$$(C_L)_{\text{cruise}} = q \left(\frac{W}{S} \right)_{\text{cruise}}. \tag{4.4}$$

As had been pointed out in the previous chapter, with long-range aircraft, the weight of the aircraft can change significantly from the start of cruise to the end of cruise. Assuming that the cruise altitude and speed remain the same, this means that the required lift coefficient changes during cruise. The average of the required lift coefficient during cruise is referred to as the "design C_L."

In order to have the maximum range during cruise, the drag needs to be a minimum. The airfoil shape is then selected based on two criteria:

1. that it can provide the "design C_L" in level flight and
2. that the range of C_L values from the start of cruise to the end of cruise is within the drag bucket.

TABLE 4.1: NACA Airfoil Nomenclature.

4-digit series	
NACA 4415	4415: maximum camber = 0.04c
	4415: position of maximum camber = 0.4c
	4415: thickness/chord ratio = 0.15

5-digit series	
NACA 23012	23012: maximum camber = 0.02c
	23012: design $C_l = 0.15(2) = 0.30$
	23012: position of maximum camber = 0.30/2 = 0.15c
	23012: thickness/chord ratio = 0.12

Series 6	
NACA $65_3 - 421$	65_3-421: series designation
	65_3-421: position of minimum pressure = 0.5c
	65_3-421: design $C_l = 0.4$
	65_3-421: extent of drag bucket = design $C_l \pm 0.3$
	65_3-421: thickness/chord ratio = 0.21

Series 7	
NACA 747A315	747A315: series designation
	747A315: upper surface position of minimum pressure = 0.4c
	747A315: lower surface position of minimum pressure = 0.7c
	747A315: design $C_l = 0.3$
	747A315: thickness/chord ratio = 0.15
	747A315: letter to indicate different sections having the same numerical designation

As an aid to this selection process, the NACA Series 6 family of airfoils provide a subscript that denotes the range of the drag bucket in tenths of $C_{l_{minD}}$. For example, the airfoil section designated as NASA 65_3-618 gives $0.3 \leq C_{l_{minD}} \leq 0.9$. The general nomenclature for NACA series airfoil sections is given in Table 4.1. Most of the cross-section shapes, and C_l versus α, and C_l versus C_d plots for these airfoils can be viewed in "Theory of Wing Sections" by Abbott and von Doenhoff or in their 1945 NACA Report 824, which is available at the NASA Internet site. Samples from each of the groups of airfoils are shown in Figure 4.7 to illustrate differences in their basic shapes.

The choice of airfoil section type that is most suitable for a new design depends on the general type of aircraft. For example, smaller low subsonic speed ($M_{\text{cruise}} \leq 0.4$) aircraft such as personal/utility or regional commuter aircraft would likely use the NACA

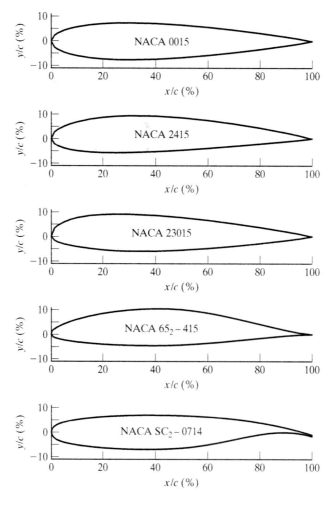

FIGURE 4.7: Drawings of NACA airfoil cross sections, which illustrate differences in their basic shapes.

FIGURE 4.8: Photograph of NASA aircraft used to perform the first flight tests of supercritical airfoils.

five-digit series with a thickness-to-chord ratio near the optimum 14 percent. These provide a relatively large maximum lift coefficient.

High subsonic speed aircraft ($M_{cruise} \simeq 0.8$), such as commercial jet transports and longer range business jets, would likely use the NACA supercritical (SC) airfoils. This airfoil, one of which is shown at the bottom in Figure 4.7, was developed to give higher critical Mach numbers compared to other shapes with comparable thickness-to-chord ratios. Figure 4.8 shows a photograph of the NASA Langley aircraft, which was used for flight tests of the supercritical airfoil. Currently, this airfoil is used on the Boeing C-17 (pictured at the start of Chapter 2) and 777, and the Airbus A-330/340.

Supersonic military aircraft would likely use the NACA Series 6 airfoils since they have good characteristics at both subsonic and supersonic Mach numbers. In order to reduce wave drag in supersonic flight, the thickness-to-chord ratio should be relatively small, of the order of 4 percent.

Following the selection of the airfoil section shape, the conversion from the 2-D lift coefficient, C_l, to the 3-D coefficient, C_L, depends on the wing planform design. This is discussed in the next two sections, which involve the wing taper ratio and sweep angle.

4.2 TAPER RATIO SELECTION

The taper ratio is denoted as $\lambda = c_t/c_r$, where the subscripts t and r refer to the wing "tip" and "root" locations, respectively. This is illustrated in the drawing of a trapezoidal wing planform in the bottom part of Figure 4.1.

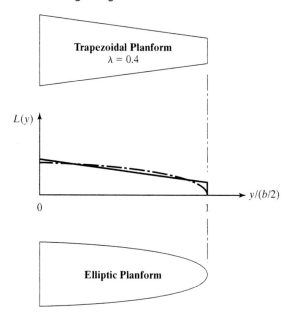

FIGURE 4.9: Schematic showing trapezoidal wing planform shape that gives an approximate elliptic spanwise lift distribution.

One of the key design drivers for the selection of the taper ratio is to minimize the amount of lift-induced drag. From lifting line theory for an unswept, untwisted wing, an ellipse-shaped wing planform gives the minimum drag. Such a wing gives an elliptic spanwise (y) variation in the lift. This can be approximated reasonably well with a more simply constructed trapezoidal wing with a taper ratio of 0.4. This is illustrated in Figure 4.9.

The taper ratio for a trapezoidal wing, which gives the minimum lift-induced drag, is slightly dependent on the aspect ratio and more significantly dependent on the wing sweep angle. This is shown in Figure 4.10, which plots the taper ratio of trapezoidal wings that best approximate an elliptic spanwise lift distribution, as a function of sweep angle. These results are based on NACA wind tunnel tests and include both rearward (positive) and forward (negative) sweep angles. Also, the sweep angle refers to the quarter chord position on the wing.

4.3   SWEEP ANGLE SELECTION

Wing sweep is defined as the angle between a line perpendicular to the aircraft centerline and a line parallel to the leading edge, Λ_{LE}, or a line parallel to the maximum thickness point along the span of the wing, $\Lambda_{t/c}$, as illustrated in Figure 4.1.

The primary reason for adding a sweep angle to a wing is to increase its section critical Mach number. As shown in Figure 4.11, sweep reduces the effective Mach number at the leading edge as

$$\text{M}_{\text{effective}} = \text{M}_\infty \cos(\Lambda_{LE}). \qquad (4.5)$$

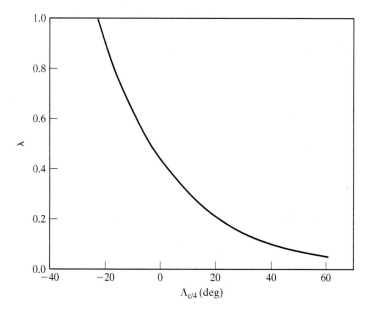

FIGURE 4.10: Taper ratio of trapezoidal wings that minimize lift-induced drag, as a function of sweep angle.

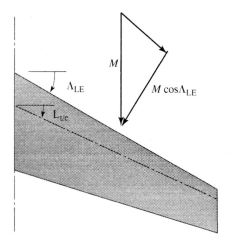

FIGURE 4.11: Schematic showing lower component leading-edge Mach number resulting from sweep.

The critical Mach number in this case is increased as

$$M_{\text{critical}} \propto \frac{1}{\cos^m(\Lambda)}, \tag{4.6}$$

where m is a function of the lift coefficient, C_L.

Adding sweep, however, has some disadvantages. These include lowering the lift through a lower effective dynamic pressure according to

$$q_{\text{effective}} = q_\infty \cos^2(\Lambda), \tag{4.7}$$

and an increase in the wing weight following

$$W_{\text{wing}} \propto [\tan(\Lambda)]^2. \tag{4.8}$$

In addition, large sweep angles generally result in poorer take-off and landing characteristics because enhanced lift devices are less effective.

Figure 4.12 shows the historic trend for sweep angle as a function of the cruise Mach number. Also included is the minimum sweep angle for the leading edge to be inside the Mach cone for supersonic flight,

$$\Lambda = 90° - \sin^{-1}(1/M_\infty). \tag{4.9}$$

It is evident that for aircraft with cruise Mach numbers below 0.4, the wings are historically designed without sweep. For transonic cruise Mach numbers, which includes most of the commercial aviation fleet, sweep angles of approximately 30 degrees are used. With cruise Mach numbers between 1 and approximately 2.4, the sweep angles are close to the angle of the Mach line.

At Mach numbers greater than 2.4, the sweep angles are historically less than needed to have subsonic flow at the wing leading edge. This is primarily the result of the poor subsonic flight characteristics and the excessive structure weight that come from having excessively large wing sweep.

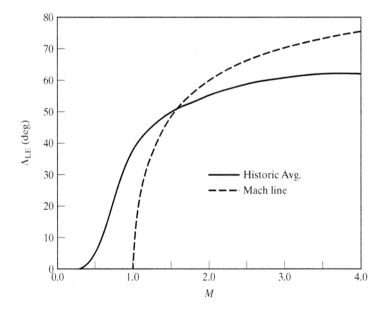

FIGURE 4.12: Historic data (solid curve) showing values of leading-edge sweep angle as a function of design Mach number (data from Raymer, 1992). Dashed curve corresponds to Mach angle versus Mach number.

4.4 3-D LIFT COEFFICIENT

Experimental airfoil data provide 2-D section coefficients. For the aircraft design, these need to be converted to 3-D coefficients.

The effect of a finite swing span on the 2-D lift coefficient, C_l, is to reduce the slope $dC_l/d\alpha$. This is illustrated in Figure 4.13 for a cambered airfoil. It amounts to a clockwise rotation of the C_l versus α line about the α_{0_L} point. The result is that at the same angle of attack, the lift generated by the 3-D wing will be less than that based on the 2-D lift coefficient.

The slope, $dC_L/d\alpha$, is derived from theory. For subsonic Mach numbers, it is given as

$$\frac{dC_L}{d\alpha} = \frac{2\pi A}{2 + \sqrt{4 + (A\beta)^2 \left(1 + \frac{\tan^2(\Lambda_{t/c})}{\beta^2}\right)}}, \tag{4.10}$$

where

$$\beta = \sqrt{1 - M_{\text{eff}}^2} \tag{4.11}$$

and

$$M_{\text{eff}} = M_\infty \cos(\Lambda_{\text{LE}}), \tag{4.12}$$

which is less than one for subsonic flows. As before, $\Lambda_{t/c}$ is the sweep angle of the wing maximum thickness line, and A is the wing aspect ratio. The units of C_{L_α} are $\circ^{-1}$.

Using the slope, 3-D lift coefficient versus angle of attack is

$$C_L = \frac{dC_L}{d\alpha}\alpha + C_{L_{\alpha=0}}, \tag{4.13}$$

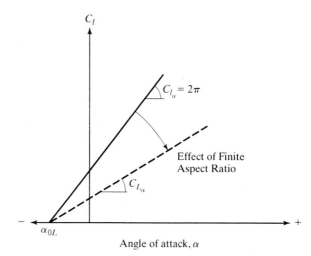

FIGURE 4.13: Schematic showing the effect of a finite aspect ratio on the 2-D section lift coefficient.

where

$$C_{L_{\alpha=0}} = -\frac{dC_L}{d\alpha}\alpha_{0_L}. \tag{4.14}$$

For an uncambered airfoil, $\alpha_{0_L} = 0$. Otherwise, it will be negative for a positive cambered airfoil. If the airfoil section was selected based on providing the design C_l at a zero angle of attack, achieving the design C_L will require placing the airfoil at a slight positive angle of attack. Otherwise the process requires selecting an airfoil section which provides a higher C_l such that the lower 3-D lift coefficient matches the design C_L at zero angle of attack.

4.5   WING DRAG ESTIMATION

In the initial estimate of the take-off weight in Chapter 2, the lift-to-drag ratio, L/D, was based on empirical data from historic aircraft. Using the value of C_L obtained in this chapter, it is now possible to more accurately calculate the total drag on the wing. This is the first step in obtaining a more accurate estimate of the total aircraft drag, which is needed for sizing the engines.

The drag coefficient for the wing corresponds to the base drag, the lift-induced drag, and any additional drag that results from viscous losses such as produced by flow separations. This is expressed in the equation

$$C_D = \underbrace{C_{D_0}}_{base} + \underbrace{kC_L^2}_{lift\ induced} + \underbrace{k'(C_L - C_{L_{minD}})}_{losses}, \tag{4.15}$$

where again,

$$k = \frac{1}{\pi Ae} \tag{4.16}$$

and e is a wing efficiency factor that accounts for taper ratio and fuselage effects on the wing.

In Chapter 2 for simplicity, a value of $e \simeq 0.8$ was recommended. As a further refinement,

$$e = e'\left[1 - \left(\frac{d}{b}\right)^2\right], \tag{4.17}$$

where d/b is the ratio of the fuselage diameter to wing span and

$$e' \simeq 0.98 \tag{4.18}$$

is a reasonable estimate for a large range of taper ratios and sweep angles.

The variable k' is dependent on the leading-edge radius and taper ratio and has values in the range

$$0.02 \leq k' \leq 0.16, \tag{4.19}$$

with the largest corresponding to sharp leading edge Δ-planform wings and the smallest corresponding to blunt rectangular-planform wings.

If the airfoil section was chosen so that the drag bucket encompasses the C_L range throughout cruise, then $(C_L - C_{L_{minD}}) = 0$, and Eq. [4.15] becomes

$$C_D = C_{D_0} + k C_L^2. \tag{4.20}$$

This in fact is the form of the equation that was presented in estimating the drag-to-weight ratio in Eq. [3.12].

4.5.1 Base Drag Estimation

The base drag coefficient, C_{D_0} in Eq. [4.20], is the zero-lift drag coefficient that corresponds to viscous skin friction and flow separations. In the best case, C_{D_0} will be equal to C_{d_0}, which is the minimum drag coefficient in the drag bucket for the 2-D wing section. In practice, $C_{D_0} > C_{d_0}$ as a result of flow disturbances caused by the wing attachments to the fuselage, wing-mounted external store such as fuel tanks, and surface imperfections such as hinge gaps at movable surfaces.

In order to estimate the 3-D base drag coefficient for the aircraft design, the viscous drag coefficient, C_f, is estimated, assuming that the flow behaves in the same fashion as over a flat plate. On a smooth wing, with a minimum amount of flow separations, $C_{D_0} = C_f$. However, to account for the imperfections in the simple 2-D wing behavior, the friction coefficient will be multiplied by two factors corresponding to a "form factor," $\mathcal{F}$, and an "interference factor," $\mathcal{Q}$.

The form factor is intended to estimate the increase in the friction coefficient, above the viscous drag, which is due to flow separations. For the main wing,

$$\mathcal{F} = \left[1 + \frac{0.6}{(x/c)_m} \left[\frac{t}{c} \right] + 100 \left[\frac{t}{c} \right]^4 \right] \left[1.34 \, M^{0.18} (\cos(\Lambda_{t/c_{max}}))^{0.28} \right], \tag{4.21}$$

where $\Lambda_{t/c_{max}}$ is the sweep angle of the maximum thickness line and $(x/c)_m$ is the chordwise location of the maximum thickness point of the airfoil section. For the tail surface drag estimates in Chapter 6, the form factor will be increased by 10 percent to account for the extra drag that occurs because of the hinge gaps by the elevator and rudder.

The interference factor, $\mathcal{Q}$, is intended to estimate the increase in the base drag due to interference effects caused by the fuselage or wing attachments. These include nacelles or external stores mounted on or near the wing, and filleted wing attachments. The value of the interference factor varies with the type and location of these components. Table 4.2 lists the recommended values of $\mathcal{Q}$ for these situations.

TABLE 4.2: Values of Interference Factor, $\mathcal{Q}$, for different arrangements.

	$\mathcal{Q}$
Wing-Mounted Nacelle or Store	1.5
Wing-Tip Missile	1.25
High-Wing, Mid-Wing	1
Well Filleted Low-Wing	1
Unfilletted Low-Wing	1.25

The wing base drag coefficient is taken as the product of the skin friction coefficient, the form factor, and interference factor, or

$$C_{D_0} = C_f \mathcal{F} \mathcal{Q} \frac{S_{\text{wet}}}{S}. \tag{4.22}$$

Here, S_{wet} corresponds to the wetted or surface area of the wing. This is the appropriate area to use with the skin friction coefficient. However, since C_D in Eq. [4.20] is based on the wing planform area, S is used to normalize C_{D_0}.

If the wing section were infinitely thin, the wetted area would be equal to twice the wing planform area. As an approximation of that for thin airfoil sections,

$$S_{\text{wet}} \simeq 2.003S \quad \text{for} \quad (t/c)_{\text{max}} \leq 0.05. \tag{4.23}$$

Otherwise, for thicker airfoil sections,

$$S_{\text{wet}} \simeq S[1.977 + 0.52(t/c)_{\text{max}})] \quad \text{for} \quad (t/c)_{\text{max}} > 0.05. \tag{4.24}$$

The friction coefficient, C_f, used in determining the drag on the wing comes from flat plate studies. If the flow is laminar,

$$C_f = \frac{1.328}{\sqrt{\text{Re}_x}} \quad \text{:Laminar,} \tag{4.25}$$

where Re_x is the Reynolds number based on the x-development length of the boundary layer.

The Reynolds number in this instance is defined as

$$\text{Re}_x = V_o x / v, \tag{4.26}$$

where V_o is the velocity at the outer edge of the boundary layer and v is the kinematic viscosity.

The flow is likely to be turbulent when $\sqrt{\text{Re}_x} \geq 1000$. In this event, the friction coefficient for a hydraulically smooth surface is

$$C_f = \frac{0.455}{(\log_{10} \text{Re}_x)^{2.58}(1 + 0.144\, M^2)^{0.65}} \quad \text{:Turbulent.} \tag{4.27}$$

The term $(1 + 0.144\, M^2)^{0.65}$ is a Mach number correction that approaches 1.0 for low Mach numbers.

The x-distance used in determining Re_x should be the local chord. Because of the wing taper, the chord length changes with the spanwise position. In order to estimate an average Reynolds number, the mean aerodynamic chord, $\bar{c}$, should be used. Therefore,

$$\bar{\text{Re}}_x = V_o \bar{c} / v. \tag{4.28}$$

Also if wing sweep is used, the velocity component normal to the leading edge,

$$V_{o_N} = V_o \cos(\Lambda_{\text{LE}}), \tag{4.29}$$

should be used in determining the Reynolds number.

Following these steps to determine C_L and C_D, the wing lift-to-drag ratio is simply

$$\frac{L}{D} = \frac{C_L}{C_D}. \tag{4.30}$$

4.6 PLANFORM GEOMETRIC RELATIONS

In laying out the planform shape of the wing there are a number of useful relations that apply to a trapezoidal shape. These are based on knowing the wing area, aspect ratio, taper ratio, and leading-edge sweep angle, which are determined based on the approaches discussed in the previous sections of this chapter.

The wing span, b, is determined from the wing area, S, and the aspect ratio, A. The relation between these is

$$b = \sqrt{SA}. \tag{4.31}$$

The wing aspect ratio is defined as

$$A = \frac{2b}{c_r(1 + \lambda)}. \tag{4.32}$$

Therefore, the root chord is found from

$$c_r = \frac{2b}{A(1 + \lambda)}. \tag{4.33}$$

The taper ratio is defined as

$$\lambda = \frac{c_t}{c_r}. \tag{4.34}$$

Therefore, the tip chord is found from

$$c_t = \lambda c_r. \tag{4.35}$$

The mean aerodynamic chord length corresponds to

$$\bar{c} = \frac{2c_r}{3} \frac{(1 + \lambda + \lambda^2)}{(1 + \lambda)}. \tag{4.36}$$

The normalized spanwise location of the mean aerodynamic chord from the center span of the wing is

$$\frac{\bar{y}}{b} = \frac{1}{6} \frac{1 + 2\lambda}{1 + \lambda}. \tag{4.37}$$

Finally, starting with the leading-edge sweep angle, the sweep angle at any x/c location on the wing is given by the following relation.

$$\Lambda_{x/c} = \tan^{-1}\left[\tan \Lambda_{\mathrm{LE}} - \frac{x}{c}\frac{2c_r}{b}(1 - \lambda)\right]. \tag{4.38}$$

Therefore, the sweep angle of the quarter-chord line is

$$\Lambda_{c/4} = \tan^{-1}\left[\tan \Lambda_{\mathrm{LE}} - \frac{1}{4}\frac{2c_r}{b}(1 - \lambda)\right]. \tag{4.39}$$

4.7   SPREADSHEET FOR WING DESIGN

The relations used for the design of the main wing have been incorporated into a spreadsheet named **wing.xls**. The format allows easy input and modification of the design parameters and provides a graphical view of the wing planform, and the 2-D and 3-D lift coefficients (C_l and C_L) versus angle of attack based on the design parameters.

A sample of the spreadsheet is shown in Figures 4.14 and 4.15. This contains the parameters for the conceptual SSBJ proposed in Chapter 1.

In the spreadsheet, there are two areas where the input parameters are placed. These are designated "Design Parameters" and "Airfoil Data."

The important airfoil data that are needed for the wing design calculations include

1. $C_{l_{max}}$;
2. $C_{l_\alpha} = dC_l/d\alpha$;
3. location of the aerodynamic center;
4. zero-lift angle of attack, α_{0L};
5. base (minimum) drag coefficient, C_{d_0};
6. leading-edge radius, r_{LE};
7. extent of the drag bucket, $C_{l_{minD}}$.

The wing design parameters based on approaches presented in this chapter that are needed as input include

1. wing area, S;
2. wing aspect ratio, A;
3. leading-edge sweep angle, Λ_{LE};
4. maximum t/c;
5. taper ratio, λ.

In addition, the following information has already been specified or determined in the design and is needed as input:

1. cruise Mach number, M;
2. cruise altitude, H;
3. weight at the start and end of cruise;
4. dynamic pressure, q, at the start and end of cruise.

The cruise Mach number and aspect ratio were specified in the first step of the design and used to determine the weight of the aircraft at each flight phase. This was done in the spreadsheet **itertow.xls**. Therefore, these values need to remain the same.

Chapter 3 dealt with the determination of the wing loading, W/S, for different flight phases. The spreadsheet associated with the wing loading calculations is **wingld.xls**. The wing area comes from this spreadsheet, based on the weight and wing loading at the start of cruise. The design C_l comes from Eq. [4.4]. This needs to be calculated based on the wing loading at the start and end of cruise. In the selection of the airfoil section shape, the drag bucket needs to be wide enough to encompass the change in C_l values.

Design Parameters				Airfoil Data					
M	2.10			Name	NACA 64A204				
S	519	ft^2		Cl_{max}	1.03				
A	2.0			Cl_α	0.11	1/deg			
Λ_{LE}	62	deg		a.c.	0.25	c			
t/c	0.04			α_{0L}	−1.33	deg			
λ	0.00			Cd0	0				
W c-start	88,817	lbf/f^2		r_{le}	0	c			
W c-end	53,976	lbf/f^2		Cl_{minD}	0.1 − 0.3				
q c-start	531.07	lbf/f^2		(t/c)max	0.40	c			
q c-end	347.41	lbf/f^2							
Cl c-start	0.32								
Cl c-end	0.20								
Calculations				**Sweep Angles**					
b	32.2	ft			x/c	$\Lambda_{x/c}$ (deg)			
M_{eff}	0.99			LE	0.00	62.0			
c_r	32.2	ft		1/4C	0.25	54.1			
c_t	0.0	ft		a.c	0.25	54.0			
m.a.c.	**21.5**	ft		(t/c)max	0.40	47.2			
				TE	1.00	−6.8			
β	0.17								
$C_{L\alpha}$	**0.044**	1/deg		**Viscous Drag**					
C_{Lo}	0.06			V_eff	904.06	f/s			
α_{trim}	**6.0**	deg		q_eff	117.05	lbf/f^2			
C_{Ltrim}	0.322			Re_mac	1.67E+07				
k	0.2			sqrt(Re)	4091.11				
C_D	**0.028**			Cf	2.54E−03				
L/D	**11.48**			S_wet	1039.56	ft^2			
				F	1.46				
				Q	1				
Total Drag	**7739.81**	lbf		C_{D0}	**0.0074**				
Plotting:									
Spanwise View									
x	y								
0	0								
32.2	0								
30.3	16.11								
30.3	16.11								
0	0								
Lift Curves									
α	Cl	α	CL						
−1.33	0	−1.33	0						
8.03	1.03	8.03	0.41						

FIGURE 4.14: WING spreadsheet for SSBJ case study (Part 1).

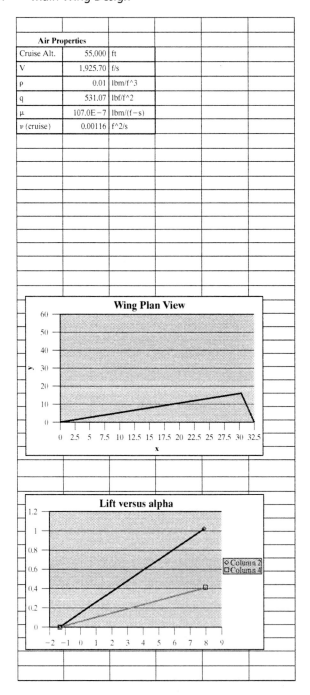

Air Properties		
Cruise Alt.	55,000	ft
V	1,925.70	f/s
ρ	0.01	lbm/f^3
q	531.07	lbf/f^2
μ	107.0E−7	lbm/(f−s)
v (cruise)	0.00116	f^2/s

FIGURE 4.15: WING spreadsheet for SSBJ case study (Part 2).

New parameters that are first used in this spreadsheet are the leading-edge sweep angle, Λ, and the taper ratio, λ. These are selected based on historical trends given in Figures 4.8 and 4.10.

Also needed as input to the spreadsheet is the cruise altitude, H. This is placed in the section denoted as "Air Properties," which is used to calculate the properties needed for the determination of the chord Reynolds number. Relations that are identical to those used in previous spreadsheets are used to determine the velocity, V; density, ρ; and dynamic pressure, q, at the cruise altitude. In addition, the kinematic viscosity, υ, is estimated. Here a constant average viscosity with altitude, μ, is assumed, whereby $\nu = \mu/\rho$. In fact, μ is a weak function of altitude. However, the dependence of ν on altitude is primarily due to changes in ρ.

The calculations are intended to determine the 3-D lift coefficient from the 2-D lift characteristics of the selected wing section, the overall 3-D drag coefficient of the wing, and the lift-to-drag ratio. In addition, planform geometry of the wing is generated.

The effective Mach number for the wing is the component that is normal to the leading edge. This is referred to as M_{eff} and was defined in Eq. [4.12]. The parameter β is a function of M_{eff}, which was defined in Eq. [4.11].

In order to calculate C_{L_α}, it is first necessary to calculate the sweep angle of the wing maximum thickness line, $\Lambda_{t/c}$. This is found using Eq. [4.38].

C_{L_α} is then calculated based on Eq. [4.9]. Next $C_{L_{\alpha=0}}$ is found from the relation given in Eq. [4.14]. If the airfoil shape was selected based on $C_l = \text{Design}C_L$, then after accounting for the 3-D wing effects, $C_{L_{\alpha=0}}$ will be less than the $\text{Design}C_L$. As a result, the wing section will have to be placed at a slightly positive angle of attack. This is referred to as α_{trim} in the spreadsheet. This is found from

$$\alpha_{\text{trim}} = \left(C_{L_{\text{Design}}} - C_{L_{\alpha=0}}\right)/C_{L_\alpha}. \tag{4.40}$$

The drag coefficient for the wing is found by solving Eq. [4.20]. This relation also involves $k = 1/(\pi A e)$. A value of $e = 0.8$ was again assumed since a more refined estimate requires knowing the fuselage diameter (Eq. [4.17]), which is the topic covered in the next chapter.

The calculation of the base drag coefficient, C_{D_0}, is found following Eq. [4.22]. This is performed in the spreadsheet in the portion marked "Viscous Drag." The Reynolds number that is used to determine C_f is found based on the conditions at cruise. The form factor, $\mathcal{F}$, is found for the main wing based on Eq. [4.21]. The wetted area is determined based on either Eq. [4.23] or Eq. [4.24] depending on the value of the maximum t/c.

The interference factor, $\mathcal{Q}$, is input into the spreadsheet in the "Viscous Drag" section. These values are based on the wing placement and on the arrangement of wing-mounted stores. Values for $\mathcal{Q}$ were given in Table 4.2.

We note that C_{D_0} is normalized by the main wing planform area, S, which is the same for the lift-induced drag coefficient. The total drag coefficient, C_D, is then found from Eq. [4.20]. The total drag is then

$$D = C_D q S. \tag{4.41}$$

In the first spreadsheet, L/D was estimated based on historic aircraft. The value used there can now be compared to the value that is calculated in this spreadsheet. Here, L/D is simply calculated as the ratio of the lift and drag coefficients, C_L/C_D.

The leading and trailing edge sweep angles are used to calculate the plan view of the symmetric half of the wing. This is plotted on the spreadsheet.

Also plotted are the 2-D (series 1) and 3-D (series 2) lift curves. This shows the effect of the finite aspect ratio on lowering C_{L_α}. These curves will form the starting point when enhanced lift devices, such as leading- and trailing-edge flaps are discussed in Chapter 9.

4.7.1 Case Study: Wing Design

The parameters that are displayed in the wing design spreadsheet correspond to the conceptual SSBJ. One of the principle design drivers is efficient cruise. The weight estimates obtained from the "itertow" spreadsheet (Figure 2.6) at the start and end of cruise were 81,817 and 53,976 pounds, respectively. The wing loading analysis in the "wingld" spreadsheet (Figure 3.13) gave $W/S = 157.51$ lb/f^2 for the most efficient cruise. This was based on the start of cruise so that using the corresponding weight, a required wing area of $S = 519.4$ f^2 was obtained. With this wing area, the decrease in weight at the end of cruise gives a wing loading of 103.01 lb/f^2. These values, along with the dynamic pressure, q, at cruise altitude were used in the list of input design parameters in the present "wing" spreadsheet.

A Series-6 airfoil was chosen because they generally exhibit good performance at both subsonic and supersonic Mach numbers. Based on the wing loading and dynamic pressure, the C_l during cruise had the range of $0.20 \leq C_l \leq 0.32$. The NACA 64A204 airfoil section provides a drag bucket which encompassed this range of lift coefficients. The 2-D lift and drag coefficient characteristics are plotted in Figure 4.16. Other characteristics of this airfoil are recorded in the spreadsheet.

The leading-edge sweep angle of $\Lambda_{LE} = 62°$ was set so that the effective Mach number at the leading edge would be slightly subsonic. This was verified in the calculations portion of the spreadsheet where $M_{eff} = 0.99$.

The taper ratio of $\lambda = 0$ was chosen to be consistent with historic aircraft with highly swept wings, as shown in Figure 4.10. The value of $t/c = 0.04$ was based on historic aircraft that have a similar cruise Mach number, as shown in Figure 4.5.

The calculations gave a wing span and root chord of 32.2 ft. For this, the mean aerodynamic chord was 21.5 ft.

The finite wing effects reduced C_{L_α} by approximately 40 percent compared to the 2-D wing. In order to produce the required lift at cruise, the wing would need to be placed at a positive angle of attack of $\alpha_{trim} = 6.0°$.

For this design, the drag coefficient for the wing at cruise was determined to be $C_D = 0.028$. In determining the base drag coefficient, C_{D_0}, the form factor, $\mathcal{F}$, was found to be 1.457. A value of $Q = 1.0$ was used for the interference factor because the design was expected to have a low wing mount on the fuselage, which is well filletted, and without any wing-mounted stores. The value for C_{D_0} of 0.00716 in this case was not much above the 2-D base drag value given for this wing section of $C_{d_0} = 0.004$. The overall drag force at cruise Mach number and altitude was found to be 7739 lbs.

The total wing L/D for this design was found to be 11.48. This is better than the value of 7.59 that was estimated from empirical data and used in the initial estimate

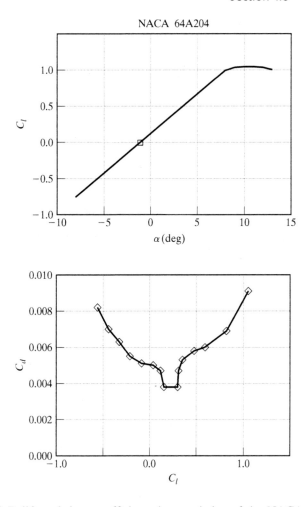

FIGURE 4.16: 2-D lift and drag coefficient characteristics of the NACA 64A204 airfoil section. (Source: NACA Report 824).

of the take-off weight in the "itertow" spreadsheet. This will improve the initial range estimate or allow more payload for the same range.

Finally, the shape of the wing planform, which is seen in the graph in the spreadsheet, is similar to other Δ-wing aircraft with similar flight characteristics, as well as to the artists view of the proposed Dassault SSBJ in Figure 1.10.

4.8 PROBLEMS

4.1. For the conditions of the supersonic business jet case study, investigate the effect of the wing aspect ratio on
   1. C_{L_α};
   2. C_D;

    3. k;

    4. L/D.

    Plot these as a function of aspect ratio and discuss how it impacts L/D.

4.2. For the conditions of the supersonic business jet case study, investigate the effect the wing taper ratio from $0 \leq \gamma \leq 1$ on

    1. C_{L_α};

    2. α_{trim};

    3. $\Gamma_{t/c}$;

    4. mean aerodynamic center.

    Plot these as a function of taper ratio. What is the mechanism by which the taper ratio affects C_{L_α}?

4.3. For the conditions of the supersonic business jet case study, investigate the effect the wing area from $700 \text{ f}^2 \leq S \leq 1500 \text{ f}^2$ on

    1. DesignC_L;

    2. C_{L_α};

    3. L/D;

    4. drag on the wing, $D = qSC_D$.

    Plot these as a function of wing area. Is the original airfoil section used in the case study suitable for all of these wing areas? Explain.

4.4. Would the airfoil section NACA 66-206 be suitable for the case study aircraft? Would it have any advantages? Explain.

4.5. For the characteristics of the jet-powered aircraft in Problem 2.3, do the following:

    1. Determine the design C_L for most efficient cruise.

    2. Select a wing section that has suitable characteristics for this design. Justify your choice.

    3. Determine the wing plan view, listing wing span; root and tip chord lengths; and leading, trailing, and maximum thickness sweep angles. Justify your choices.

4.6. Consider a small business or commuter jet aircraft with the following characteristics:

    — cruise Mach number = 0.80;

    — cruise altitude = 40,000 f;

    — wing area = 200 f^2;

    — weight at start of cruise = 15,300 lbs;

    — weight at end of cruise = 11,500 lbs.

    Design the main wing that would be suitable for this aircraft by specifying the following:

    1. wing section type;

    2. aspect ratio;

    3. taper ratio;

    4. sweep angle.

    Justify all of your selections.

4.7. Consider a private four-place aircraft with the following characteristics:

    — cruise Mach number = 0.2;

    — cruise altitude = 10,000 f;

    — wing loading = 20 lb/f^2;

    — take-off weight = 15,000 lbs.

Design the main wing that would be suitable for this aircraft by specifying the following:
1. wing section type;
2. aspect ratio;
3. taper ratio;
4. sweep angle.
Justify all of your selections.

CHAPTER 5

Fuselage Design

Photograph of the Aero Spacelines B-377PG "Pregnant Guppy," which was designed to transport outsized cargo for the NASA Apollo Program. The upper fuselage was more than 20 feet in diameter to accommodate portions of the Saturn V rocket. (NASA Dryden Research Center Photo Collection.)

Following the main wing, the next logical step in the conceptual design involves the design of the fuselage. The fuselage has a number of functions that vary depending on the type and mission of the aircraft. These include accommodating crew, passengers, baggage or other payload, as well as possibly housing internal engines. Other considerations for the fuselage design include possible fuel storage, the structure for wing attachments and accommodations for retractable landing gear.

All of these generally aim at setting the internal volume, height, width and length of the fuselage. For example, the primary mission of the "Pregnant Guppy" pictured on the previous page, was to carry 20-foot diameter sections of the Apollo Saturn V rocket. This specifically dictated the size and shape of the fuselage, and resulted it its distinctive appearance.

Depending on the Mach number regime, the optimum shape of the fuselage, or more specifically the length-to-diameter ratio, may be determined on the basis of minimizing the aerodynamic drag. With subsonic aircraft, this ratio is historically far from the optimum and dictated more by function. With supersonic aircraft, however, the penalty for deviating from the optimum leaves little room for compromise.

5.1   VOLUME CONSIDERATIONS

5.1.1   Passenger Requirements

The size and shape of subsonic commercial aircraft are generally determined by the number of passengers, seating arrangements and cargo requirements. Table 5.1 gives typical dimensions for the passenger compartments. These are generally based on an average passenger who is assumed to weigh 180 lbs. Figure 5.1 accompanies the table and provides graphical definitions of the nomenclature used.

Seating arrangements on commercial passenger aircraft vary depending on the size and range. Examples of different aisle seating arrangements for a variety of aircraft is illustrated in Figure 5.2. Table 5.2 lists additional seating information for these aircraft.

The seating arrangements in Figure 5.2 generally correspond to "coach class." In large commercial aircraft, there is additional variability in seating arrangements based on other passenger classes. This is illustrated in Table 5.3 with a comparison between coach and first class for two typical jet transports.

As indicated in Table 5.1, passengers are allocated from 40–60 lbs of baggage. On average, this is expected to occupy a volume of 15–25 f^3. In addition, passengers are allocated 3 f^3 for onboard overhead baggage storage.

In modern aircraft, checked baggage and other cargo are carried in standard cargo containers. Examples of these are shown in Figure 5.3.

TABLE 5.1: Passenger compartment requirements.

	Long-Range	Short-Range
Seat Width (in)	17–28	16–18
Seat Pitch (in)	34–40	30–32
Headroom (in)	>65	—
Aisle Width (in)	20–28	>15
Aisle Height (in)	>76	>60
Passengers/Cabin	10–36	≤50
Lavatories/Passenger	1/(10–20)	1/(40–50)
Galley Volume/Passenger (f^3)	1–8	0–1
Baggage/Passenger (lbs)	40–60	40

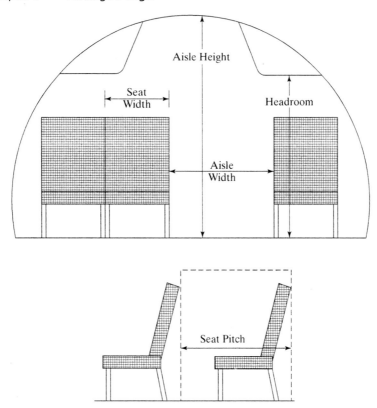

FIGURE 5.1: Schematic drawing of a passenger seating arrangement defining parameters.

TABLE 5.2: Passenger aircraft seating arrangements.

Passenger No.	Fuselage Diam. (in.)	Aisle Seating	Examples
4–9	64	1 + 1	Citation V
10–20	58	1 + 1	Beech 1900
	94	2 + 1	Gulfstream II
20–50	91	2 + 1	Saab 340
50–75	106	2 + 2	DHC-8/300
75–190	130	2 + 3	MD-80
	148	3 + 3	Boeing 757
190–270	198	2 + 3 + 2	Boeing 767
	222	2 + 4 + 2	Airbus A300
270–360	222	2 + 4 + 2	Airbus A330
	236	2 + 5 + 2	DC-10, L-1011, Boeing 777
360–450	256	3 + 4 + 3	Boeing 747

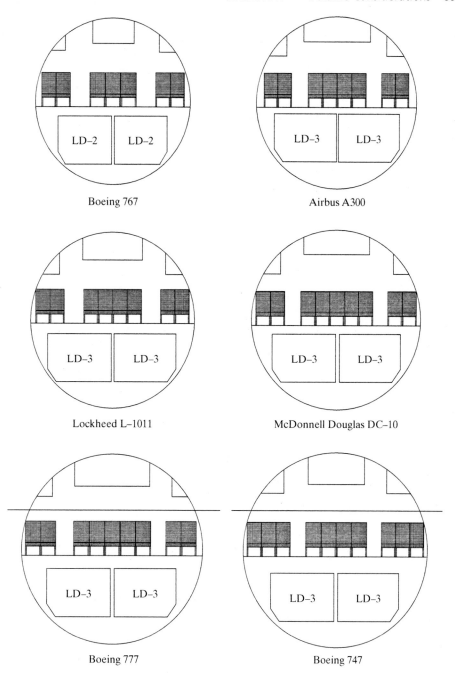

FIGURE 5.2: Schematic drawings of coach compartment cross-sections that are typical of different commercial passenger aircraft. Also shown are types of lower deck cargo containers used on these aircraft.

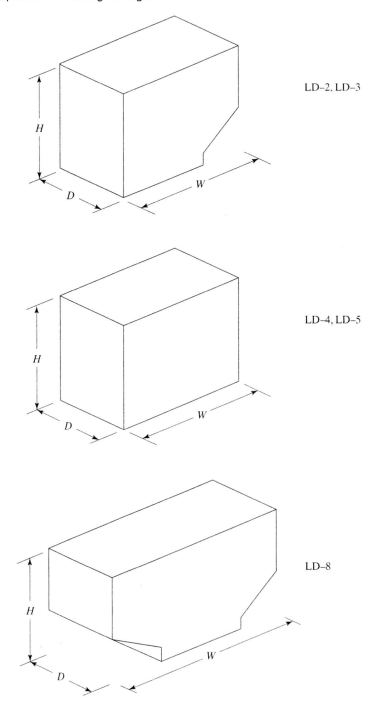

FIGURE 5.3: Schematic drawing of different styles of lower deck containers listed in Table 5.4

TABLE 5.3: Typical passenger accommodations for large jet transports.

		B757	DC-10
Seats	Total	178	292
	First Class	16 (9%)	24 (8.2%)
	Coach	162 (91%)	268 (91.8%)
	First-Class Pitch	38 in.	38 in.
	Coach Pitch	34 in.	34 in.
	First-Class Aisle	2 + 2	3 + 3
	Coach Aisle	3 + 3	2 + 5 + 2
Lavatories	First Class	1	1
	Coach	3	6
	First-Class Pass./Lav.	16	24
	Coach Pass./Lav.	54	45
Galleys	First-Class Volume	70 f^3	120 f^3
	First-Class f^3/Pass.	4.4	5.0
	Coach Volume	231 f^3	450 f^3
	Coach f^3/Pass.	1.4	1.7

TABLE 5.4: Cargo container dimensions.

Type	Height (in.)	Width (in.)	Depth (in.)	Volume (f^3)	Gross Wt. (lbs)
LD-2*	64.0	61.5	60.4	120	2700
LD-3*	64.0	79.0	60.4	156	3500
LD-4	64.0	96.0	60.4	195	5400
LD-5	64.0	125.0	60.4	279	5400
LD-8*	64.0	125.0	60.4	245	5400

* Trapezoidal shape.

Table 5.4 lists the dimensions of the more widely used containers. The smaller containers are suitable for smaller commercial/transport aircraft, such as the Boeing 727. The larger ones, for example the LD-3, are commonly used on larger commercial aircraft, such as shown in Figure 5.2. Table 5.5 lists the number of LD-3 containers that can be carried on these transports.

Smaller, short-range aircraft do not use cargo containers, but rather have space only for bulk cargo with a volume that is based on 6–8 f^3 per passenger.

Passenger aircraft also have requirements for the number, placement and type of emergency exits in the event of a survivable accident. These are based on the number of passenger seats installed on the aircraft. The requirements are summarized in Table 5.6, with the description of the exit types given in Table 5.7. They indicate a fairly straightforward criteria for passenger numbers up to 179. After which, the number and type of exits is based on the arrangement that gives sufficient "seat credits." These credits need

TABLE 5.5: Large-body aircraft cargo compartment arrangements.

	Number of LD-3 Containers	Bulk Cargo Volume (f^3)
B-747	30	1000
L-1011	16	700
DC-10	14	800
A-300	10	600

TABLE 5.6: Number and type of emergency exits required for passenger transport aircraft by FAR 25.807.

No. Pass.	Type I	Type II	Type III	Type IV
1–9				1
10–19			1	
20–39		1	1	
40–79	1		1	
80–109	1		2	
110–139	2		1	
140–179	2		2	

180–299   Add exits so that 179 plus "seat credits" $\geq$ passenger number.

Seat Credit	Exit Type
12	Single Ventral
15	Single Tailcone
35	Pair Type III
40	Pair Type II
45	Pair Type I
110	Pair Type A

≥ 300 Use pairs of Type A or Type I with the sum of "seat credits" $\geq$ passenger number.

to correspond to the number of passenger seats in excess of 179 on the aircraft. This procedure holds for up to 299 passenger seats. Above this number, the designs use pairs of Type A or Type I exits with the sum of the credits equal to the passenger seat total.

5.1.2 Crew Requirements

The size of the crew compartment will vary depending on the aircraft. With long-range military/commercial transport and passenger aircraft, the crew compartment should be designed to accommodate from two to four crew members. Recommendations suggest

TABLE 5.7: Types of emergency exits for passenger transport aircraft defined by FAR 25.807.

Type	Location	Min. Dimensions Width × Height (in.)	Min. Step Height Inside:Outside (in.)
Type I	Floor Level	24 × 48	—
Type II	Floor Level	20 × 44	—
	Overwing	20 × 44	10:17
Type III	Overwing	20 × 36	24:27
Type IV	Overwing	19 × 26	29:36
Tailcone	Aft of Pressure Hull	20 × 60	24:27
Ventral	Bottom of Fuselage	Equiv. Type I	—
Type A	Floor Level	42 × 72	—

that the crew compartment have a length of approximately 150 inches for four crew members, 130 inches for three crew members and 100 inches for two crew members.

An important factor that impacts the shape of the forward section of the fuselage is the requirement that the pilot have an unobstructed forward view. A critical need in achieving this is obtaining the proper amount of over-nose angle. This is especially important for the landing phase for all aircraft, and during the combat phase of military-fighter aircraft. The Concorde and Russian Tu-144 supersonic passenger jets have a nose section that deflects downward in order to give the necessary over-nose angle for landing. Figure 5.4 shows a photograph of the Tu-144 with the nose deflected during landing.

The over-nose angle, α_{overnose}, is defined as the angle between a horizontal line through the pilot's eye, down to the point of the highest visual obstruction. A schematic representation is shown in Figure 5.5. The proper over-nose angle depends on the landing approach angle, γ_{approach}, and the landing approach velocity, V_{50}. In the landing analysis in Chapter 8, the approach angle is found from

$$\gamma_{\text{approach}} = \sin^{-1}\left(\frac{-D}{W}\right), \tag{5.1}$$

where D is the drag and W is the weight at landing. The approach velocity, V_{50}, refers to the velocity at an elevation of 50 f, which starts the landing phase with

$$V_{50} = 1.3 V_s, \tag{5.2}$$

and V_s is the stall velocity that will include an enhanced lift configuration of the main wing, which is covered in Chapter 9. If these quantities are known, a reasonable empirical relation for the over-nose angle is

$$\alpha_{\text{overnose}} = \gamma_{\text{approach}} + 0.07 V_{50}, \tag{5.3}$$

where V_{50} has units of knots, and α and γ have units of degrees. Table 5.8 gives some typical values for a variety of aircraft.

A somewhat less critical visual requirement for the crew compartment is the over side vision angle, α_{overside}. This is the unobstructed viewing angle from a horizontal line

FIGURE 5.4: Photograph of Tu-144 with nose deflected to give necessary over-nose angle during landing. (NASA Dryden Research Center Photo Collection.)

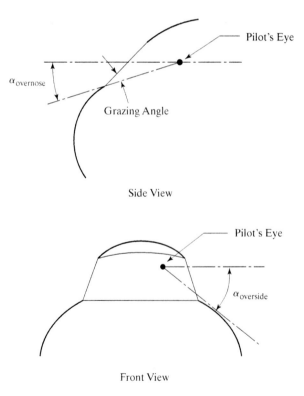

FIGURE 5.5: Illustration defining crew compartment vision parameters.

TABLE 5.8: Values of over-nose and over-side angles for different aircraft.

	α_{overnose}	α_{overside}
Military Transports/Bombers	17°	35°
Military Fighter	11°–15°	40°
General Aviation	5°–10°	35°
Commercial Transport	11°–20°	35°

through the pilot's eye down to the highest visual obstruction formed by the side of the fuselage. Typical values for a variety of aircraft are also given in Table 5.8.

The upward vision angle is also important. Military/commercial transport and passenger aircraft should have an unobstructed viewing angle of at least 20° above the horizon. Military combat aircraft should have at least a 120° unobstructed upward viewing angle.

Another factor that impacts pilot vision, which is mostly only an issue with combat and high-speed aircraft, is the transparency grazing angle. (See Figure 5.5.) This corresponds to the smallest angle between a line of vision of the pilot and the cockpit window or windscreen. If this angle becomes too small, the visibility through the window can become substantially reduced or distorted. The recommended minimum grazing angle is 30°.

5.1.3   Fuel Storage Requirements

In long-range aircraft, a large percentage of the weight at take-off is due to the weight of the fuel. The volume required to hold this fuel can be allocated to the fuselage or wing, or most likely both. The decision on where to store the fuel depends on a number of factors. These include the location of the center of mass with respect to the center of lift, thus affecting the static stability, the vulnerability of crew and passengers in the event of an uncontrolled landing, and the vulnerability of the fuel in combat aircraft caused by enemy fire.

In the case of the static stability, the placement of the fuel is also important in defining how the location of the center of mass might change as a result of the changing fuel weight from the beginning to the end of cruise. In order to maintain static stability in the pitch direction, the center of mass must always be forward of the center of lift. As a result, if any fuel is stored in the fuselage, it should be located at or slightly forward of the wing attachment point. A detailed analysis of the static stability will be done later in Chapter 11.

The volume needed to accommodate the fuel is based on the maximum weight of fuel at take-off and the density of the fuel. The specific volumes of various fuels are given in Table 5.9.

The actual volume that is available in locations in the fuselage or wing depends on the type of fuel tank used. There are generally three types: discrete, bladder and integral. The choice of these determines what percentage of the available volume is capable of holding fuel.

TABLE 5.9: Specific volumes for different aviation fuels (f^3/lb).

	0°F	100°F	Mil-spec
AV-gas	0.0219	0.0235	0.0223
JP-4	0.0199	0.0208	0.0206
JP-5	0.0186	0.0196	0.0197
JP-8	—	—	0.0199

Discrete tanks are generally only used for small general aviation aircraft. They consist of separately manufactured containers that mount in the aircraft. In the wing, these are often mounted at the inboard span portion, near the leading edge. In the fuselage, they are generally placed just behind the engine and above the pilots feet.

Bladder fuel tanks are thick rubber bags that are placed into cavities in the wing or fuselage. The advantage of bladder tanks is that they can be made to be self-sealing. This feature improves the aircraft survivability in the event of an uncontrolled landing or enemy fire. The thickness of the rubber bladder walls reduces the available volume of the cavity. As a general rule, 77 percent of a cavity volume in the wing, and 83 percent of a cavity volume in the fuselage, is available with bladder tanks.

Integral tanks are cavities within the airframe structure that are sealed to form fuel tanks. Examples are the wing box areas formed between wing spars and the area between bulkheads in the fuselage. Because integral tanks are more prone to leaking compared to the other two types, they should not be located near air inlet ducts or engines. The fire hazard of integral tanks can be reduced by filling the tank with a porous foam material. This, however, reduces the volume capacity by approximately 5 percent. As a general rule, 85 percent of the volume measured to the external skin of the wing, and 92 percent measured to the external skin of the fuselage, is available with integral tanks.

5.1.4 Internal Engines and Air Inlets

Engines may be mounted internal to the fuselage. This is often the practice in combat aircraft and small general aviation aircraft, but sometimes has been done with long-range commercial passenger aircraft, such as the B-727 and L-1011. At this stage of the design, the total drag is not yet known so that the thrust requirements of the engine (size and number) are not yet determined. This will be covered in Chapter 7. As a result, the internal arrangement of the engine and air delivery system is difficult to assess at this point in the conceptual design. Therefore, if the propulsion system is likely to be internal to the fuselage, the volume to enclose it needs to be accounted for in the design. At this stage, the best approach is to rely on suitable comparison aircraft.

For internally mounted jet engines, the air delivery system is an integral element. The type and geometry of the inlet will determine the pressure loss and uniformity of the air supplied to the engine. These in turn affect the installed thrust of the engine and fuel consumption.

The types of air inlets depend on the operating Mach number. In general, the objective of the air inlet system for turbojet and turbofan engines is to reduce the Mach number of the air at the compressor face to between 0.4 and 0.5. In subsonic aircraft, this is accomplished using a subsonic diffuser. In supersonic aircraft, this is done through area changes at the inlet that result in the formation of one or more compressive shocks.

At this stage of the design, the objective is to size the fuselage. Therefore in addition to basing the design on suitable comparison aircraft, a first estimate can come from empirical data that indicate that the diameter of the air inlet be the same as that of the engine compressor face and that the length of the inlet be 60 percent of the engine length.

5.1.5 Wing Attachments

The manner in which the main wing attaches to the fuselage is an important element in the fuselage design. For structural reasons, the wing is generally constructed as an integral unit. The portion that passes through the fuselage is referred to as the wing carry-through. The root-span portion of the wing has the largest thickness in order to withstand the large bending moment in the wing. As a result, the wing carry-through occupies a large volume where it passes through the fuselage. An illustration of the wing carry-through is shown in Figure 5.6. A photograph of the partially assembled fuselage section with the main wing attachment for the Boeing 777 is shown in Figure 5.7.

Since the details of the main wing are known from the analysis of the preceding chapter, the volume requirements for the carry-through structure can be directly applied to the design of the fuselage.

5.1.6 Landing Gear Placement

In most aircraft, the fuselage needs to accommodate all or some parts of the landing gear when it is retracted. Therefore, its placement and volume requirements need to be considered in the design of the fuselage.

The size and location of the landing gear will vary depending on the aircraft. Again, a good first estimate can come by examining suitable comparison aircraft. In selecting comparisons for the basis of the landing gear size and placement, it is important to select aircraft that have a comparable take-off weight. Figure 5.8 shows an illustration of some of different landing gear arrangements.

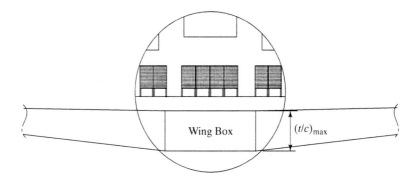

FIGURE 5.6: Illustration of a typical fuselage wing carry-through arrangement.

FIGURE 5.7: Photograph of a Boeing 777 fuselage wing attachment section during assembly. (Courtesy of the Boeing Company.)

Most commercial passenger/transport and combat aircraft use retractable tricycle-type landing gear. With these, the nose wheel is mounted on and retracts into the fuselage. The main wheels are usually mounted on the main wing. Depending on their size, they retract into the main wing, or into the main wing/fuselage junction.

Heavier aircraft use a multi-bogeyed tricycle arrangement that have multiple sets of wheels at three points. Two-wheel bogeys are typically used on aircraft with take-off weights in the range of 50,000 to 150,000 lbs. Four-wheel bogeys are generally used on aircraft with higher take-off weights between 200,000 to 400,000 lbs.

A fewer number of military long-range aircraft such as the B-52, use a quadricycle landing gear arrangement in which the wheels are on either side of and retract into the fuselage. An extreme variation of this is used on the Russian built An-225, which has a take-off weight of 1.3 million pounds and uses 14 two-wheel bogeys on either side of the fuselage.

Another variation on the quadricycle landing gear arrangement can be seen on the MD-11. This uses a fourth set of wheels, which are located further aft, on the fuselage centerline. These are revealed in the photograph of an MD-11 during take-off in Figure 5.9

The largest portion of the landing gear for which space has to be allotted in the fuselage is the landing gear wheels. Here wheels refer to the hub and tire. The size of the wheels is proportional to the percentage of the aircraft weight that they hold. Most typically, the tires on the main landing gear carry approximately 90 percent of the aircraft weight. The other 10 percent is carried by the nose gear.

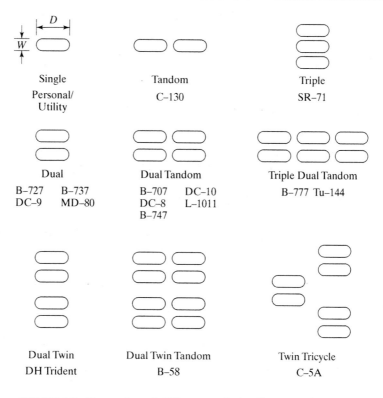

FIGURE 5.8: Illustration of different main landing gear footprints.

At this point in the design, the size of the main landing gear wheels can be estimated using a statistical fit of historic data. This gives the diameter and width of the wheels, with units of inches, as

$$\text{Main Wheel Diameter or Width (in.)} = A W_{\text{main}}^{B}, \tag{5.4}$$

with A and B as given in Table 5.10. The weight on each wheel of the main landing gear depends on the number of wheels, so that

$$W_{\text{main}} = \frac{0.9 W_{\text{TO}}}{N_{\text{wheels}}}. \tag{5.5}$$

After determining the size of the wheels on the main gear, the size of the nose wheel can be assumed to be approximately 40 percent smaller. An exception is for the quadricycle arrangement, where the nose wheel is typically the same size as the main wheels. The calculated values for the diameter and width of all the wheels should be increased by 30 percent if the aircraft is intended to be operated on unpaved runways.

5.1.7 Armament Placement

With combat aircraft, the number and size of bombs and armament are generally decided in the initial design proposal when the mission requirements are set. At this stage, when

FIGURE 5.9: Photograph of MD-11 during take-off, which shows quadricycle landing gear arrangement, with one landing gear on the aft fuselage centerline. (NASA Dryden Research Center Photo Collection.)

TABLE 5.10: Main landing gear wheel sizing coefficients for Eq. [5.4].

	Diameter		Width	
	A	B	A	B
General Aviation	1.510	0.349	0.715	0.312
Business Twin jet	2.690	0.251	1.170	0.216
Transport/Bomber	1.630	0.315	0.104	0.480
Jet Fighter/Trainer	1.590	0.302	0.098	0.467

the design of the fuselage is being considered, the arrangements for storage, positioning and release of weapons need to be examined.

If weapons are carried externally, they add a considerable amount of aerodynamic drag. In some cases, this drag can be as much as that on the total aircraft. A compromise with externally mounted weapons is the use of semi-submerged or conformally carried designs. In these cases, the weapons are partially or fully recessed in the underside of the wing or fuselage.

The lowest drag design has the weapons internally mounted. In most cases, this requires them to be located inside the fuselage, in a weapons bay. In this instance, provisions need to be made in the design of the fuselage to have the necessary volume and exterior access.

5.2   AERODYNAMIC CONSIDERATIONS

5.2.1   Fuselage Fineness Ratio

Once the required internal arrangement and volume of the fuselage have been determined, the exterior shape (length and diameter) are decided.

The fuselage is generally considered to be made up of circular conic leading and trailing sections, and a central cylindrical section. The total length of the fuselage, l, is the sum of the lengths of these three sections. The largest diameter, d, is usually that of the central section.

The ratio of the fuselage maximum diameter to its length is referred to as the fineness ratio, d/l. In terms of the aerodynamic drag, there exists an optimum fineness ratio, although it is different for subsonic and supersonic flight.

Subsonic Aircraft.   For subsonic aircraft, the overall drag coefficient, C_{D_0}, is made up of contributions from the viscous drag, with coefficient C_F, and pressure or form drag, with coefficient $C_{D_{P-\min}}$. The percentage that each contributes to the total depends on the fineness ratio. An example of this is shown in Figure 5.10.

In this figure, the solid curve corresponds to the overall drag. The dashed curve corresponds to the viscous drag. At any value of the fineness ratio, the difference between the values of C_{D_0} and C_F corresponds to the pressure drag. In one extreme, an infinitely slender body ($d/l \simeq 0$) has a drag that is dominated by viscous forces. Conversely, a spherical body ($d/l = 1$) has a drag that is nearly all pressure drag, due to flow

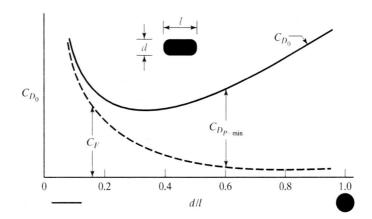

FIGURE 5.10: Effect of fineness ratio on the contributions of viscous and pressure drag to the overall drag of a subsonic body.

TABLE 5.11: Fineness ratio for different passenger transport aircraft.

Subsonic		d/l
	757-200	0.08
	767-200	0.10
	777	0.10
	MD-11	0.10
	A330-300	0.08
Supersonic		d/l
	Concorde	0.05
	Tu-144	0.05

separations. In between, an optimum exists. This optimum, which minimizes the total drag coefficient, occurs at a fineness ratio (d/l) of approximately 0.3.

In a majority of subsonic aircraft, a fineness ratio of 0.3 is not practical and is never used. In most cases, the payload requirements are a more important design driver, and smaller fineness ratios from 0.08 to 0.125 ($8 \leq l/d \leq 12.5$) are used. This is illustrated in the top part of Table 5.11, which gives fineness ratios for a number of subsonic passenger transport aircraft.

Supersonic Aircraft.   For supersonic aircraft, the overall drag coefficient, C_{D_0}, is made up of contributions from the viscous drag, with coefficient C_F, and the supersonic wave drag, with coefficient C_{D_W}. The percentage that each contributes to the total drag, as a function of the fineness ratio, is shown in Figure 5.11.

In general for supersonic flight, the overall drag coefficient on a slender body is 2–3 times higher than for subsonic flight. For a blunt body, approaching a sphere ($d/l = 1$), the overall drag is predominantly wave (bow shock) drag. The rise in wave drag as d/l increases is particularly severe. Only with extremely slender bodies ($d/l \simeq 0$) is the wave drag negligible to the point that the viscous drag becomes dominant.

Again an optimum fineness ratio exists. In this case, it occurs at $d/l = 0.07$ ($l/d \simeq 14$). In contrast to subsonic aircraft, minimizing the aerodynamic drag **is the design driver** for long-range supersonic aircraft, and their fuselage designs use the optimum fineness ratio.

For short or intermediate range supersonic aircraft that operate about 50 percent of the time at subsonic speeds, it is necessary to compromise on the fuselage fineness ratio. However, the severe drag penalty imposed by supersonic flight skews the selection more towards the supersonic optimum. As such, fineness ratios of 0.125 to 0.1 ($8 \leq l/d \leq 10$) are most typically used. This range of values would be representative of supersonic fighter aircraft, such as the F-15. Two examples of supersonic passenger aircraft are given in the lower part of Table 5.11.

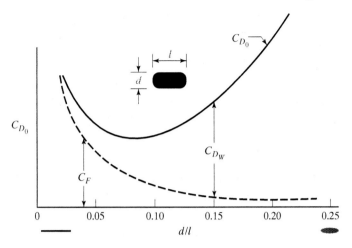

FIGURE 5.11: Effect of fineness ratio on the contributions of viscous and wave drag to the overall drag of a supersonic body.

5.2.2 Fuselage Shapes

Within the design constraints imposed by the volume requirements, the fuselage shape should be aerodynamic, with smooth and gradual dimension changes and blended curves. Of particular concern are large divergence angles that can cause the flow over the fuselage to separate. This would lead to a higher base drag of the fuselage and a reduction in cruise efficiency.

One area that could be prone to separation is the furthest aft portion, where the fuselage ends. The aft body usually has an upward slope to allow ground clearance during pitch-up in the "rotation" portion of take-off. (See Chapter 8.)

Figure 5.12 illustrates some of the different geometries. As a general rule, the total divergence angle should be less than 24 degrees. If this is divided around the fuselage, the local angle should be less than 12 degrees.

5.3 DRAG ESTIMATION

On subsonic aircraft, minimizing the surface or wetted area of the fuselage is one of the most powerful considerations in reducing the drag. The drag force due to viscous drag is simply

$$F_f = q S C_f, \tag{5.6}$$

where q is the dynamic pressure, for example based on cruise conditions, S is the surface area and C_f is the friction coefficient.

The surface area can be estimated in a number of ways. For elliptic cross-sections, the local perimeter, is

$$P(x) \simeq \pi \left[\frac{h(x) + w(x)}{2} \right], \tag{5.7}$$

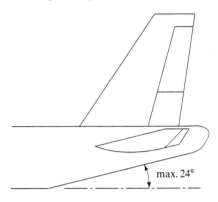

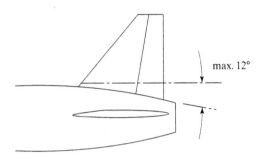

FIGURE 5.12: Schematic drawings showing the divergence angle limits for different types of aft-fuselage designs.

where $h(x)$ and $w(x)$ are the local height and width, respectively, corresponding to the major and minor axes at the streamwise, x, location. In the family of ellipses, the relation gives the exact perimeter length when $w = h$ (circle). To find the total surface area, the perimeter is integrated over the total streamwise extent, namely,

$$S = \int_0^L P(x)dx. \tag{5.8}$$

The integral can be approximated by dividing the fuselage shape into N piecewise-linear x-portions with constant dimensions. In that case, the integral is estimated to be

$$S = \sum_1^N P_i x_i. \tag{5.9}$$

This approximation obviously improves as the number of constant dimension portions increases.

For an elliptic cross-section with constant dimensions (h and w) and length L, Eqs. [5.7 & 5.9] give the total surface area as

$$S \simeq \pi \left[\frac{(hL) + (wL)}{2} \right] \tag{5.10}$$

or

$$S \simeq \pi \left[\frac{A_{\text{side}} + A_{\text{top}}}{2} \right], \tag{5.11}$$

where A_{side} and A_{top} are the areas of the respective side and top projected views of the fuselage.

Thus, a simple method for estimating the total surface area of the fuselage can come by taking the projected areas from the two-view drawings of the fuselage and substituting those into Eq. [5.11]. The projected areas can be determined graphically by drawing to-scale views onto graph paper and counting the number of square cells within the interior of the fuselage boundary. Alternatively, the determination of surface area is a standard function on most computer drafting software.

The friction coefficient used in determining the drag on the fuselage comes from flat plate studies. If the flow is laminar,

$$C_f = \frac{1.328}{\sqrt{\text{Re}_x}} \quad \text{:Laminar,} \tag{5.12}$$

where Re_x is the Reynolds number based on the x-development length of the boundary layer. Reynolds number in this instance is defined as

$$\text{Re}_x = U_o x / v, \tag{5.13}$$

with U_o being the velocity at the outer edge of the boundary layer and v being the kinematic viscosity, at a given flight condition.

The flow is likely to be turbulent when $\sqrt{\text{Re}_x} \geq 1000$. In this event, the friction coefficient for a hydraulically smooth surface is

$$C_f = \frac{0.455}{\left(\log_{10} \text{Re}_x \right)^{2.58} \left(1 + 0.144 M^2 \right)^{0.65}} \quad \text{:Turbulent.} \tag{5.14}$$

The term $\left(1 + 0.144 M^2 \right)^{0.65}$ is a Mach number correction that approaches 1.0 for low Mach numbers.

Surface roughness affects the Reynolds number at which the flow becomes turbulent and increases the friction coefficient. The effect of roughness on boundary layers can be expressed in terms of an effective Reynolds number, which is a function of the roughness height, k, with respect to the boundary layer thickness. The following relations correlate empirical data from flat plate experiments:

$$\text{Re}_{\text{effective}} = 38.21 \left(\frac{x}{k} \right)^{1.053} \quad (M < 1) \tag{5.15}$$

TABLE 5.12: Roughness heights for different aircraft surface conditions.

Surface Type	k $(10^{-5}$ ft$)$
Smooth Molded Composite	0.17
Polished Sheet Metal	0.50
Production Sheet Metal	1.33
Smooth Paint	2.08

and

$$\text{Re}_{\text{effective}} = 44.62 \left(\frac{x}{k}\right)^{1.053} M^{1.16} \quad : (M \geq 1). \tag{5.16}$$

Values of roughness height, k, for different aircraft surface conditions are given in Table 5.12.

The value of Reynolds number used in Eqs. [5.12 & 5.14] should be the one that gives the higher friction coefficient. Thus, if $\text{Re}_{\text{effective}} > \text{Re}_x$, then the effective Reynolds number should be used. In addition, the friction coefficient for turbulent flow will be larger than that for laminar flow. Therefore, an upper (conservative) estimate comes from assuming that the flow is everywhere turbulent on the fuselage.

Even if the flow is considered turbulent everywhere over the fuselage, the friction coefficient is a function of the streamwise location, x. Therefore, the friction force on an elemental streamwise segment of the fuselage is

$$F(x) = q P(x) C_f(x). \tag{5.17}$$

The total force can be determined by integrating Eq. [5.17] over the length of the fuselage. The integral can again be approximated by dividing the fuselage into N piecewise-linear x-portions. In that case, the total viscous drag force can be estimated by

$$F = q \sum_{1}^{N} P_i x_i C_f^i. \tag{5.18}$$

This approximation again improves as the number of fuselage segments increases.

In the area of the nose of the aircraft up to the near-constant diameter fuselage section, the favorable pressure gradient results in a slightly higher friction coefficient compared to the flat plate equivalent. An estimate of the friction coefficient on the nose is then

$$C_{f_{\text{nose}}} \simeq 1.16 C_{f_{\text{flatplate}}}. \tag{5.19}$$

The pressure gradient on the nose virtually assures that the boundary layer is laminar. Therefore, $C_{f_{\text{flatplate}}}$ is determined from Eq. [5.12].

Form Factor.   As with the estimation of the base drag coefficient on the main wing, the estimate of the friction drag coefficient of the fuselage makes use of a form factor, $\mathcal{F}$. For the fuselage, this is given by

$$\mathcal{F} = 1 + \frac{60}{f^3} + \frac{f}{400}, \tag{5.20}$$

where f corresponds to the fuselage inverse fineness ratio, l/d.

In most cases, the fuselage has a negligible interference factor. Therefore, $\mathcal{Q} = 1$ is appropriate.

The total viscous drag on the fuselage is then

$$F_f = q S C_f \mathcal{F} \mathcal{Q}, \tag{5.21}$$

where again S is the fuselage surface (wetted) area.

Quantitative Shapes.   There are a number of quantitative fuselage shapes for which drag data are available. These are useful as a starting point in laying out the fuselage since their analytic form allows direct integration for determining the surface area and volume. The following is a list of the most useful of these as illustrated in Figure 5.13.

1. Cone-cylinder. In this shape, the nose of the fuselage is a right circular cone whose base matches onto a constant diameter cylindrical fuselage section. With this shape, at subsonic Mach numbers, the total drag force on the fuselage is primarily due to viscous drag, with viscous drag coefficients that are determined using the relations presented in the previous section. At supersonic Mach numbers, the wave drag coefficient, C_{Dw} is equal to the pressure coefficient, C_p, due to the conical shock wave that will form at the leading nose. The C_p values are easily determined from conical shock charts that are available in most compressible flow handbooks.

2. Power series-cylinder. This shape is similar to the cone-cylinder except that the nose section is derived from the following relation

$$\frac{r(x)}{r(l_c)} = \left(\frac{x}{l_c}\right)^n, \tag{5.22}$$

where l_c is the length of the nose section, $r(l_c)$ is the radius at the base ($x = l_c$) and $r(x)$ is the local radius. When $n = 1$, Eq. [5.22] describes a right-circular cone. A good amount of drag data are available for a variety of values of n. For this family of shapes, $n = 3/4$ gives the minimum wave drag.

3. Von Karman Ogive. This is a symmetric body of revolution that is described by the following relation:

$$\left[\frac{r(x)}{r(0)}\right]^2 = \frac{1}{\pi}\left[\frac{2x}{l}\sqrt{1 - \left(\frac{2x}{l}\right)^2} + \cos^{-1}\left(\frac{-2x}{l}\right)\right] \quad (-l/2 \le x \le l/2). \tag{5.23}$$

For this, $r(0)$ is the maximum radius, which occurs at $x = 0$ and l is the overall length.

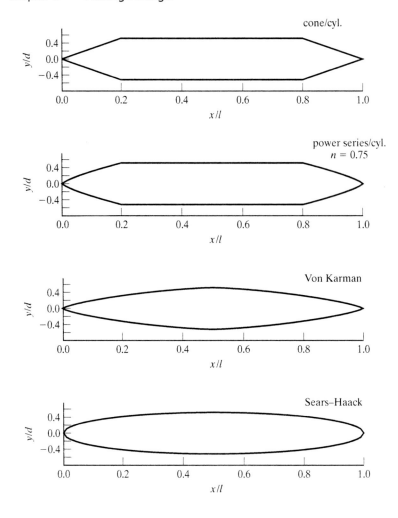

FIGURE 5.13: Schematic drawings of different quantitative fuselage shapes.

In a fuselage design, Eq. [5.23] will form the leading half, namely, from the leading point to the point of largest diameter. Thus, if the fuselage length is intended to be L, and the maximum diameter is D, then in Eq. [5.23], $l = L/4$, $r(0) = D/2$ and $x = x' - L/4$, where x' is the streamwise position along the fuselage starting from the most leading point ($0 \leq x' \leq L$). The downstream half of fuselage can be made as a mirror image of the leading half. An example of this is presented for the case study supersonic business jet. The overall volume of the body is

$$\text{Volume} = \frac{l}{2} A_{\max} = \frac{l}{2} \pi r(0)^2. \tag{5.24}$$

This shape is primarily suited to supersonic aircraft, where its wave drag coefficient is the lowest of the group of shapes listed:

$$C_{DW} = \frac{4A_{\max}}{\pi l^2} = 4\left[\frac{r(0)}{l}\right]^2. \tag{5.25}$$

4. Sears–Haack. This is a symmetric body of revolution that also has a relatively low wave drag compared to the other shapes. The profile is described by the following relation:

$$\left[\frac{r(x)}{r(0)}\right]^2 = \left[1 - \left(\frac{2x}{l}\right)^2\right]^{3/2} \quad (-l/2 \le x \le l/2). \tag{5.26}$$

Note that in contrast to Eq. [5.23], which only described the leading half of the fuselage, Eq. [5.26] describes the complete fuselage, from leading to trailing points. The overall volume of the body in this case is

$$\text{Volume} = \frac{3}{16}\pi l A_{\max} = \frac{3}{16}l[\pi r(0)]^2. \tag{5.27}$$

The surface (wetted) area is

$$S = 1.8667[(\text{Volume})(l)]^{1/2} = 0.8083\pi l r(0). \tag{5.28}$$

The wave drag coefficient is given as

$$C_{DW} = \frac{9}{2}\frac{\pi}{l^2}A_{\max} = \frac{9}{2}\left[\frac{\pi r(0)}{l}\right]^2. \tag{5.29}$$

It is evident from the formulas for the wave drag coefficients given in Eqs. [5.25 & 5.29], that the wave drag depends on the cross-sectional area. This applies not only to the fuselage, but to the fuselage and wing together. As a result, the cross-section of the fuselage is often indented in the vicinity of the wing attachment location in order to keep a nearly constant and smooth wing-fuselage cross-section area distribution along the length of the aircraft. This process is called "area ruling." A properly area-ruled fuselage design can reduce the wave drag by as much as 50 percent over a non-area-ruled design. An example of an area-ruled fuselage design is shown in Figure 5.14.

5.4 SPREADSHEET FOR FUSELAGE DESIGN

The relations used for the design of the fuselage have been incorporated into a spreadsheet file named **fuse.xls**. The format allows easy input and modification of the design parameters and provides a graphical view of fuselage perimeter dimension, the value of the total drag and the equivalent drag coefficient based on the reference main wing area. A sample of the spreadsheet is shown in Figure 5.15. This contains the parameters for the conceptual supersonic business jet that was proposed in Chapter 1.

FIGURE 5.14: Photograph of F-106 that demonstrates the thinning of the fuselage in the region of the wing according to area ruling used to minimize wave drag in supersonic aircraft. (NASA Dryden Research Center Photo Collection.)

In the spreadsheet, there are two areas where the input parameters are placed. These correspond to the flight regime data, and the dimension data. The flight regime would correspond to that phase of the flight plan that is the most important design driver. Generally, these calculations are intended to determine the drag on the fuselage, which when combined with the other components, is used to size the engines. Often this is done for cruise conditions.

In the spreadsheet, the input parameters are the cruise Mach number and cruise altitude. Relations that are identical to those used in previous spreadsheets are used to determine the velocity, V; density, ρ; and dynamic pressure, q, at the cruise altitude, H. In addition, the kinematic viscosity, ν is determined. Here a constant average viscosity with altitude, μ, is assumed, whereby $\nu = \mu/\rho$. Actually, μ is a weak function of altitude; however, this is primarily due to changes in ρ.

The dimension data come from filling the requirements of having the necessary volume to enclose crew, passengers, payload, etc., as defined by the mission requirements. This generally starts with specifying a maximum diameter (or equivalent diameter for non-circular cross-sections). The length of the fuselage is specified by specifying the fineness ratio, d/l. As discussed earlier, for subsonic aircraft, the choice of the fineness ratio is not critical. However, for supersonic aircraft, in order to minimize the wave drag, the fineness ratio should be near the optimum, $d/l = 0.07$ ($l/d = 14$).

			Fuselage Design					
Flight Regime Data:								
Cruise Mach	2.1							
Cruise Alt. (ft)	55,000							
V (f/s)	1,925.70							
ρ (lbm/f^3)	0.01							
q (lbf/f^2)	531.07							
μ (lbm/(f–s))	0							
ν (cruise) (f^2/s)	0							
Dimension Data:			**Form Factors:**					
D–max (ft)	9		F	1.06				
L/D	14		Q	1				
L (ft)	126		F*Q	1.06				
S (f^2)	519							
Viscous Drag Calculations:			*Von–Karman Ogive Fuselage Shape*					
x/L	x (ft)	x−L/4 (ft)	D (ft)	P (ft)	Sw(ft^2)	Re_x	C_F	Drag (lbf)
0.00	0.00	−31.50	0	0.0				
0.10	12.60	−18.90	3.4	10.7	134.4	2.1E+07	1.94E−03	147
0.20	25.20	−6.30	5.5	17.3	217.7	4.2E+07	1.75E−03	214
0.30	37.80	6.30	7.12	22.4	282.0	6.3E+07	1.65E−03	261
0.40	50.40	18.90	8.33	26.2	329.9	8.4E+07	1.59E−03	294
0.50	63.00	31.50	9	28.3	356.3	1.0E+08	1.54E−03	307
0.60	75.60	−	8.33	26.2	329.9	1.3E+08	1.50E−03	277
0.70	88.20	−	7.12	22.4	282.0	1.5E+08	1.47E−03	232
0.80	100.80	−	5.5	17.3	217.7	1.7E+08	1.44E−03	176
0.90	113.40	−	3.4	10.7	134.4	1.9E+08	1.42E−03	107
1	126.00	−	0	0.0	0.0	2.1E+08	1.40E−03	0
Totals:					2284.4			2015
Wave Drag Calculations:								
A_max	63.62							
CDW	0.02							
Drag (lbf)	**689.5**							
Total Drag:	**2705**							
(lbf)								
Equiv. CD	0.0098							

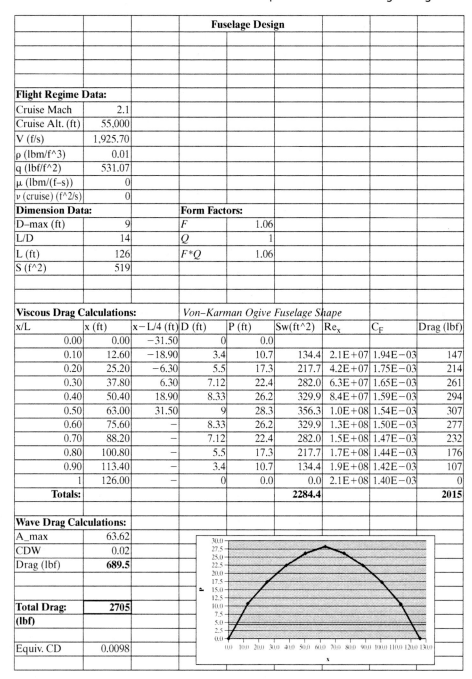

FIGURE 5.15: Spreadsheet for fuselage design (FUSE) showing results for conceptual supersonic business jet.

Included in the dimensions input is the wing area. This is not necessary in the calculation of the fuselage drag, but is used to provide a reference equivalent drag coefficient that can be compared to the wing value.

The drag calculations consider viscous drag and wave drag (in the case of supersonic aircraft). The procedure for the viscous drag is to divide up the fuselage length into 10 equal elements. The first column then shows x/l in 10 percent increments. The equivalent locations along the fuselage are given in the second column. As given in Eq. [5.6], the viscous drag corresponds to the product of the dynamic pressure, surface (wetted) area and the local friction coefficient. The product of the form factor, $\mathcal{F}$, and interference factor, $\mathcal{Q}$, is then multiplied by the viscous drag to satisfy Eq. [5.21]. The formula for $\mathcal{F}$ is taken from Eq. [5.20], and $\mathcal{Q} = 1$.

The diameter at each x/l station can be either input by hand, or as in the case study, calculated from a given analytic function (Von Karman Ogive). The local perimeter is based on the dimension and shape at each x/l-location. In the case study, the cross-section is circular. An estimate of the perimeter for an elliptic cross-section was given in Eq. [5.7]. The product of the local perimeter and segment length gives the surface area. Equation [5.10] is used for estimating the surface area of a segment with an elliptic cross-section.

The friction coefficient is inversely proportional to Reynolds number. The equation to estimate the friction coefficient depends on if the flow is laminar or turbulent. The spreadsheet uses the following logic:

If $\sqrt{\mathrm{Re}_x} < 1000$, the flow is presumed to be laminar, and C_f is based on Eq. [5.12];

otherwise, the flow is presumed to be turbulent, and C_f is based on Eq. [5.14].

The drag corresponding to each segment is calculated based on the local surface area and friction coefficient. These are then summed up to obtain the total viscous drag.

In the case of a supersonic aircraft, the other important drag component is wave drag. In some shapes, such as the Von Karman Ogive and the Sears–Haack, equations exist for determining the wave drag. These are functions of the maximum cross-section area, $A_{\max}$, such as given in Eqs. [5.23 & 5.27]. The Von Karman Ogive can be used as the shape of the leading portion of the fuselage and, therefore, offers an easy method for determining the wave drag coefficient. Alternatively, the leading portion can be approximated by a right circular cone. In that case, $C_{D_W} = C_p$, and C_p can be found from shock tables.

The spreadsheet is configured to calculate the wave drag coefficient, C_{D_W}, for a Von Karman Ogive shape, with a length that corresponds to half of the overall fuselage length. Other formulas can be used or values taken from shock tables can be input as is appropriate to the fuselage shape. The wave drag is defined as

$$F_W = q\, A_{\max} C_{D_W}. \qquad (5.30)$$

The total drag force is defined as the sum of the viscous drag and wave drag forces. For subsonic aircraft, the total drag should only be taken to be the viscous drag. The trend in the design of subsonic aircraft is to use smaller fineness ratios (larger l/d) than is optimum for the overall drag. As a result, in a good design that minimizes large flow separations, a majority the total drag is due to the viscous drag.

Finally, an equivalent drag coefficient to that of the main wing is calculated at the bottom of the spread sheet. This is defined as

$$C_{D_0} = \frac{F_f + F_W}{q S},$$

(5.31)

where S corresponds to the area of the main wing. The value given in the spreadsheet can then be compared to C_D for the wing, which was derived in Chapter 4, in order to see the relative contributions of each to the overall drag on the aircraft.

5.4.1 Case Study: Wing Design

The passenger compartment was designed to comfortably seat from 12 to 15 passengers. The diameter of the fuselage was based on having two seats that are separated by a center aisle. This arrangement is comparable to aircraft with a similar number of passengers (Table 5.2). Based on the guidelines for passenger comfort requirements given in Table 5.1, Table 5.13 lists the proposed seating arrangement for the conceptual SSBJ.

This arrangement is equivalent to first-class seating on a commercial passenger aircraft. The diameter of the fuselage is based on the sum of the seat and aisle widths plus a 4-inch fuselage wall thickness, which is common for aircraft of this type. This gives a fuselage diameter of 9 feet. This is illustrated in Figure 5.16.

The length of the fuselage was stipulated by having an optimum fineness ratio of $d/l = 0.07$ ($l/d = 14$). Therefore, based on $d = 9$ f, $l = 126$ f.

Because long cruise range is a principle design driver in this aircraft, the Von Karman Ogive shape was chosen for the fuselage, because it has the lowest wave drag among the well-known and documented shapes. In order to use Eq. [5.21] to define the local diameters of the fuselage, the length, l, used in the equation, corresponds to half the total length of the fuselage. In addition, to accommodate Eq. [5.21], the range of the

TABLE 5.13: Conceptual SSBJ passenger compartment data.

Passengers/Cabin	12–15
Seats Across	2
Number of Aisles	1
Seat Width (in.)	35
Seat Pitch (in.)	40
Headroom (in.)	66
Aisle Width (in.)	30
Aisle Height (in.)	78
Lavatories	1
Galley Volume (f^3)	160
Baggage/Passenger (lbs)	40

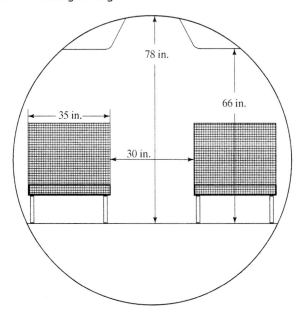

FIGURE 5.16: Cross-section drawing of fuselage design for conceptual supersonic business jet.

streamwise locations on the fuselage is $-l/4 \leq x \leq l/4$. This requires that the original x-locations, up to $x = l/2$, be shifted by $-l/4$. This will define the first half of the fuselage length. The second half of the fuselage is taken to be the mirror image of the first half.

The cross-section of the fuselage is circular, so that the perimeter is $P = \pi D$. The values are based on the diameter at the end of each segment. The surface area is then the product of the perimeter and segment length.

The local Reynolds numbers are based on the x-locations at the end of each segment. In all cases, $\sqrt{\mathrm{Re}_x} > 1000$ so that the flow is assumed to be turbulent everywhere. Equation [5.14] is then used to determine the local drag coefficient. Finally, the drag force corresponding to each segment is calculated and summed up at the bottom of the column. In this case, the total viscous drag force was estimated to be 2015 lbs. The form factor corresponded to $\mathcal{F} = 1.056$, which added 108 lbs (approximately 5 percent) to the viscous drag force.

Equation [5.23] was used to determine the wave drag coefficient on the fuselage. This was based on the maximum cross-section area, which occurs at $x/l = 0.5$. The total wave drag force was found to be 689 lbs. Summing the viscous and wave drag forces, the total drag on the fuselage was estimated to be 2705 lbs. Using the total drag force on the fuselage, the equivalent drag coefficient, with the reference main wing area was found to be approximately 0.0061. This turned out to be approximately four times smaller than C_D for the main wing (given in **wing.xls**).

The local perimeter at the end of each x-station is automatically plotted at the bottom of the spreadsheet. This is useful to highlight any regions where sharp changes in the fuselage dimensions might exist. Attempts should be made to eliminate any jumps in the fuselage dimensions since these can lead to flow separations and an increase in the drag force.

In this design, the main wing will be mounted downstream of the largest diameter of the fuselage. As a result, the added cross-section area of the wing will be compensated for by the decrease in the cross-section area of the fuselage. The Ogive shape selected for this design is convenient for achieving this area-ruling.

5.5 PROBLEMS

5.1. Using an aircraft reference book such as "Jane's All the World Aircraft," tabulate and plot the fineness ratio, d/l, for different aircraft as a function of their design cruise Mach numbers. Discuss the historical trend in context to Figures 5.10 and 5.11.

5.2. For the conditions of the supersonic business jet case study, keeping the fuselage diameter fixed, investigate the effect of the inverse fineness ratio in the range, $3 \leq l/d \leq 16$ on
  1. the drag due to skin friction;
  2. the wave drag;
  3. the ratio of the skin friction drag to wave drag;
  4. the overall drag.
  Plot these as a function of d/l and compare the results to Figure 5.11.

5.3. For the conditions of the supersonic business jet case study, keeping the fuselage fineness ratio fixed, investigate the effect of the maximum diameter in the range, $4\,f \leq d \leq 12\,f$ on
  1. the drag due to skin friction;
  2. the wave drag;
  3. the ratio of the skin friction drag to wave drag;
  4. the overall drag.
  Plot these as a function of d. How does this relate to "area-ruling"?

5.4. For the conditions of the supersonic business jet case study, investigate the effect of the cruise altitude in the range, $30{,}000\,f \leq H \leq 65{,}000\,f$, on:
  1. the drag due to skin friction
  2. the wave drag
  3. the ratio of the skin friction drag to wave drag
  4. the overall drag
  Plot these as a function of H. Discuss the physics behind the trends.

5.5. Consider a private 4-place aircraft with the following characteristics:
  — Cruise Mach number = 0.2
  — Cruise altitude = 10,000 f
  Design the fuselage which would be suitable for this aircraft by specifying the following:
  1. Maximum fuselage diameter
  2. Fineness ratio
  3. Fuselage shape
  Determine the overall fuselage drag for your design.

5.6. Consider a jet-powered shuttle aircraft that is to have the following characteristics:
  — 6–9 passengers, 2 crew, side-by-side seating,
  — Cruise Mach number of 0.8,
  — Cruise altitude of 40,000 f

The fuselage has a circular cross-section with a maximum diameter of 80 inches and a length of 54 feet. The leading nose and trailing cone of the fuselage are formed as power-series cylinders with $n = 3/4$. The length of these cone sections correspond to 20 percent each, of the total fuselage length. For this, determine the total drag on the fuselage.

5.7. For the aircraft in Problem [5.5], how much will the fuselage drag increase if the constant diameter section of the fuselage is lengthened by 40 percent in order to carry more passengers?

CHAPTER 6

Horizontal and Vertical Tail Design

Photograph of Lockheed F-22 Raptor, which illustrates an intricate tail design that addresses issues of maneuverability, survivability, and stealth. (Lockheed Martin photograph.)

This chapter deals with the design and placement of the horizontal and vertical tail surfaces. Although these surfaces are more traditionally located at the aft portion of the fuselage, forward horizontal (canard) surfaces are also discussed.

A large variety of horizontal and vertical tail designs have been used in the past. A sampling of these is shown in the aircraft photographs in Figure 6.1. Their choice and

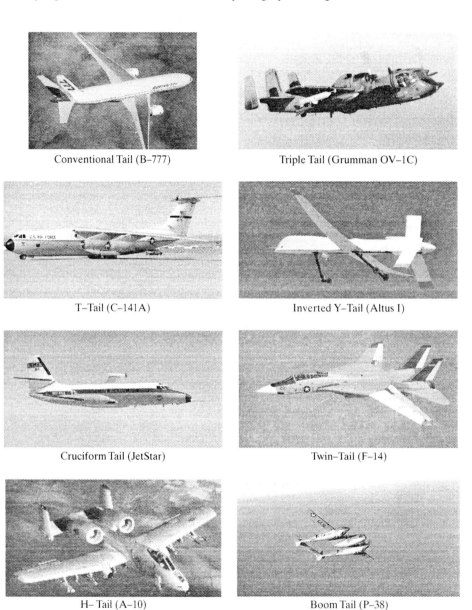

Conventional Tail (B–777)   Triple Tail (Grumman OV–1C)

T–Tail (C–141A)   Inverted Y–Tail (Altus I)

Cruciform Tail (JetStar)   Twin–Tail (F–14)

H– Tail (A–10)   Boom Tail (P–38)

FIGURE 6.1: Photographs of aircraft with different tail designs.

placement depends on a number of factors including weight, stability and control, spin recovery, survivability, and combat stealth.

Empirical relations will be used in sizing the vertical and horizontal tail surfaces. These are based on coefficients that correlate the tail surface areas and their locations on the fuselage, with the wing area, chord, and span of historic aircraft.

The selection of the tail airfoil section types will be based on the same procedures that were used for the design of the main wing, and covered in Chapter 4. The focus will be on the use of symmetric airfoil sections, which have a low base drag coefficient. The choice of the planform shape of the tail surfaces will also be based on historic trends.

At this stage of the design, no attempt will be made to design the tail control surfaces. This step is left for Chapter 11, which performs a detailed analysis of the static stability and control.

The final objective of this chapter is to obtain a quantitative estimate of the 3-D lift characteristics and total drag on the horizontal and vertical tail surfaces. This follows the same procedures used for the design of the main wing. As a result, many of the formulas and spreadsheet elements that were used in Chapter 4 have been duplicated in this chapter. The drag produced by the tail surfaces is then summed with those from the main wing and fuselage as the final step towards sizing the engine(s) for the aircraft, which is presented in the next chapter.

6.1 TAIL ARRANGEMENTS

A variety of tail designs have been used on past aircraft. All of these are intended to provide certain benefits to a design, and the selection of one over another depends on which of these best meets the overall mission requirements for the aircraft. The following list details some of these tail types and their characteristics and benefits.

1. Conventional tail. A majority of commercial and general purpose aircraft use this tail design. An example is the Boeing 777 aircraft shown in Figure 6.1. This design places the horizontal stabilizer at or near the fuselage vertical centerline. The advantages of this tail design are that it provides sufficient stability and control, and it has the lowest tail weight. A large tail weight is a particular problem since static stability requires that the center of gravity be forward of the center of lift. A tail that is too heavy can force a redistribution of other weight or a change in the position of the main wing, which sometimes can be difficult.

2. T-tail. The T-tail is also a relatively popular design. Examples are the Boeing 727, Douglas YC-15 (shown in Figure 1.4), and the C-141 transport shown in Figure 6.1. This design places the horizontal tail high on the end of the vertical tail. It has two main advantages. The first is that the vertical tail can be smaller than on a conventional tail because the placement of the horizontal stabilizer acts as a winglet and increases the effective aspect ratio. The second advantage is that the horizontal stabilizer can also be made smaller because it is placed high, out of the wake of the main wing. The main disadvantage of the T-tail is that it is heavier than the conventional tail design, since the vertical tail structure needs to be made stronger in order to carry the load of the horizontal tail.

3. <u>Cruciform tail</u>. The Cruciform tail is a compromise between the conventional and T-tail designs, where the horizontal tail is at the approximate mid-span of the vertical tail. An example is the JetStar shown in Figure 6.1. Its advantages are that it raises the horizontal stabilizer out of the wake of the main wing, with less of a weight penalty compared to the T-tail. However, because the horizontal stabilizer is not at the end of the vertical stabilizer, there is no reduction in the vertical tail aspect-ratio requirement that comes with the T-tail.

4. <u>H-tail</u>. The H-tail is a popular design for some combat aircraft. An example is the YA-10 shown in Figure 6.1 (and Figure 1.1). The advantages of the H-tail design are that it positions the vertical stabilizers in air, which is not disturbed by the fuselage, and that it reduces the required size of the horizontal stabilizer because of the winglet effect of the vertical tail surfaces. Another particular advantage is that it lowers the required height of the vertical tail. This is particularly important on aircraft that must have a low clearance height or on combat aircraft where it reduces the projected area of this vulnerable component. The required added strength of the horizontal stabilizer makes the H-tail heavier than the conventional tail.

5. <u>V-tail</u>. A V-tail is designed to reduce the surface (wetted) area by combining the vertical and horizontal tail surfaces. Control in this case is through "ruddervators." For example, a downward deflection of the right elevator and an upward deflection of the left elevator will push the tail to the left, and thereby the nose to the right. Unfortunately, this same maneuver produces a roll moment toward the left, which opposes the turn. This effect is called an "adverse yaw-roll." The solution to this is an inverted V-tail.

6. <u>Inverted V-tail</u>. An inverted V-tail avoids the adverse yaw-roll coupling of the V-tail. In this case, the elevator deflections produce a complimentary roll moment, which enhances a coordinated turn maneuver. This design also reduces spiral tendencies in the aircraft. The only disadvantage of the inverted V-tail is the need for extra ground clearance.

7. <u>Y-tail</u>. The Y-tail is similar to the V-tail except that a vertical tail surface and vertical rudder are used for directional control. This eliminates the complexity of the "ruddervators" on the V-tail, but still retains a lower surface area compared to the conventional tail design. An inverted Y-tail was used on the F-4 as a means of keeping the horizontal surfaces out of the wake of the main wing at high angles of attack. A photograph is shown in Figure 6.2. Another example of an inverted Y-tail is on the Altus I high-altitude long-duration surveillance drone, which is shown in Figure 6.1.

8. <u>Twin-tail</u>. The twin-tail is a common design on highly maneuverable combat aircraft. Examples include the F-14, shown in Figure 6.1, and the F-15, F-18, Mig-25, and F-22 shown at the beginning of the chapter. The purpose of the twin-tail is to position the vertical tail surfaces and rudders away from the fuselage centerline, where it can be affected by the fuselage wake at high angles of attack.

9. <u>Canard</u>. This is a horizontal stabilizer that is located forward of the main wing, on the fuselage. The canard can be designed to provide very little lift, compared to the main wing, or up to 15–25 percent of the total lift. The former is called a *control* canard; the latter is called a *lifting* canard. The control canard provides the same

FIGURE 6.2: Photograph of F-4 that illustrates an inverted Y-tail, which was designed to keep the control surfaces out of the wake of the main wing at high angle-of-attack flight. (Courtesy of the USAF Museum Archives.)

function as the aft horizontal stabilizer by introducing a moment that changes the angle of attack of the fuselage and main wing. Examples of control canards can be found on the Concorde and Tu-155, which was shown in Figure 5.4.

The lifting canard carries a larger portion of the lift compared to the control canard and, therefore, reduces the lift on the main wing. The lifting canard is designed to stall at a lower angle of attack than the main wing. As a result, the nose of the aircraft will drop before the main wing can stall and, therefore, make it statically stable.

In principle, the lifting canard design lowers the overall drag on the aircraft by reducing the lift, and thereby reducing the lift-induced drag on the main wing. In addition, in contrast to an aft tail, a canard uses a positive (downward) elevator to offset the moment produced by the main wing in level flight. This produces an upward lift component, which augments the main wing and further reduces the lift and lift-induced drag on the main wing. An example of an aircraft that uses a lifting canard is shown in Figure 6.3.

FIGURE 6.3: Photograph of Q-200 "Quickie," which demonstrates a lifting canard design. The canard has a plain elevator and also doubles as the main landing gear spring.

6.2 HORIZONTAL AND VERTICAL TAIL SIZING

In the conceptual design, the sizing of the vertical and horizontal tail surfaces is based on historical trends. This is done through coefficients that correlate features of different aircraft that are relevant to the tail design. The aircraft used in obtaining these coefficients are usually grouped according to their general mission requirements, such as range, cruise Mach number, high maneuverability, etc.

6.2.1 Vertical Tail Sizing

The coefficient that is used in scaling the vertical stabilizer is referred to as, C_{VT}. The area of the vertical stabilizer is found from the equation

$$S_{VT} = C_{VT} \frac{b_W S_W}{l_{VT}}, \tag{6.1}$$

where b_W and S_W are the span and area of the main wing, respectively, and l_{VT} is the distance between the quarter-chord locations of the mean-aerodynamic-chords (m.a.c.) of the main wing and vertical stabilizer. This is illustrated in Figure 6.4. Note that the area of the vertical stabilizer includes only the portion that is exposed above the fuselage. Values of C_{VT} for different types of aircraft are listed in Table 6.1.

The coefficient C_{VT}, should be taken from aircraft with similar mission requirements. At this stage of the design, the main wing is designed so that S_W and b_W are known. In addition, the fuselage is designed so that the vertical tail can be placed on the fuselage with respect to the main wing position. The distance, l_{VT}, is in effect the moment arm upon which the aerodynamic force generated by the vertical stabilizer acts on the fuselage. Equation [6.1] indicates that a larger distance requires a smaller vertical tail area. Therefore, this length is a useful parameter in the design of the tail.

6.2.2 Aft-Horizontal Tail Sizing

The coefficient that is used in scaling the aft-horizontal stabilizer is C_{HT}. This is referred to as an <u>aft</u> stabilizer in order to distinguish it from a forward canard.

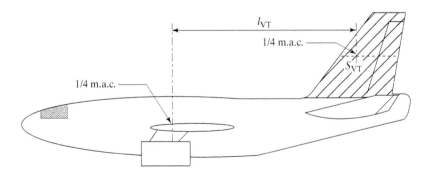

FIGURE 6.4: Schematic illustrating the distance l_{VT} used in the vertical tail sizing.

TABLE 6.1: Vertical and aft-horizontal tail coefficients.

	C_{VT}	C_{HT}
Sail Plane	0.02	0.50
Homebuilt	0.04	0.50
General Aviation (single engine)	0.04	0.70
General Aviation (twin engine)	0.07	0.80
Twin Turboprop	0.08	0.90
Combat Jet Trainer	0.06	0.70
Combat Jet Fighter	0.07	0.40
Military Transport/Bomber	0.08	1.00
Commercial Jet Transport	0.09	1.00

The area of the aft stabilizer is given by

$$S_{\mathrm{HT}} = C_{\mathrm{HT}} \frac{\bar{c}_W S_W}{l_{\mathrm{HT}}}, \tag{6.2}$$

where $\bar{c}_W$ is the m.a.c. of the main wing and l_{HT} is the distance between the quarter-chord locations of the mean-aerodynamic-chords (m.a.c.) of the main wing and horizontal stabilizer. These parameters are illustrated in Figure 6.5. Values of C_{HT} based on different types of historic aircraft are also listed in Table 6.1.

Note that in contrast to the vertical stabilizer, S_{HT} *includes* the portion that runs through the fuselage and, therefore, is not exposed. This is consistent with the definition of the main wing area. As with the vertical tail design, C_{HT} should be taken from aircraft with similar mission requirements.

6.2.3 Canard Sizing

The coefficient used in scaling a forward-horizontal stabilizer (canard) is C_C. The area of the canard is given by

$$S_C = C_C \frac{\bar{C}_W S_W}{l_C}, \tag{6.3}$$

where l_C is the distance between the quarter-chord locations of the mean-aerodynamic-chords (m.a.c.) of the main wing and canard. The distance l_C is illustrated in Figure 6.6. Values of C_C based on different types of aircraft are listed in Table 6.2.

TABLE 6.2: Control-canard sizing coefficient.

	C_C	Cruise Mach No.
B-70	0.104	2+
CL-408	0.12	3
NAA-M3.0	0.10	3
F-108	0.11	2+

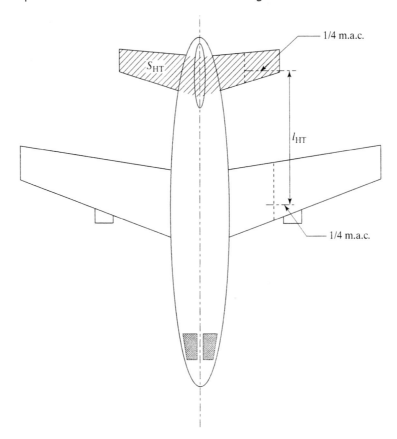

FIGURE 6.5: Schematic illustrating the distance l_{HT} used in the aft-horizontal tail sizing.

The values of the sizing coefficient in Table 6.2 are only relevant for control canards. For lifting canards, the area is primarily based on the percentage of the total lift that the canard is designed to produce. This is most typically from 15 to 25 percent.

In contrast to the aft-horizontal stabilizer, the area of the canard, S_C, includes only the exposed portion, outside of the fuselage. This is consistent with the vertical stabilizer.

6.2.4   Scaling for Different Tail Types

As a general trend, for T-tail designs, the vertical and horizontal tail coefficients can be reduced by 5 percent compared to a conventional tail. For an H-tail design, the horizontal tail coefficient can be reduced by 5 percent. Also with an H-tail, the vertical tail area on each side will be one-half of the required total area corresponding to a conventional tail.

For a V-tail design, the area should be the same as the combined horizontal and vertical surface areas of an equivalent conventional tail design. In addition, the dihedral angle of the two surfaces should be the arc-tangent of the square root of the ratio between the required vertical and horizontal tail areas. This is illustrated in Figure 6.7 and should give an angle of approximately 45°.

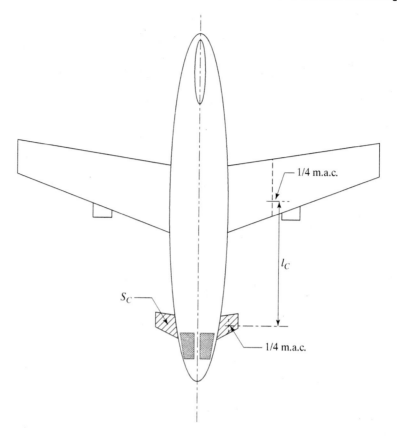

FIGURE 6.6: Schematic illustrating the distance l_C used in the canard sizing.

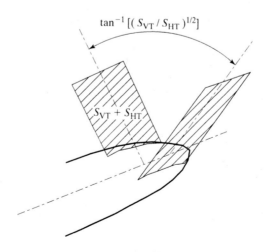

FIGURE 6.7: Schematic illustrating the area and dihedral angle for a V-tail, where S_{VT} and S_{HT} are based on a conventional tail.

TABLE 6.3: Coefficient scaling for different tail types.

Type	Equivalent C_{VT}	Equivalent C_{HT}
T-Tail	0.95	—
H-Tail	0.50	0.95
V-Tail	1.00	1.00

TABLE 6.4: Typical lengths, l_{VT}, l_{HT}, and l_C.

Type	$l_{Tail}/l_{Fuselage}$
Front-Mounted Prop.	0.60
Wing-Mounted Engines	0.50–0.55
Fuselage-Mounted Engines	0.45–0.50
Canard	0.30–0.50

The lengths, l_{VT} and l_{HT}, will vary somewhat depending on the type of aircraft. For an aircraft with a front-mounted propeller engine, these lengths are approximately 60 percent of the fuselage length. With an aircraft with wing-mounted engines, these lengths are approximately 50–55 percent of the fuselage length. For engines that are mounted on the aft portion of the fuselage, these lengths are from 45–50 percent of the fuselage length.

With control canards, l_C, varies from 30 to 50 percent of the fuselage length. All these are summarized in Tables 6.3 and 6.4.

6.3   TAIL PLANFORM SHAPE

Once the required areas of the horizontal and vertical surfaces are found, the planform shapes are next determined. As with the main wing design, the planform shape is defined by the aspect ratio, A, and the taper ratio, λ. The aspect ratio, given in Chapter 4, relates the area, S, and the span, b, as

$$A = \frac{b^2}{S}. \tag{6.4}$$

The taper ratio then defines the root chord, C_r, as

$$C_r = \frac{2S}{b(1 + \lambda)} \tag{6.5}$$

and the tip chord as

$$C_t = \lambda C_r. \tag{6.6}$$

Historic values of aspect and taper ratios of aft-horizontal and vertical tail surfaces are given in Table 6.5.

The leading edge sweep angle, Λ_{LE}, of the aft-horizontal stabilizer is typically set to be a few degrees more than the sweep angle of the main wing. This gives the aft-tail

TABLE 6.5: Aft-horizontal and vertical tail aspect and taper ratios based on historic aircraft.

	Aft-horizontal		Vertical	
	A	λ	A	λ
Combat	3–4	0.2–0.4	0.6–1.4	0.2–0.4
Sail Plane	6–10	0.3–0.5	1.5–2.0	0.4–0.6
Other	3–5	0.3–0.6	1.3–2.0	0.3–0.6
T-Tail	—	—	0.7–1.2	0.6–1.0

a higher critical Mach number than the main wing. This also helps to avoid the loss of elevator effectiveness due to shock formation. With the same sweep angle as the main wing, the same benefits can be accomplished by reducing $(t/c)_{max}$ of the horizontal stabilizer compared to the main wing.

The sweep angle of the vertical stabilizer generally varies between 35 and 55°. For supersonic aircraft, higher sweep angles may be used if the leading-edge Mach number is intended to be subsonic.

6.4   AIRFOIL SECTION TYPE

The selection of the airfoil section type used for the horizontal and vertical stabilizers should be based on

1. being a symmetric airfoil and
2. having a low base drag coefficient, C_{D_0}.

The tail surfaces do not produce lift except with the deflection of control surfaces, which are the elevator and rudder for the horizontal and vertical stabilizers, respectively. As a result, the stabilizers should be symmetric airfoils that are not placed at an angle of attack. When the control surfaces are deflected, the effect is equivalent to adding camber to the section shape. Only then is lift produced.

Both of the vertical stabilizers can be considered to be wings. As such, the analysis that was done in Chapter 4 for the design of the main wing is also used here in order to determine the 3-D characteristics, $dC_L/d\alpha$ and C_{D_0}.

In general, we would like to maximize $dC_L/d\alpha$, since this will maximize the effect of deflecting the control surfaces. Recall from the main wing design (Eq. (4.9)), that $dC_L/d\alpha$ increases with increasing aspect ratio, A, and decreases with increasing leading-edge sweep angle, Λ_{LE}. Tail designs such as the T-tail or H-tail, which produce winglet-type effects on the ends of the vertical or horizontal stabilizers, increase the effective aspect ratio and, therefore, improve $dC_L/d\alpha$.

Because the stabilizers are symmetric sections, at 0° angle of attack, they do not produce lift or lift-induced drag. Therefore, the only drag component is the base drag, C_{D_0}. As a result, wing sections that have a lower base drag are preferable for cruise efficiency.

TABLE 6.6: Values of interference factor, Q, for different tail arrangements.

	Q
Conventional Tail	1.05
V-Tail	1.03
H-Tail	1.08

Aircraft often fly with a small amount of "trim" deflection on the elevator and rudder. As discussed with the topic of main-wing loading, long-range aircraft tend to climb to higher altitudes as the fuel weight decreases. In order to maintain a constant altitude, a small amount of elevator trim is needed. Rudder trim is necessary when flying in a cross-wind. Because of the need for control trim in these instances, the choice of the airfoil section should be one with as wide a drag bucket as possible in order to minimize the trim drag.

As with the main wing, C_{D_0} relates the sum of the drag due to viscous skin friction and flow separations. The viscous drag coefficient, C_f, is again estimated, assuming that the flow behaves in the same manner as over a flat plate. As with the main wing, in order to account for imperfections over the simple 2-D wing behavior, the friction coefficient will be multiplied by the form and interference factors.

The equation for the form factor, $\mathcal{F}$, is the same as for the main wing, which was given in Eq. [4.21]. However, this equation slightly under-predicts the effect of the hinge gaps that occur by the elevator and rudder. As a result, the form factor should be increased by approximately 10 percent.

Values for the interference factor vary with the tail design. Table 6.6 gives some typical values.

It is important that the horizontal and vertical stabilizers have a higher critical Mach number than the main wing. This can be achieved by choosing a slightly smaller section $(t/c)_{max}$. However, care should be taken to be sure that the stall angle, α_s, of the tail surfaces are not reduced too much by reducing $(t/c)_{max}$. This is essential with the horizontal stabilizer for maintaining pitch-up control.

6.5 TAIL PLACEMENT

6.5.1 Stall Control

The placement of the aft-horizontal and vertical stabilizers affects the stall and spin characteristics of an aircraft. Stall characteristics are affected by the location of the horizontal stabilizer with respect to the main wing. If the horizontal stabilizer is in the wake of the main wing at the stall angle of attack, α_s, elevator control will be lost, and further pitch-up may occur.

The solution to this potential problem is to locate the horizontal stabilizer in one of two regions:

1. near the mean chord line of the main wing or
2. above the wake of the main wing at the stall angle of attack.

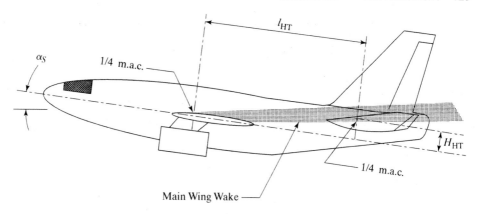

FIGURE 6.8: Schematic illustrating the influence of the wake of the main wing on the horizontal stabilizer at stall.

To determine whether the horizontal tail will be in the wake of the main wing, the relative positions of the main wing and horizontal stabilizer on the fuselage need to be drawn while at the stall angle of the main wing, α_s. This is illustrated in Figure 6.8. When the main wing stalls, the airflow will separate from the leading and trailing edges. The wake of the main wing will spread with a total angle of approximately $30°$.

Figure 6.9 illustrates a safe region for the vertical placement of a horizontal stabilizer in a conventional tail. Both the downstream location, l_{HT}, and height above the main wing centerline, H_{HT}, are normalized by the main wing m.a.c.

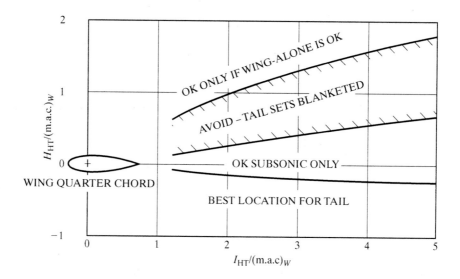

FIGURE 6.9: Recommendation for the placement of the horizontal stabilizer in a conventional tail for maximum stall control. (From NACA-TMX-26.)

This indicates that the best location for the horizontal tail is below the wing centerline. However, a higher position, such as with Cruciform or T-tails, is possible if they are set high enough above the wing.

For a T-tail, all but the trailing edge of the elevator needs to be outside the wake of the main wing. Having the elevator just inside the wake produces an unsteady buffeting on the pitch control that signals the pilot of an imminent stall.

6.5.2   Spin Control

Spin characteristics are affected by the vertical tail. During an uncontrolled spin, an aircraft is falling vertically and rotating about its vertical axis. Recovery from the spin requires having a sufficient amount of rudder control.

As illustrated in Figure 6.10 for a conventional tail, the vertical stabilizer is caught in the wake of the horizontal stabilizer during an uncontrolled spin. This makes the rudder ineffective. The solution for a conventional tail design is to move the horizontal stabilizer either forward or aft of the vertical stabilizer position. This is evident in the tail design for the Boeing 777, shown in Figure 6.1.

Alternatively, the horizontal stabilizer can be positioned higher on the vertical stabilizer. This is an advantage of the Cruciform or T-tail designs.

In either approach, a good design should have approximately 30 percent of the rudder outside of the wake of the horizontal stabilizer during a spin for proper recovery.

6.6   SPREADSHEET FOR TAIL DESIGN

A spreadsheet file named **tail.xls** is used in designing the horizontal and vertical stabilizers. The horizontal stabilizer can be either aft or canard designs. Many of the formulas that are used for the design of the tail are taken from the spreadsheet that was used to design the main wing (wing.xls). The format of the spreadsheet allows easy input and modification of the design parameters and provides a graphical view of the vertical and horizontal stabilizer plan shapes. In addition, the 3-D characteristics, $dC_L/d\alpha$ and C_{D_0}, are calculated. A sample of the spreadsheet is shown in Figures 6.11 and 6.12. This contains the parameters for the conceptual supersonic business jet case study that was proposed in Chapter 1.

The top of the spreadsheet contains input that represents general characteristics of the main wing design. This includes the wing span, b; wing mean aerodynamic chord (MAC), $\bar{c}$, wing area, S; cruise Mach number, M; leading-edge sweep angle, Λ_{LE}; the maximum thickness-to-chord, t/c_{max}, and taper ratio, λ. These can all be found in the wing design spreadsheet, **wing.xls**, and copied into this spreadsheet.

Also needed as input to the spreadsheet is the cruise altitude, H. This is placed in the section denoted as "Air Properties," which is used to calculate the properties needed for the determination of the chord Reynolds numbers. Relations that are identical to those used in previous spreadsheets are used to determine the velocity, V; density, ρ; and dynamic pressure, q; at the cruise altitude. In addition, the kinematic viscosity, ν, is estimated. Again a constant average viscosity with altitude, μ, is assumed, whereby $\nu = \mu/\rho$, and the dependence of ν on altitude is due to changes in ρ.

The tail design spreadsheet is divided into two parts. The top part deals with the vertical stabilizer. It ends with a graphical representation of the plan view of the total

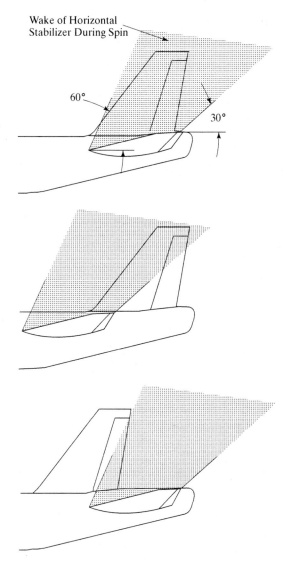

Wake of Horizontal
Stabilizer During Spin

60°

30°

FIGURE 6.10: Illustration of possible locations of the aft-horizontal stabilizer with respect to the vertical stabilizer rudder for spin recovery with a conventional tail.

stabilizer. The bottom part deals with the horizontal stabilizer. At the end, a plan view of one-half of the surface is shown. The other half is a mirror image.

For the vertical tail design, there are two sets of input parameters. The left set includes the vertical tail coefficient, C_{VT}; the distance between the quarter-chord locations of the mean-aerodynamic-chords (m.a.c.) of the main wing and vertical stabilizer, L_{VT}; the sweep angle of the vertical stabilizer, Λ_{LE}; the $(t/c)_{max}$ for the vertical stabilizer; and the vertical stabilizer taper and aspect ratios, λ and A_{VT}. The leading-edge sweep

Main Wing Reference			Air Properties					
b	32.2	ft	Cruise Alt.	55.000	ft			
m.a.c.	21.5	ft	V	1.925.70	f/s			
S	519	ft^2	ρ	0.00922	lbm/f^3			
M	2.10		q	531.07	lbf/f^2			
Λ_{LE}	62	deg	μ	107.0E-7	lbm/(f-s)			
t/c	0.04		ν (cruise)	116.0E-5	f^2/s			
λ	0.00							

Vertical Tail

Design Parameters			Airfoil Data					
Cvt	0.07		Name	NACA 64-	004			
Lvt	40.0	ft	Cl_{max}	0.8				
Λ_{LE}	63	deg	Cl_α	0.11	1/deg			
t/c	0.04		a.c.	0.26	c			
λ	0.30		α_{0L}	0	deg			
Avt	1.10		Cd	0				

Calculations			Sweep Angles			Viscous Drag		
Svt	29	ft^2		x/c	$\Lambda_{x/c}$(deg)	V_eff	874.25	f/s
b	5.7	ft	LE	0.00	63.0	q_eff	109.46	lbf/f^2
c_r	7.9	ft	1/4 chord	0.25	55.8	M_eff	0.95	
c_t	2.4	ft	(t/c)max	0.35	51.9	Re_mac	426.1E+4	
m.a.c.	5.7	ft	TE	1.00	0.3	sqrt(Re)	2064.22	
β	1.85					Cf	3.19E−03	
$C_{L\alpha}$	0.023	1/deg				S_wet	58.58	ft^2
						F	1.57	
						Q	1.05	
Total Drag	163.880	lbf				C_{D0}	0.0106	

Spanwise View

x	y
0	0
7.9	0
7.95	5.7
5.57	5.7
0	0

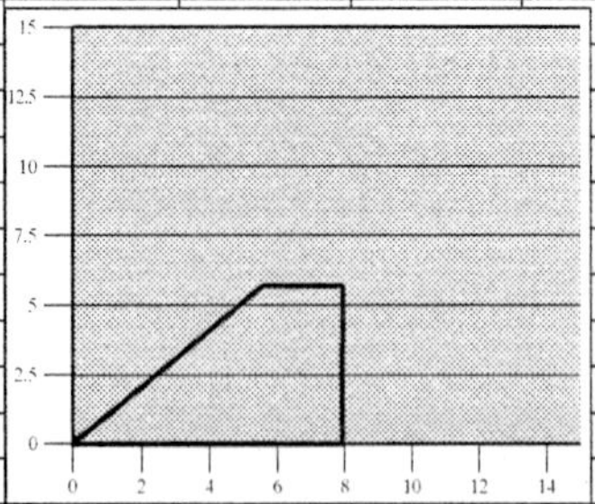

FIGURE 6.11: Spreadsheet for tail design (TAIL) showing results for conceptual supersonic business jet (Part 1).

Horizontal Tail									
Design Parameters			**Airfoil Data**						
Cht	0.11		Name	NACA 64–	004				
Lht	50.0	ft	Cl_{max}	0.8					
Λ_{LE}	63	deg	Cl_α	0.11	1/deg				
t/c	0.04		a.c.	0.26	c				
λ	0.35		α_{0L}	0	deg				
Aht	2.00		Cd	0.0040					
Calculations			**Sweep Angles**				**Viscous Drag**		
Sht	25	ft^2		x/c	$\Lambda_{x/c}$(deg)		V_eff	874.25	f/s
b	7.0	ft	LE	0.00	63.0		q_eff	109.46	lbf/f^2
c_r	5.2	ft	1/4 chord	0.25	59.9		M_eff	0.95	
c_t	1.8	ft	(t/c)max	0.35	58.4		Re_mac	284.4E+4	
m.a.c.	3.8	ft	TE	1.00	45.0		sqrt(Re)	1686.43	
β	1.85						Cf	3.42E–03	
$C_{L\alpha}$	**0.030**	1/deg					S_wet	49.17	ft^2
							F	1.5	
							Q	1.05	
Total Drag	**140.864**	lbf					C_{D0}	**0.0108**	

Spanwise View		
x	y	
0	0	
5.2	0	
8.69	3.5	
6.88	3.5	
0	0	

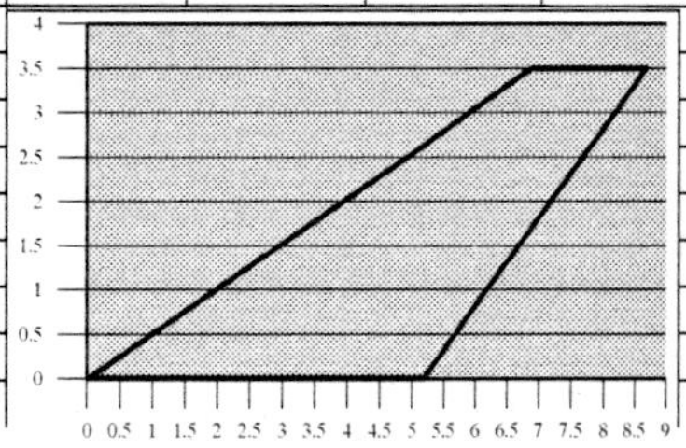

FIGURE 6.12: Spreadsheet for tail design (TAIL) showing results for conceptual supersonic business jet (Part 2).

angle of the vertical (and horizontal) stabilizer is set by a formula in the spreadsheet to be 1° larger than that of the main wing.

The coefficient, C_{VT}, is selected from Table 6.1 for aircraft with similar mission requirements. The distance, L_{VT}, is measured from a drawing of the aircraft in which the relative placement of the main wing and tail on the fuselage are shown. The value of the m.a.c. for the vertical stabilizer depends on the input parameters and is calculated in the spreadsheet. Reference values can be taken from Table 6.4.

Based on the input values, the projected area of the vertical stabilizer, S_{VT}, is calculated using Eq. [6.1]. The span, or height, of the vertical tail, b, is calculated based on the area and aspect ratio through the relation given in Eq. [6.4]. Through these, the root and tip chord lengths are calculated using Eqs. [6.5 & 6.6]. The mean-aerodynamic-chord is then determined using

$$\bar{c} = \frac{2C_r}{3}\left[\frac{1+\lambda+\lambda^2}{1+\lambda}\right]. \tag{6.7}$$

The 3-D wing property, $dC_L/d\alpha$, is based on the 2-D characteristic, $dC_l/d\alpha$, which is given in the airfoil data, as well as the planform shape according to Eq. [4.9]. In this, the quantity, β, has two definitions based on whether $M > 1$ or $M < 1$. This is accounted for by an IF statement in the formula for β.

The 3-D drag coefficient, C_{D_0}, is found from the product of the skin friction coefficient and the form and interference factors, such as given in Eq. [4.19] for the main wing design. This is based on the wetted area. Based on $(t/c)_{max}$, this is estimated using Eq. [4.23] or Eq. [4.24].

Again as with the main wing, the friction coefficient, C_f, is a function of Re_x, and whether the flow is laminar or turbulent. The two equations for C_f that are used here are the same as given by Eq. [4.25] (laminar) and Eq. [4.27] (turbulent). Re_x is based on the mean-aerodynamic-chord, $\bar{c}$, and the velocity component that is normal to the leading edge, such as given by Eq. [4.28]. The flow is considered to be turbulent when $\sqrt{\text{Re}_x} > 1000$.

The value for the form factor, $\mathcal{F}$, is calculated directly following Eq. [4.21] and then increased by 10 percent. The interference factor, $\mathcal{Q}$, must be input directly at the labeled locations in the spreadsheet. The values are based on Table 6.6.

The value of the base drag coefficient, C_{D_0}, is calculated following Eq. [4.22]. Note that this coefficient is normalized by the respective planform area of the vertical tail, S_{VT}, or horizontal tail, S_{HT}. The total drag for the vertical tail is then found as

$$D = C_{D_0} q S_{VT}. \tag{6.8}$$

The drag on the horizontal tail is determined using the respective value of C_{D_0} and S_{HT}.

The leading-edge and trailing-edge sweep angles are calculated for the vertical tail based on the planform shape. These are then used in constructing a graph of the plan view of the vertical stabilizer. Care should be taken in interpreting the view since the two axis scales may not be the same. To insure that they are, the axis limits need to be set manually.

The calculations for the horizontal stabilizer are carried out in similar fashion in the spreadsheet. The area, S_{HT}, is found by solving Eq. [6.2]. Note that the equation is identical to that used for determining the area of a canard (Eq. [6.3]), so that only the coefficient value (C_{HT}) needs to be changed, based on whether it represents an aft-horizontal stabilizer or canard. Also note that for the horizontal stabilizer, the planform area represents the sum of the two symmetric halves, as with the main wing. This is in contrast to the vertical stabilizer in which there is no symmetric half.

The planform view of one-half of the horizontal stabilizer is shown below the calculations portion of the spreadsheet. This again provides some visual feedback as to how the parameters affect the planform shape. As before, care must be taken to set the axis limits so that shape of the graphed view is not distorted.

6.6.1 Case Study: Tail Design

The parameters that are initially set in the tail design spreadsheet, **tail.xls**, correspond to the conceptual supersonic business jet. Recall that this has a design cruise Mach number of 2.1, and a cruise altitude of 55,000 feet. As a result of the relatively high supersonic cruise Mach number, the design uses a control canard for the horizontal stabilizer. The vertical stabilizer design, however, is relatively standard.

The characteristics of the main wing have been input in the reference table at the top of the spreadsheet. An important characteristic is the leading-edge sweep angle, which is 62° in this design. As a result of this value, the leading-edge sweep angles of the vertical stabilizer and canard were made to be 1° larger, or 63°.

For the vertical stabilizer, a size coefficient, $C_{VT} = 0.07$, was chosen. This was based on Table 6.1, with the closest comparison aircraft being combat jet fighters. The length, L_{VT}, was based on the placement of the main wing on the fuselage. A top-view drawing of the aircraft showing the fuselage, main wing, vertical tail, and canard, is shown in Figure 6.13. For this placement, $L_{VT} = 40$ f.

A taper ratio of $\lambda = 0.30$ and an aspect ratio of $A_{VT} = 1.10$ were chosen based on Table 6.3. These were in the approximate middle range for combat aircraft, which are the closest comparison aircraft.

A NACA 64-004 airfoil section was selected for both the vertical and horizontal stabilizers. This is a symmetric airfoil, which is in the same family as that used for the main wing. It has a relatively low base drag coefficient, $C_{D_0} = 0.004$, and a drag bucket that extends to $C_l = \pm 0.1$. Because the leading-edge sweep angles of the two stabilizers are slightly larger than that of the main wing, the same $(t/c)_{max}$ as the main wing was used. The larger sweep angle will insure that the critical Mach number is higher for the stabilizer sections and, as with the main wing, that the leading-edge Mach number will be less than one.

Based on these input values, the calculations gave the following design for the vertical tail:

$S_{VT} = 29$ f^2,

$b = 5.7$ ft, which is the height of the vertical stabilizer;

$c_r = 7.9$ ft;

$c_t = 2.4$ ft;

$\bar{c} = 5.7$ ft;

$dC_L/d\alpha = 0.023$ per degree;

$C_{D_0} = 0.004$.

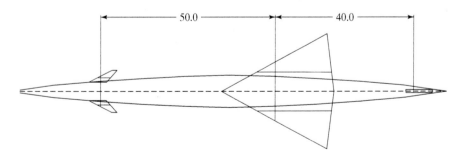

FIGURE 6.13: Top-view drawing showing the locations of the main wing, vertical stabilizer, and canard on the fuselage of the conceptual supersonic business jet.

The drawing of the vertical stabilizer in the spreadsheet illustrates the plan-view shape. Of particular importance is the trailing-edge sweep angle. With the vertical stabilizer, this angle should not be too large, because it can reduce the effectiveness of the rudder. In this design, $\Lambda_{TE} = 0.3°$. This value can be easily adjusted by changing the aspect ratio.

For the canard, a size coefficient, $C_C = 0.11$, was chosen. This was based on Table 6.2 and represents an approximate average of the different aircraft cited. The length, $L_{HT} = l_C = 50.0$, was again based on the placement of the main wing on the fuselage. This value was consistent with those listed in Table 6.4.

A taper ratio of $\lambda = 0.35$, and an aspect ratio of $A_{HT} = 2.00$ was chosen based on Table 6.3. Because there are not a great number of comparison aircraft with canards, these values were the most difficult to estimate.

Based on these input values, the calculations gave the following design for the canard:

$S_{HT} = 25$ f^2;

$b = 7.0$ ft, which is the tip-to-tip span;

$c_r = 5.2$ ft;

$c_t = 1.8$ ft;

$\bar{c} = 3.8$ ft;

$dC_L/d\alpha = 0.030$ per degree;

$C_{D_0} = 0.004$.

The ratio of the planform areas of the canard to the main wing is $S_C/S_W = 0.071$. Thus, the canard plan area is approximately 7 percent of the area of the main wing. This is very typical of a control canard. The area of a lifting canard would be a larger percentage (15–25 percent) of the main wing area.

The drawing of the plan-view of the canard in the spreadsheet demonstrates its shape. In this design, instead of an elevator, the lift generated by the canard will be varied by changing its angle of attack. When used as an aft stabilizer, this arrangement is called an "all flying tail" design. In an aft tail, this arrangement has been used on many aircraft, ranging from the B-52 and B-727 to combat fighter aircraft such as the F-14, F-15, and F-18.

Without an aft-horizontal stabilizer, the vertical stabilizer offers excellent spin recovery. The rudder is placed well aft of the main wing so that it is well clear of the wake of the main wing during an uncontrolled spin.

One of the benefits of the canard is the excellent pitch-up control, without any risk of being in the wake of the main wing. In order that the wake of the canard does not affect the main wing in level flight, it should be located either slightly above or below the main wing mean chord line.

The base drag coefficient for the vertical tail was found to be $C_{D_0} = 0.0099$. This is approximately 2.5 times larger than C_{d_0} based on an ideal 2-D wing section. To obtain this, a value of $Q = 1.05$ was used. This was based on Table 6.4 for a conventional tail design.

Because the canard is an "all flying tail" design, it does not have a hinge gap for an elevator. Therefore, it is not likely necessary to increase the value of $\mathcal{F}$ by 10 percent, as otherwise recommended. The value of Q could also likely be smaller than the conventional 1.05 magnitude used.

The total drag on the vertical stabilizer was found to be 163.8 lbs. The drag on the canard was a slightly lower value of 140.6 lbs. This gave a total drag associated with the tail of 304.4 lbs. This was approximately one-seventh the drag on the fuselage and approximately 20-times smaller than that on the main wing.

As Eqs. [6.1–6.3] indicate, the area of the horizontal and vertical stabilizers varies with the distances they are from the 1/4-m.a.c. point on the main wing. This has an ultimate effect on the drag. For example, Figure 6.14 demonstrates how the drag changes on the canard as the distance, l_C, changes. This is a consideration in the design and led to the choice of $l_C = 50$ that was used.

The complete design of the horizontal and vertical stabilizers is done with regard to the static stability and control of the aircraft. The actual stability analysis will be completed in Chapter 11. At that point, the sizing of the control surfaces, such as the rudder and elevator, will also be done. However, the tail design at this point is sufficient to estimate the drag at cruise conditions. We then have all the elements (main wing, fuselage, and tail) from which the total drag can be estimated for use in selecting the engine(s) for the aircraft. This is the topic of the next chapter.

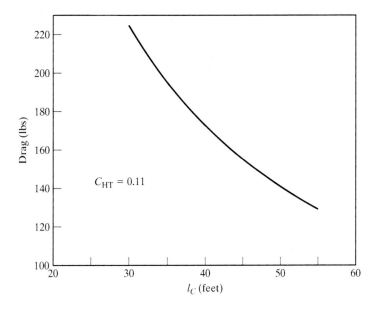

FIGURE 6.14: Effect of l_C on the drag produced by the canard used in the conceptual supersonic business jet.

6.7 PROBLEMS

6.1. Using an aircraft reference book such as "Jane's All the World Aircraft," tabulate the following properties for a variety of commercial transport aircraft:
1. l_{VT};
2. S_{VT};
3. b_W;
4. S_W.

Plot the product $l_{VT} S_{VT}$ versus the product $b_W S_W$. Compare your results to Eq. [6.1] and the value of C_{VT} in Table 6.1.

6.2. Using an aircraft reference book such as "Jane's All the World Aircraft," tabulate the following properties for a variety of commercial transport aircraft:
1. l_{HT};
2. S_{HT};
3. $\bar{c}_W$;
4. S_W.

Plot the product $l_{HT} S_{HT}$ versus the product $\bar{c}_W S_W$. Compare your results to Eq. [6.2] and the value of C_{VT} in Table 6.1.

6.3. Repeat Problem [6.2], but substitute the main wing span, b_W, for $\bar{c}_W$. Do the results correlate as well?

6.4. Repeat Problem [6.1] for a variety of general aviation single-engine aircraft. Do the results correlate as well as with commercial aircraft? Explain reasons for any differences.

6.5. Repeat Problem [6.2] for a variety of general aviation single-engine aircraft. Why do you think that there is such a large difference in the values of C_{HT} in Table 6.1 between these and commercial transport aircraft?

6.6. Examine the following aircraft, which have T-tail designs:
1. B-727;
2. C-17;
3. MD-80;
4. Gulfstream V;
5. Lear jet 31A.

Tabulate the height of the horizontal stabilizer above the mean chord line of the main wing (h_{HT}) and the distance l_{HT}. Plot $h_{HT}/\bar{c}$ versus $l_{HT}/\bar{c}$. Can this plot be useful in specifying the height of the horizontal stabilizer in a T-tail design? Explain.

6.7. Examine the tail design of the Grumman E-2C Hawkeye early warning control aircraft. Discuss the various features and possible motivation of that design.

6.8. Examine the tail design of the Antonov AN-225 heavy-transport aircraft. Discuss the various features and possible motivation of that design.

6.9. The B-747 aircraft has a conventional tail design. Input the characteristics of B-747 aircraft into the tail design spreadsheet and compare the values of S_{VT} and S_{HT} to the actual values.

6.10. The Cessna Citation VI has a T-tail design. Input the characteristics of this aircraft into the tail design spreadsheet and compare the values of S_{VT} and S_{HT} to the actual values.

CHAPTER 7

Engine Selection

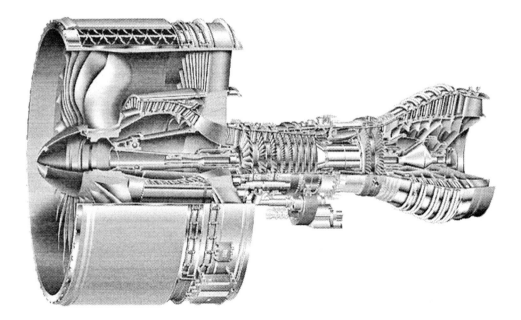

Cutaway drawing of General Electric and Pratt & Whitney alliance GP7000 turbo-jet engine, which is designed for the Airbus A380. This engine has a fan tip diameter of 116 in. and a length of 187 in. The take-off thrust is 70,000 lbs. The A380 is scheduled for its first flight in January 2006. (Courtesy of the GE- Pratt & Whitney Engine Alliance.)

At this point in the design, the total drag on the main wing, fuselage, and tail surfaces have been determined. This chapter deals with the scaling of available engines to provide the thrust necessary to overcome the drag based on the different mission requirements.

TABLE 7.1: Thrust-to-weight ratios based on mission requirements for different historic aircraft.

Primary Mission Requirement	T/W
Long Range	0.20–0.35
Short and Intermediate Range	0.30–0.45
STOL	0.40–0.60
Combat: Close-Air Support	0.40–0.60
Combat: Air-to-Air	0.80–1.30
Combat: High-Speed Intercept	0.55–0.80

For long-range aircraft where efficient cruise is the main design driver, the engines are selected based on the drag at the cruise Mach number and altitude. These aircraft tend to have a lower thrust-to-weight ratio, T/W. Combat aircraft that are designed for maneuverability tend to have higher T/W. However, they generally have poor cruise efficiency and, therefore, shorter range. Aircraft that are designed for short take-off and landing, generally have T/W ratios that fall between these two extremes. Thus, the selection of an extreme engine thrust to meet one mission requirement may limit another, so that compromises may be necessary. Sample T/W values for different aircraft are given in Table 7.1.

With new designs, there is an advantage to using engines that are already available, since this generally leads to fewer early development problems. However, in many cases, an engine having the *exact* characteristics for the design does not exist. In that case, the method of engine scaling is used whereby the ratio of the required thrust to that of a reference engine is used to determine the weight, length, and diameter of the required engine.

7.1 PROPULSION SELECTION

The appropriate propulsion system for an aircraft depends on a number of factors. These include the design Mach number and altitude, fuel efficiency, and cost. Figure 7.1 presents choices for propulsion systems based on the aircraft design Mach number.

The piston-engine driven propeller was the first form of propulsion for historic aircraft. Modern designs have the advantage of providing the lowest fuel consumption and the lowest cost. Their disadvantages are that they have a low T/W ratio, and produce higher noise and vibration. The maximum altitude for piston-engine aircraft is limited by a decrease in engine horsepower with altitude, due to decreasing atmospheric pressure. This can be overcome to some extent through a turbo-charger, which increases the air intake manifold pressure. Turbo-charged piston engines can maintain a constant horsepower up to an altitude of approximately 20,000 feet. Presently, most piston-engine designs are used with smaller, lighter-weight aircraft.

Turbo-jet-driven propeller aircraft are an improvement on piston-engine aircraft. With these, a majority of the energy in the exhaust is extracted by a turbine stage, which is used to turn a propeller. The jet exhaust retains some of the thrust capability and can

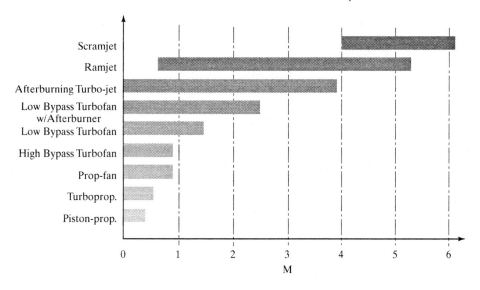

FIGURE 7.1: Useful range of flight Mach numbers for different types of engines.

contribute as much as 20 percent more to the total thrust. The advantages of this system are an increase in the engine T/W, lower vibration, and an increase in the maximum operational altitude.

Turboprop designs have a higher efficiency than piston-prop designs at Mach numbers greater than 0.5 because of the residual jet thrust. However, all propeller-driven aircraft have a limited maximum cruise Mach number because the propeller tip Mach number cannot exceed approximately 0.7. Because of their high efficiency, turboprop designs are popular for mid-range commercial (commuter) passenger aircraft.

At higher subsonic Mach numbers, the propulsion system that gives the highest efficiency is the turbofan-jet engine. In these designs, the incoming air is ducted into a fan stage. Part of the air in the fan stage is ducted through a compressor stage, combustor, and turbine stage before it exhausts at the jet exit. This air is referred to as the "primary air." Energy from the primary air turbine stage powers the fan stage. The remaining part of the air in the fan stage that does not go through compression and combustion is called the "bypass air." The bypass ratio is the ratio of the volumetric mass flow of the bypass air to primary air. Bypass ratios range from as much as 6 to as low as 0.25. High bypass-ratio engines have better fuel efficiency, but less thrust. Figure 7.2 shows two turbofan engines with low and moderately high bypass ratios.

Turbo-jet engines use air-fuel ratios that are higher than stoichiometric in order to keep internal temperatures below the property limits of the turbine blades. As a result, only about 25 percent of the primary air is actually used in combustion. Therefore, if fuel is added downstream of the turbine stage, it will combust. This is the basis of a turbo-jet after-burner.

When an after-burner is used, it can increase the thrust by a much as two times. However, it greatly increases the fuel consumption and is, therefore, only used over a small portion of a flight plan, where extra thrust is needed. After-burner installed turbo-jet

FIGURE 7.2: Photographs that illustrate low and high bypass ratio turbofan engines. The top engine is a P&W JT8d with a bypass ratio of 1.7:1 and $T_{SL} = 21,000$ lbs. This is used on the B-727, B-737-200, DC-9, and MD-11 aircraft. The bottom engine is a P&W JT9d with a bypass ratio of 4.8:1 and $T_{SL} = 56,000$ lbs. This is used on the B-747, B-767, A-300, A-310, and DC-10 aircraft. (Courtesy of Pratt & Whitney a United Technologies Company).

engines are most typically used with combat aircraft and operated for short periods during take-off, high-speed intercept, and combat.

As shown in Figure 7.1, a fairly wide range of flight Mach numbers are possible with turbo-jet engines. The maximum Mach numbers depend on the bypass ratio. Mach numbers nearly as high as 4.0 are possible with low bypass ratios and full after-burners.

All of the propulsion systems discussed to this point can operate down to zero velocity (static). This is not the case for a ram-jet engine. This type of engine uses the natural compression of air produced by high-speed flight to produce the pressure ratios needed for combustion. The minimum Mach number for the ram-jet effect is approximately 2.0. However, flight Mach numbers of greater than about 3.0 are needed to exceed the efficiency of low bypass turbo-jets. Ram-jets have a useful maximum Mach number of approximately 5.0.

The highest flight Mach numbers for air breathing engines are produced by scram-jet engines. The scram-jet is similar to a ram-jet with the one exception that the internal flow and combustion is supersonic. Scram-jet-powered aircraft would be used for flight Mach numbers that exceed 5.0. Currently, these designs are being considered for earth-to-orbit vehicles that might ultimately replace the U.S. space shuttle. At the moment, they are highly experimental and not ready for general use.

The basis for the choice of one of these propulsion systems over another can typically be made based on the maximum design Mach number, available thrust, fuel efficiency, cost, and reliability. The necessary data can be obtained from engine manufactures. A fairly concise list can also be found in Jane's "All the World Aircraft" and in periodicals such as *Aerospace America*. Because of the fundamental differences between the turbo-jet and propeller propulsion systems, a detailed discussion of the selection of each for a design is presented separately.

7.2   NUMBER OF ENGINES

The number of engines is often specified by the need to produce a sufficient amount of thrust based on mission requirements and the available thrust per engine. However, if possible, a design should use the fewest number of engines necessary. This generally leads to a simpler, lighter, more efficient, and less expensive aircraft.

However, for commuter and commercial passenger aircraft certified under FAA regulations, at least two engines are required. The performance of these aircraft is then demonstrated with one engine inoperative.

7.3   ENGINE RATINGS

The maximum performance of an engine under various conditions is specified by the engine rating. These ratings correspond to different thrust conditions that are specified for take-off, maximum climb, and maximum cruise.

7.3.1   Take-Off

The take-off rating is the maximum thrust that the engine is certified to produce. This is generally specified for short periods of time, of the order of five minutes, to be used only at take-off. For turbo-jet engines, the take-off rating is specified as sea-level static

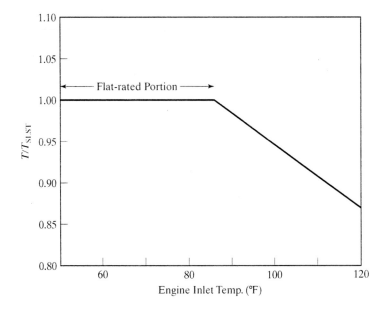

FIGURE 7.3: Typical turbo-jet thrust rating based on inlet air temperature.

thrust (SLST). There is a maximum ambient temperature for which the SLST can be maintained. This is set by temperature limits for internal engine parts. An example of the SLST thrust rating with temperature is shown in Figure 7.3. The temperature range where the engine thrust equals the rated SLST thrust is called the "flat-rated" portion. The engine thrust must decrease below the SLST value when the inlet air temperature exceeds the manufacturer's limit. In Figure 7.3, this occurs at 86°F.

The take-off rating is generally used when "sizing" an engine for a design. If an engine is equipped with after-burners, the take-off rating is also the maximum rating with after-burners.

7.3.2 Maximum Climb

The maximum climb rating is the maximum thrust that the engine is certified to produce for normal climb operation. This rating is from 90 to 93 percent of the take-off rating.

7.3.3 Maximum Cruise

The maximum cruise rating is the maximum thrust that the engine is certified to produce for normal cruise. This corresponds to 80 percent of the take-off rating. In addition, the cruise rating is for continuous operation, with no time limits.

7.4 TURBO-JET ENGINE SIZING

The ideal situation in a new design, is to find an existing turbo-jet engine that meets the mission requirements perfectly. However, in most cases, this will not happen. In this instance, the designer would start with an existing engine with characteristics that are close to those needed in the design and scale it up or down based on suitable scaling laws.

It is possible to develop scaling laws based on conservation of mass and momentum for fluid flows. The thrust force, T, is given by

$$T = \dot{m}(V_e - V_a) + A_e(P_e - P_a), \tag{7.1}$$

where $\dot{m}$ is the mass flow rate, V is velocity, P is pressure, and the subscripts a and e refer to atmospheric and jet-exit conditions, respectively.

Turbo-jet engines are designed to increase the momentum of the incoming air, but not the exit pressure. Therefore,

$$P_e \simeq P_a. \tag{7.2}$$

If we consider the static thrust, then $V_a = 0$. Therefore, Eq. [7.1] reduces to

$$T = \dot{m}V_e. \tag{7.3}$$

Hence, the thrust of a turbo-jet engine should vary with respect to that of a reference engine as

$$T = T_{\text{ref}} \frac{(\dot{m}V_e)}{(\dot{m}V_e)_{\text{ref}}}. \tag{7.4}$$

It is a reasonable assumption that the jet-exit velocity, V_e, is not a function of the thrust. Therefore, $V_e = V_{e_{\text{ref}}}$, and

$$T = T_{\text{ref}} \frac{\dot{m}}{\dot{m}_{\text{ref}}}. \tag{7.5}$$

The mass flow rate of air through the engine is

$$\dot{m} = (\rho A V)_e = \left[\rho V \frac{\pi d^2}{4}\right]_e, \tag{7.6}$$

where d is the engine diameter, V is the bulk air velocity, and the subscript e again refers to the jet exit. If we consider a reference engine with an exit diameter, d_e, then the mass flow rate through any engine is related to that of the reference engine as

$$\dot{m} = \dot{m}_{\text{ref}} \left[\frac{d}{d_{\text{ref}}}\right]_e^2. \tag{7.7}$$

Based on this, the diameter of any engine is related to the diameter of the reference engine as

$$\left[\frac{d}{d_{\text{ref}}}\right]_e = \left[\frac{\dot{m}}{\dot{m}_{\text{ref}}}\right]^{1/2}. \tag{7.8}$$

Using Eq. [7.5], the diameter ratio can be put in terms of the thrust ratio as

$$\left[\frac{d}{d_{\text{ref}}}\right]_e = \left[\frac{T}{T_{\text{ref}}}\right]^{1/2}. \tag{7.9}$$

Equations [7.5] and [7.9] have been based on conservation laws for the air flow and provide a useful scaling factor corresponding to the ratio of a desired thrust to that

TABLE 7.2: Empirical scaling relations for turbo-jet engines. (M is the maximum Mach number, BPR is the bypass ratio.)

Non-after-burning, $M < 1$, $0 \leq BPR \leq 6$:

$$W = 0.084T^{1.1}e^{(-0.045BPR)}$$
$$L = 2.22T^{0.4}M^{0.2}$$
$$D = 0.393T^{0.5}e^{0.04BPR}$$

After-burning, $1 \leq M \leq 2.5$, $0 \leq BPR \leq 1$:

$$W = 0.063T^{1.1}M^{0.25}e^{-0.81BPR}$$
$$L = 3.06T^{0.4}M^{0.2}$$
$$D = 0.288T^{0.5}e^{0.04BPR}$$

---

of a reference engine. The weight and length of a turbo-jet engine are based on *empirical* relations. These are given in Eqs. [7.10 & 7.11]:

$$W_{eng} = W_{eng_{ref}}\left[\frac{\dot{m}}{\dot{m}_{ref}}\right]^a = \left[\frac{T}{T_{ref}}\right]^a \quad \text{with } (0.8 \leq a \leq 1.3) \qquad (7.10)$$

and

$$L_{eng} = L_{eng_{ref}}\left[\frac{\dot{m}}{\dot{m}_{ref}}\right]^{(2a-1)/2} = \left[\frac{T}{T_{ref}}\right]^{(2a-1)/2} \quad \text{with } (0.8 \leq a \leq 1.3). \qquad (7.11)$$

Equations [7.8] through [7.11] are reasonably accurate if the scale factor T/T_{ref} falls in the approximate range, $0.5 \leq T/T_{ref} \leq 1.5$. Outside this range, it may be better to use completely empirical relations, such as the ones given in Table 7.2.

7.4.1   Altitude and Velocity Effects

The thrust value provided by the engine manufactures corresponds to the uninstalled sea-level static thrust (SLST). If the engine is being sized based on the thrust required for cruise, the sea-level thrust needs to be corrected for altitude. Based on Eq. [7.3], the thrust is a function of the mass flow rate, $\dot{m} = (\rho AV)_e$. The air density, ρ, is a function of altitude. Assuming an ideal gas relation,

$$\rho = \frac{P}{R\theta}, \qquad (7.12)$$

where R is the gas constant for air. The pressure, P, and temperature, θ, are functions of altitude for a standard atmosphere. Therefore, the thrust at an altitude, H, is related to the thrust at sea-level, T_{SL}, as

$$T_H = T_{SL}\frac{P_H}{P_{SL}}\frac{\theta_{SL}}{\theta_H}. \qquad (7.13)$$

As the speed of the aircraft increases, the velocity of air entering the engine, V_a, increases. The exit nozzle of turbo-jet engines is usually always close to sonic, so that

V_e is nearly independent of V_a. Therefore, the term $(V_e - V_a)$ in Eq. [7.1] decreases as the speed of the aircraft increases. By itself, this would result in a decrease in thrust with increasing airspeed. However, this effect is more than compensated for by an increase in the mass flow of air, $\dot{m}$, due to a "ram effect," which increases with velocity. The result is that the thrust *increases* with the speed of the aircraft.

At subsonic speeds, the ram effect is not very significant, so that the engine thrust can be considered to be approximately constant with velocity. At supersonic speeds, however, the ram effect can be a major factor and needs to be accounted for in determining the thrust.

The last effect of altitude on turbo-jet performance is for the thrust-specific fuel consumption, TSFC, defined as

$$\text{TSFC} = \frac{w_f}{T}, \tag{7.14}$$

where w_f is the weight of fuel used per hour for a given thrust, T.

The fuel flow depends on the engine throttle position. For a constant throttle setting, for example at military continuous power, TSFC varies with thrust. Since thrust depends on altitude (Eq. [7.13]), TSFC is a function of altitude. For subsonic turbo-jet aircraft, there is an optimum altitude where the TSFC is a minimum. This generally occurs at approximately 36,000 feet. Thus, this represents the best altitude for most efficient cruise of a turbo-jet-powered aircraft.

7.4.2 Installed Thrust Corrections

The thrust that is reported by engine manufacturers is referred to as the "uninstalled thrust." That is, it is the thrust that is produced by the engine on a test stand. The thrust that is relevant to the aircraft performance is referred to as the "installed thrust." The installed thrust is the uninstalled thrust that is corrected for installation effects, minus any drag produced by the propulsion system.

Installation effects include the following:

1. inlet pressure recovery;
2. bleed air and power extraction;
3. inlet airflow distortion effects;
4. exit nozzle performance.

Added propulsion drag can originate from the following:

1. inlet drag caused by air "spillage" that results from a mismatch in the amount of air demanded by the engine compared to the amount of air delivered to the inlet based on the flight condition.
2. exit nozzle drag;
3. trim drag that would result from added control trim that might be needed to offset moments produced by the engine thrust.

Estimates of the added propulsion drag are difficult to assess and will not be attempted for a conceptual design. The effect of the engine installation on the uninstalled thrust can be reasonably estimated for the most significant effects of the inlet pressure recovery and bleed air extraction.

The uninstalled thrust is based on having a perfect inlet pressure recovery, where $P_i/P_\infty = 1.0$. For subsonic Mach numbers, this is usually the case. Deviations can occur when $M_\infty > 1$, which can then give a lower installed thrust. An estimate of the percent thrust loss is given by

$$\text{Loss}(\%) = C_{\text{ram}}\left[\left(\frac{P_i}{P_\infty}\right)_{\text{ref}} - \left(\frac{P_i}{P_\infty}\right)_{\text{act}}\right] \times [100], \tag{7.15}$$

where

$$(P_i/P_\infty)_{\text{ref}} = 1 - 0.075\,(M_\infty - 1)^{1.35}, \tag{7.16}$$

$$(P_i/P_\infty)_{\text{act}} = 1 \text{ for } M_\infty < 2.5, \tag{7.17a}$$

$$(P_i/P_\infty)_{\text{act}} = -0.1\,M_\infty + 1.25 \text{ for } M_\infty \geq 2.5, \tag{7.17b}$$

and

$$C_{\text{ram}} = 1.35 - 0.15\,(M_\infty - 1). \tag{7.18}$$

The reference recovery ratio (Eq. [7.16]) is based on MIL-E-5008B. The actual pressure recovery ratio can vary depending on the inlet design. Equation [7.17] is based on an ideal mixed-compression isentropic spike inlet.

In commercial jet aircraft, air is bled from the engines to circulate inside the cabin. The amount of bleed air is typically from 1 to 5 percent of the total engine mass flow. The thrust is proportional to the mass flow rate (Eq. [7.5]), so that the use of bleed air reduces the engine thrust. However, it has a disproportionate effect, where the percent loss in thrust is given by

$$\text{Loss }(\%) = C_{\text{bleed}}\left[\frac{\dot{m}_{\text{bleed}}}{\dot{m}_{\text{engine}}}\right] \times [100], \tag{7.19}$$

with $C_{\text{bleed}} \simeq 2.0$.

7.4.3   Spreadsheet for Turbo-Jet Engine Sizing

A spread sheet file named **engine.xls** is used for sizing turbo-jet engines used in the design that are based on existing engine data. The requirements are based on the cruise conditions, which are defined by cruise altitude and Mach number. The sizing is based on providing thrust that equals the drag at cruise. A sample of the spreadsheet is shown in Figure 7.4. This contains the parameters for the conceptual supersonic business jet case study proposed in Chapter 1.

The top of the spreadsheet contains a summary of the drag calculations for the main aircraft components: fuselage, main wing, and horizontal and vertical tail sections. The values of the constituent drag forces are summed in the spreadsheet to yield the total drag that must be offset by the thrust.

The next portion of the spreadsheet is designed to calculate the uninstalled thrust for a given engine. This is based on the engine manufacturer's values of static thrust at

Drag:

	Drag (lbf)	% Total
Fuselage	2705.0	25.17
Wing	7739.0	72.00
Hor. Tail	140.8	1.31
Ver. Tail	163.8	1.52
Total	**10748.6**	**100.00**

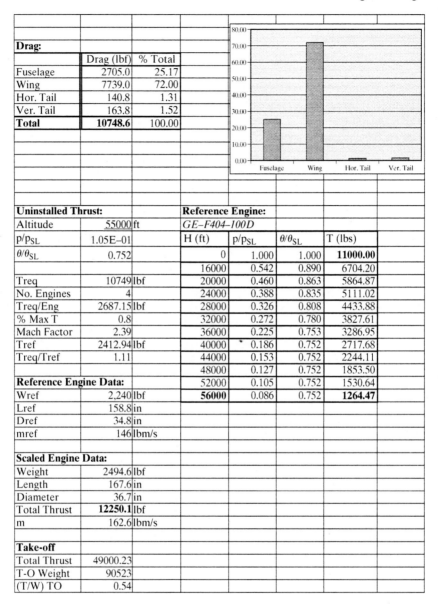

Uninstalled Thrust:

Altitude	55000	ft
p/p_{SL}	1.05E–01	
θ/θ_{SL}	0.752	
Treq	10749	lbf
No. Engines	4	
Treq/Eng	2687.15	lbf
% Max T	0.8	
Mach Factor	2.39	
Tref	2412.94	lbf
Treq/Tref	1.11	

Reference Engine Data:

Wref	2,240	lbf
Lref	158.8	in
Dref	34.8	in
mref	146	lbm/s

Scaled Engine Data:

Weight	2494.6	lbf
Length	167.6	in
Diameter	36.7	in
Total Thrust	**12250.1**	lbf
m	162.6	lbm/s

Take-off

Total Thrust	49000.23
T-O Weight	90523
(T/W) TO	0.54

Reference Engine: *GE–F404–100D*

H (ft)	p/p_{SL}	θ/θ_{SL}	T (lbs)
0	1.000	1.000	**11000.00**
16000	0.542	0.890	6704.20
20000	0.460	0.863	5864.87
24000	0.388	0.835	5111.02
28000	0.326	0.808	4433.88
32000	0.272	0.780	3827.61
36000	0.225	0.753	3286.95
40000	0.186	0.752	2717.68
44000	0.153	0.752	2244.11
48000	0.127	0.752	1853.50
52000	0.105	0.752	1530.64
56000	0.086	0.752	**1264.47**

FIGURE 7.4: Spreadsheet for turbo-jet engine sizing (ENGINE) showing results for conceptual supersonic business jet.

sea-level altitude. The values of thrust are corrected for altitude and Mach number in the spreadsheet.

The table in the spreadsheet under the heading "Reference Engine" contains data for the altitude variation in the pressure ratio, $P(H)/P_{SL}$, and temperature ratio, $\theta(H)/\theta_{SL}$, for a standard atmosphere. This is used to correct the sea-level engine thrust for altitude, which is based on Eq. [7.13]. The sea-level thrust is input into the top row of the thrust column in the table. The remainder of the column displays the altitude corrected thrust.

In the uninstalled thrust calculations, T_{req} is equal to the total drag from above. This is divided by the number of engines, n, to give the thrust required per engine, T_{req}/Eng. At cruise, the engines operate at their cruise rating, which as mentioned earlier, corresponds to 80 percent of their SLST rating. The fraction of the maximum thrust that is intended for the design is entered into the spreadsheet as the "%MaxT."

The final correction to the static engine thrust is to account for any "ram" effect produced by the flight Mach number. This effect is denoted in the spreadsheet as the "Mach Factor." In general, the Mach Factor is a number that is greater than or equal to 1, which is multiplied by the static thrust value. For cruise Mach numbers less than 1, the Factor should be 1 to express the fact that there is little ram augmentation of the thrust. Above Mach 1, the Factor will be greater than 1. The Mach Factor needs to be input to the spreadsheet, since algorithms to determine the ram effect require more lengthy calculations. Such information can be found in textbooks on propulsion. Some of these even contain packaged software to perform this and other engine-performance calculations.

The quantity T_{ref} in the spreadsheet represents the thrust for a single engine, whose sea-level thrust was supplied at the top of the table, which has been corrected for altitude and Mach number. To account for altitude, the appropriate row index from the table needs to be placed in the formula. For example, in the supplied spreadsheet, G29 refers to row 29 where the thrust value at the altitude of 52,000 feet is located in the table.

The engine scale factor corresponds to the ratio, T_{ref}/T_{req}. This is then used with the reference engine data to determine the weight, length, and diameter of the scaled engine for the design. The formulas used for these calculations correspond to Eqs. [7.8–7.11].

Finally, the bottom of the spreadsheet calculates the thrust-to-weight ratio at take-off. This is based on the total thrust at sea level per engine, times the number of engines, times the engine scale factor. The take-off weight is supplied from the first spreadsheet, **itertow**. The $(T/W)_{TO}$ is needed for the refined take-off analysis, which is covered in the next chapter.

7.5 PROPELLER PROPULSION SYSTEMS

For a propeller-driven aircraft, the important design parameters that need to be determined are the propeller diameter and the engine shaft horsepower. The propeller generates thrust in the same way that a wing generates lift. As with a wing, the propeller is designed for a particular flight condition (cruise, take-off, etc.).

The nondimensional forms for the thrust developed and power required for a propeller are given in terms of a thrust coefficient, C_T, and power coefficient, C_P, where

$$C_T = \frac{T}{n^2 D^4} \tag{7.20}$$

and

$$C_P = \frac{P}{\rho n^3 D^5}. \tag{7.21}$$

Here n is the propeller rotational velocity with units of revolutions per second, D is the propeller diameter, and ρ is the air density. Care needs to be taken with the units in these equations. C_P can be put in terms of break horsepower with the proper unit conversion. For completeness, a dimensionless form for the torque, Q, is also defined as

$$C_Q = \frac{Q}{\rho n^2 D^5}.$$ (7.22)

The propulsive, or propeller efficiency, is defined as

$$\eta_P = \frac{\text{Thrust Power Output}}{\text{Shaft Power Input}} = \frac{TV}{P},$$ (7.23)

where V is the true airspeed. The propulsive efficiency takes into account profile losses, with efficiency factor, η_o, as well as induced losses, with efficiency factor, η_i. Based on this,

$$\eta_P = \eta_o \eta_i.$$ (7.24)

The tip speed of the propeller is

$$V_{\text{tip}} = \pi n D.$$ (7.25)

The ratio of the true airspeed to the tip speed is defined as the advance ratio,

$$J = \frac{V}{nD}.$$ (7.26)

The propeller efficiency can then be expressed in terms of the advance ratio as

$$\eta_P = J \frac{C_T}{C_P}.$$ (7.27)

Combining Eqs. [7.26 & 7.27] gives an expression for the thrust

$$T = \frac{P}{nD} \frac{C_T}{C_P},$$ (7.28)

which is useful at static conditions (take-off) where the propeller efficiency equation (Eq. [7.23]) is not usable at $V = 0$.

In propeller designs, the sectional camber is defined by the design lift coefficient, C_{l_d}, at a specific radius of $r = 0.7R$. Generally, $0.4 \leq C_{l_d} \leq 0.6$.

The propeller blade plan-form is expressed by the "activity factor," AF, which represents the integrated power absorption capability of all the blade elements. The activity factor per blade is conventionally defined as

$$\text{AF} = \frac{10^5}{D^5} \int_{0.15R}^{R} cr^3 dr,$$ (7.29)

where c is the local blade chord.

Generally, $80 \leq \text{AF} \leq 200$. The actual value depends on the structural characteristics of the propeller blade. The lower values correspond to light aircraft, and the upper values to turboprop aircraft. For example, AF = 162 for a C-130 transport. The activity factor for a straight, tapered propeller blade with a taper ratio, λ, is

$$\text{AF} = \frac{10^5 c_{\text{root}}}{16D}[0.25 - 0.2(1 - \lambda)]. \tag{7.30}$$

7.5.1 Propeller Design for Cruise

The general methodology for designing a propeller propulsion system for cruise is outlined subsequently. The following steps have been implemented in the spreadsheet **prop.xls**:

1. Start with aircraft conditions at cruise. This consists of the cruise altitude and Mach number, which then specifies the air density, ρ, and velocity, V.

2. Specify a propeller diameter, D. This may be selected based on physical limits such as ground clearance.

3. Estimate the engine shaft horsepower (SHP). This is probably best done by starting with an existing engine, whereby the shaft horsepower (SHP) at a particular r.p.m., n, is specified. An estimate used in selecting the engine can come from Eq. [7.23], assuming a propulsive efficiency, $\eta \simeq 0.8$. Here the shaft power input, P, has units (British) of (f-lbf)/s, which can be converted to SHP using the conversion, 1 hp = 550 (f-lbf)/s.

4. Specify a propeller tip velocity, V_{tip}. This is best done while also calculating the propeller tip Mach number, M_{tip}. If the tip Mach number is less than approximately 0.85, compressibility effects on the propeller efficiency can be neglected. $M_{\text{tip}} \leq$ 0.85 is also a good upper limit to minimize propeller-generated sound levels.

5. Compute the advance ratio, J, based on Eq. [7.26]. This uses the n from the engine selection and will vary based on the selection of V_{tip} and D.

6. Compute the power coefficient, C_P, based on Eq. [7.21]. It is important to note that if British units are used, the proper conversion between pound-mass and pound-force is needed so that C_P is dimensionless. This can be observed in the spreadsheet by examining the formula cell for C_P.

7. The values of C_P and J are next used with propeller data to determine the propeller efficiency, η_P. A representative plot of such data is shown in Figure 7.5. This shows iso-contours of η_P and blade profile twist angle at $3/4D$, $\Theta_{3/4}$, as a function of C_P and J, for a three-bladed propeller with AF = 100. These conditions are fairly typical of a smaller private and sport-utility aircraft. The efficiency is a function of the number of propeller blades. For example, a two-bladed propeller is approximately 3 percent more efficient than an equivalent three-bladed version. Similarly, a four-bladed propeller is about 3 percent less efficient than a propeller with three blades. In the spreadsheet, the efficiency is corrected based on the difference in the blade number compared to 3 used in Figure 7.5. This is referred to as $\eta_{P_{\text{cor}}}$.

8. Calculate the thrust coefficient, C_T, based on Eq. [7.27].

9. Calculate the thrust using Eq. [7.28]. Again the value for SHP can be used for P with the proper units conversion.

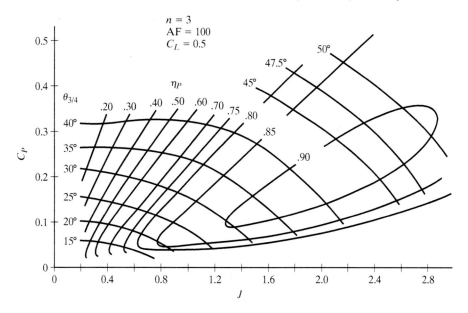

FIGURE 7.5: Sample propeller efficiency for forward flight (three-blade propeller, $C_{l_d} = 0.5$, AF $= 100$). (From Report PDB 6101A, United Aircraft Corp., 1963.)

10. The thrust from the previous step is then compared to the required thrust, which must equal the drag at cruise. If it is too large or small, the most straightforward approach is to change the SHP. If an existing engine is to be used without modification, the alternative is to vary the propeller design (V_{tip} and D) until the thrust equals the drag. This latter approach is less automated in the spreadsheet because it requires inserting values of η_P from Figure 7.5.

7.5.2 Static Thrust

The static thrust produced by the propeller is needed to determine the take-off distance. At the start of take-off, the velocity is zero. From Eq. [7.23], this would imply that the static thrust is infinite. In actuality, the static thrust is found from Eq. [7.28]. This uses the ratio, C_T/C_P. For a given propeller design, this ratio is determined for static conditions as a function of the power coefficient, C_P. A representative plot for the same propeller as in Figure 7.5, is shown in Figure 7.6.

The procedure to determining the static thrust is outlined in the steps that follow. These steps have been included in the bottom part of the **prop.exe** spreadsheet:

1. Determine the ratio C_T/C_P from Figure 7.6 based on the value of C_P found from Eq. [7.21].

2. Substitute this ratio into Eq. [7.23] to determine the static thrust. For equivalent propeller designs, the static thrust depends on the number of blades. For example a two-bladed propeller develops approximately 5 percent less thrust than a three-bladed version. Similarly, a four-bladed propeller develops about 5 percent

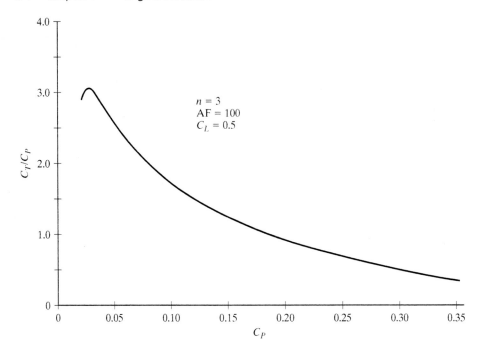

FIGURE 7.6: Sample static propeller characteristics (three-blade propeller, $C_{l_d} = 0.5$, AF = 100). (From Report PDB 6101A, United Aircraft Corp., 1963.)

more thrust than one having three blades. In the spreadsheet, the static thrust is corrected based on the difference in the blade number compared to the three used in Figure 7.6. This is referred to as T_{cor} in the spreadsheet.

7.5.3 Turboprop Propulsion

For a turboprop propulsion system, residual thrust, T_j, is generated by the exhaust jet. An equivalent shaft horsepower is used to account for this in the total power output. For this, the jet thrust is converted to thrust horsepower by the relation

$$\mathrm{THP} = \frac{T_j V}{0.8}. \qquad (7.31)$$

Here the factor, 0.8, comes from assuming a standard 80 percent propeller efficiency. Based on Eq. [7.31], the equivalent shaft horsepower (ESHP) for a turboprop system is

$$\mathrm{ESHP} = \mathrm{SHP} + \frac{T_j V}{0.8} \ \text{ for } V > 50 \text{ f/s} \qquad (7.32)$$

and

$$\mathrm{ESHP} = \mathrm{SHP} + \frac{T_j}{2.5} \ \text{ for } V \leq 50 \text{ f/s}. \qquad (7.33)$$

Equation [7.33] is relevant to static thrust conditions that would exist at take-off. The factor, 2.5, comes from a 2.5 lbs/hp thrust conversion.

TABLE 7.3: Piston and Turboprop sizing relations.

$a[SHP]^b$	Piston Engines						Turboprop	
	Opposed		In-line		Radial			
	a	b	a	b	a	b	a	b
W (lbs)	5.47	0.780	5.22	0.780	4.90	0.809	1.67	0.803
L (f)	3.86	0.424	5.83	0.424	6.27	0.310	4.14	0.373
D (f)	Width $= 32-34$ in		Width $= 17-19$ in		20.2	0.130	9.48	0.120
	Height $= 22-25$ in		Height $= 24-26$ in					

The design of the propeller follows the same steps as before with the exception of using the ESHP for the SHP. Since the ESHP is different for cruise and static conditions, separate values need to be used to determine the thrust in the cruise and take-off flight phases that are treated in the spreadsheet.

7.5.4 Piston and Turboprop Sizing

If an existing engine is available for the design, the dimensions and weight are available from the manufacturer. Otherwise, these need to be estimated. In this case, the relations and coefficients in Table 7.3 can be used. These are based on historic engine data that were compiled by Raymer (1992). They encompass three types of piston engines, and turboprop engines.

7.5.5 Propeller Spreadsheet

The steps for designing a propeller-driven engine system have been implemented in the spreadsheet **prop.xls**. A sample with values corresponding to a hypothetical single-place recreational aircraft is shown in Figure 7.7. The propeller design is based on cruise conditions consisting of a cruise altitude of 8000 feet and a cruise Mach number of 0.18. The total drag for this aircraft under these conditions is estimated to be 400 pounds.

The appropriate engine for this design would be a reciprocating type, with an SHP in the range of 150 hp. An example of an appropriate engine would be a Lycoming 0-320-A2B, which delivers 150 hp at 2700 rpm.

Based on the engine r.p.m., V_{tip} and D were selected so that $M_{tip} \leq 0.85$. In this case, a value of $M_{tip} = 0.74$ was chosen. These values gave $n = 45$ rps (2700.14 rpm).

The calculations portion of the spreadsheet determines the quantities needed to obtain the propeller efficiency, η_P. This involves using Figure 7.5. Since this design will use a two-blade propeller, η_P is corrected to reflect this difference. Using $\eta_{P_{cor}}$, C_T and the thrust, T, are calculated.

The nominal 150 SHP used as an initial estimate was found to produce a thrust of 368 lbs, which was less than the estimated 400 lbs of drag. Without any changes to the propeller design, a SHP of 165 lbs was found to produce the necessary thrust (405.1 lbs).

The static thrust is calculated in the bottom part of the spreadsheet. This starts with input of C_T/C_P based on Figure 7.6. From this, T_{static} is calculated. Since Figure 7.6 is

Cruise Conditions:				
Cruise Alt.	8,000	ft		
M	0.18			
V	193.82	f/s		
ρ	0.06	lbm/f^3		
Tot. Drag	400	lbf		
Propeller Design:				
V_tip	820.00	f/s		
M_tip	0.76			
D	5.8	f		
SHP	165	hp		
blade no.	2			
Calculations:				
n	45	rps	2700.15	rpm
J	0.74	1/rev		
C_P	0.08			
η_P	0.84	(Fig. 7.5)		
η_P (COR)	0.87			
C_T	0.09			
T_cruise	405.09	lbf		
Static Thrust:				
C_T/C_P	2.5	(Fig 7.6)		
T_static	869.21	lbf		
T_(COR)	843.13	lbf		

FIGURE 7.7: Spreadsheet for propeller-driven engine sizing (PROP) for conditions with a hypothetical single-place recreational aircraft.

based on a three-bladed propeller, the static thrust is corrected for the two-blade design. This lowers the static thrust to a value of 843 lbs.

7.6   SUPERSONIC BUSINESS JET CASE STUDY

The spreadsheet for the engine scaling based on the parameters for the conceptual supersonic business jet case study was presented in Figure 7.4. The top part of the spreadsheet contains the drag values of the fuselage, main wing, vertical tail, and canard (horizontal tail). A plot of the relative drag values is shown to the right. This shows that an overwhelming amount of the drag comes from the main wing. The fuselage is next, with the drag on the tail surfaces being almost inconsequential to the total drag. Recall that the optimum fineness ratio was used for the fuselage in order to minimize its drag.

The cruise conditions for the business jet are an altitude of 55,000 feet and a Mach number of 2.2. The required thrust is the drag at cruise, which is equal to 10,748 lbs. The aircraft will have four engines, so that the required thrust per engine is 2687 lbs. The percentage of the maximum thrust at cruise was selected to be a standard maximum cruise rated value of 80 percent. At the cruise Mach number, there will be a fairly substantial increase in the thrust due to the ram effect. The value of 3.25 for the Mach Factor reflects this.

The existing engine that was selected as a reference engine for the conceptual aircraft design is the GE-F404-100D. A photograph of the engine is shown in Figure 7.8.

FIGURE 7.8: Photograph of GE-F404, which is selected as the reference engine for the conceptual SSBJ. (Courtesy of General Electric).

TABLE 7.4: Data for GE-F404-100D, selected as the reference engine for the conceptual SSBJ.

Thrust	11,000 lbs SLST
Bypass Ratio	0.27
Pressure Ratio	26:1
Weight	2,240 lbs
Length	158.8 in.
Diameter (Max)	34.8 in.
$\dot{m}$	146 lb/s

The manufacturer's data on the engine are given in Table 7.4. The F404 engine has been used to power the F/A-18, Grumman X-29, and Rockwell/MBB X-31. The 100D is a non-after-burning version of this engine.

The 11,000-lb static sea-level thrust for the reference engine was input in the reference engine thrust table in the spreadsheet. Near the cruise altitude (52,000 f), the thrust is reduced to approximately 900 lbs. However, with the Mach Factor, the reference engine should be able to supply approximately 2700 lbs of thrust at cruise. As a result, the scaling factor is $T_{req}/T_{ref} = 1.11$. This is within the acceptable range where the empirical scaling coefficients are valid. Based on this scaling factor, the weight, length, and maximum diameter of the required engine are determined.

The scale factor was used to determine the length and diameter of the engines for the SSBJ. These have been added to the drawing of the aircraft, as shown in Figure 7.9.

Finally, the thrust-to-weight ratio is determined in the bottom of the spreadsheet. The take-off weight of 90,000 lbs was taken from the first spreadsheet, **itertow**, for the parameters of the design. Using the scaling factor, times the reference engine T_{SL}, times four engines, $(T/W)_{SL} \simeq 0.54$. This value is slightly high for a long-range aircraft (see Table 7.1), but is reasonable based on the high cruise Mach number.

The value of $(T/W)_{SL}$ will be used in the refined analysis of the take-off and landing distances. This will be covered in the next chapter.

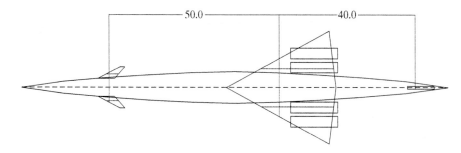

FIGURE 7.9: Top-view drawing showing the locations of the engines on the conceptual supersonic business jet.

7.7 PROBLEMS

7.1. Using a reference book such as Jane's "All the World Aircraft," select turbo-jet engines with a range of bypass ratios, and plot their thrust-specific fuel consumption, TSFC, as a function of BPR.

7.2. Using a reference book such as Jane's "All the World Aircraft," tabulate the thrust-to-weight ratio and bypass ratio for combat aircraft such as the F-15, F-16, F/A-18, and F-22. Compare these to values for a range of commercial jet aircraft, such as the B-747, B-757, and MD-80.

7.3. The Pratt & Whitney JT9D-3A turbofan engine, which powers the B-747, is rated at a sea-level thrust of 43,600 lbs. The manufacturer rates the maximum thrust at cruise conditions (Mach 0.85, 35,000-f altitude) to be 10,200 lbs. How does this compare to the value obtained from the spreadsheet?

7.4. Using the JT9D-3A as a reference engine, if a design requires 20 percent more thrust at cruise than the manufacturer specified value, what will be the new engine weight, length, and maximum diameter?

7.5. Select the engine/propeller combination for a single-place sport aircraft given the following information:
1. engine shaft height = 83 in.;
2. weight = 1500 lbs;
3. thrust-to-weight ratio = 0.6;
4. cruise speed = 130 mph, at 21,000 f;
5. two-blade propeller.

7.6. For the conditions in the spreadsheet **prop.xls**, plot how the thrust, T_{cruise}, changes with the following input values: SHP, blade number, and V_{tip}. Be careful that you update any changes in the values read from Figure 7.7.

7.7. A small homebuilt aircraft is designed to cruise at a velocity of 115 knots, at an altitude of 8000 feet. It is proposed to use a 150-hp engine, which will turn a two-bladed, 70-in. diameter propeller, at 2700 rpm. For this, determine the thrust generated at cruise and the static thrust.

7.8. If a propeller is a variable pitch design, its pitch is adjusted to the optimum blade angle at a given flight condition to produce a constant engine RPM regardless of the power being produced. In this instance, the advance ratio, J, and the power coefficient, C_P, are independent variables.

7.9. Using Figure 7.5, plot C_P versus J for an efficiency, $\eta_P = 0.9$. How is this plot useful?

7.10. For the homebuilt aircraft in Problem 7.7, with a fixed pitch angle of $\theta_{3/4} = 35°$, plot how the engine RPM will change with the cruise velocity? What happens to the thrust as a function of cruise velocity?

7.11. The designers of a conceptual aircraft determined the total thrust at a cruise Mach number of 0.8 and cruise altitude of 40,000 feet to be 900 lbs. They propose to use two Allied Signal TFE 731-40R turbofan engines, which has a sea-level static thrust of 4250 lbs. Determine the scaling factor using the Allied Signal engine as a reference. Propose at least one other engine type that would be appropriate for this design.

CHAPTER 8

Take-Off and Landing

8.1   TAKE-OFF
8.2   LANDING
8.3   SPREADSHEET APPROACH FOR TAKE-OFF AND LANDING ANALYSIS
8.4   PROBLEMS

Boeing 767-300 at the rotation portion of take-off. (Courtesy of The Boeing Company.)

In Chapter 3, simple empirical formulas were used to estimate the effect of wing loading on take-off and landing distances. In this chapter, a more rigorous analysis of the take-off and landing flight phases are made. This makes use of the fact that by this point, the aerodynamic surfaces have been designed, and the engine(s) have been selected. This information will be used to supply the necessary parameters including the weight, wing loading, and thrust-to-weight ratio at take-off and landing.

160

The maximum lift coefficient, $C_{L_{max}}$, is another important parameter used in the take-off and landing analysis. This involves the use of lift-enhancing devices, such as trailing-edge flaps and leading-edge slats, which will be discussed in the next chapter. As such in this chapter, take-off and landing distance estimates will be treated in one of two ways:

1. by estimating $C_{L_{max}}$ based on historical values, and using it to determine the take-off and landing distances or

2. specifying the take-off and landing distances and then finding the necessary value of $C_{L_{max}}$.

The latter approach occurs when an important design feature is take-off and landing distances, such as with short take-off and landing (STOL) and ultra-STOL aircraft. In those cases, the methods for enhancing the maximum lift coefficient strongly drives the design.

8.1 TAKE-OFF

The take-off flight phase consists of accelerating from rest to a take-off velocity, V_{TO}, and climbing to an altitude, which is greater than a reference obstacle height, $H_{obstacle}$. The distance required to accomplish this is the take-off distance, s_{TO}. The take-off velocity was first defined in Eq. [3.2], which is reproduced here:

$$V_{TO} = 1.2V_s = 1.2\left[\left(\frac{W}{S}\right)_{TO}\frac{2}{\rho C_{L_{max}}}\right]^{0.5}. \tag{8.1}$$

For analysis, the take-off phase is divided into four parts consisting of

1. ground roll,

2. rotation,

3. transition, and

4. climb.

A schematic representation of these four parts is shown in Figure 8.1.

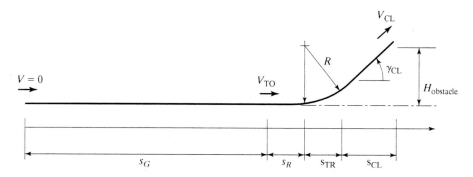

FIGURE 8.1: Schematic illustration of the take-off flight phase, which is made up of four parts.

The ground roll is the portion in which the aircraft accelerates from rest to the take-off velocity, V_{TO}. The distance of the ground roll is designated s_G. The rotation portion consists of a rotation maneuver in which the aircraft is pitched up to increase the angle of attack. The lift coefficient in the rotation portion is defined as C_{L_R}, where

$$C_{L_R} = 0.8 C_{L_{max}}. \tag{8.2}$$

The distance for the rotation is designated s_R.

The transition portion of take-off is where the aircraft first leaves the ground and flies at constant velocity along a circular arc of radius, R. This portion ends when the aircraft reaches the climb angle, γ_{CL}. The ground distance for the transition portion is designated s_{TR}.

The climb portion begins where the aircraft first reaches its climb angle and ends when an altitude of a reference obstacle is reached. The ground distance for this portion is designated s_{CL}.

The total take-off distance is the sum of the distances of each portion, namely,

$$s_{TO} = s_G + s_R + s_{TR} + s_{CL}. \tag{8.3}$$

Depending on the climb angle, it is sometimes possible to reach $H_{obstacle}$ in the transition portion of take-off. This most often happens with low wing loading designs, which are specifically designed to have short take-off distances. In that case, s_{TO} would not include s_{CL}. Each of the parts of take-off will now be discussed in detail in the following sections. Table 8.1 lists various minimum take-off parameters for three classes of aircraft.

TABLE 8.1: Minimum take-off specifications.

	MIL-C5011A Military	FAR Part 23 Civil	FAR Part 25 Commercial
Velocities:			
	$V_{TO} \geq 1.1 V_s$	$V_{TO} \geq 1.1 V_s$	$V_{TO} \geq 1.1 V_s$
	$V_{CL} \geq 1.2 V_s$	$V_{CL} \geq 1.1 V_s$	$V_{CL} \geq 1.2 V_s$
Climb Gradient:			
Gear Up, AEO	500 fpm at SL	300 fpm at SL	—
Gear Up, OEI	100 fpm at SL	—	3% at V_{CL}
Gear Down, OEI	—	—	0.5% at V_{CL}
Rolling Friction Coefficient:			
	0.025	—	—
Field Length Definition:			
	Distance needed to clear 50-f obstacle	Distance needed to clear 50-f obstacle	115% of distance needed to clear 35-f obstacle

AEO = all engines operating.
OEI = one engine inoperative.

8.1.1 Ground Roll

In the ground-roll portion of take-off, the aircraft accelerates from rest, until it reaches the take-off velocity given by Eq. [8.1]. The ground distance required for this portion of take-off is

$$s_G = \int_0^{V_{TO}} \left(\frac{V}{a}\right) dV = \frac{1}{2} \int_0^{V_{TO}} \frac{dV^2}{da}. \tag{8.4}$$

The acceleration, a, is found from

$$a = \frac{g}{W_{TO}} \sum F_x = \frac{g}{W_{TO}} \left[T - D - F_f\right], \tag{8.5}$$

where F_f is the rolling friction force given by

$$F_f = \mu \left[W_{TO} - L_G\right]. \tag{8.6}$$

A schematic that indicates the forces acting on an aircraft during ground roll is shown in Figure 8.2.

The rolling friction coefficient, μ, varies with the runway conditions. Examples are given in Table 8.2. The lift generated during ground roll, L_G, is based on an enhanced lift configuration for the wing and the angle of attack produced when the aircraft is on the ground.

The drag force consists of the base drag and lift-induced drag as before (Eq [3.12] for example) plus additional base drag produced by extended flaps and landing gear. That is,

$$D = qS \left[C_{D_0} + kC_{L_G}^2 + \Delta C_{D_{0\text{flap}}} + \Delta C_{D_{0\text{LG}}}\right]. \tag{8.7}$$

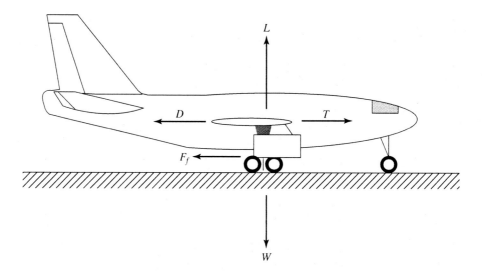

FIGURE 8.2: Schematic illustration of forces acting on aircraft during the ground-roll portion of take-off.

TABLE 8.2: Rolling friction coefficients for different runway surfaces.

Surface Type	μ
Concrete (wet or dry)	0.03–0.05
Hard Turf	0.05
Firm & Dry Dirt	0.04
Soft Turf	0.07
Wet Grass	0.10
Snow or Ice-Covered	0.02

TABLE 8.3: Added base drag due to flaps.

Flap Type (60% span, 25% chord)	δ_f	$\Delta C_{D_{0_{\text{flap}}}}$
Fowler	30	0.032
Fowler	50	0.083
Split or Plain	30	0.05
Split or Plain	50	0.10
Slotted	30	0.02
Slotted	50	0.05

For the flaps, the added base drag depends on the type of flap and deflection angle, δ_f. This will be discussed in detail in the next chapter on enhanced lift devices. In order to make an estimate at this stage of the design, empirical data are given in Table 8.3.

The additional base drag coefficient, which is due to the extended landing gear, can be estimated from the following empirical relation

$$\Delta C_{D_{0_{\text{LG}}}} = f_{\text{LG}} \frac{A_{\text{LG}}}{S}, \tag{8.8}$$

where A_{LG} is the frontal (projected) area of the landing gear. f_{LG} is a correlation function that is based on the take-off weight of the aircraft. It is given by the following relation

$$f_{\text{LG}} = 3.23 \sqrt{\frac{W_{\text{TO}}}{1000}}. \tag{8.9}$$

In Eq. [8.6], L_G is

$$L_G = qS C_{L_G}. \tag{8.10}$$

For take-off and landing, where the aircraft is in close proximity to the ground, the effective aspect ratio of the wing is larger by an amount

$$\frac{A}{A_{\text{effective}}} = \sqrt{\frac{2H}{b}}, \tag{8.11}$$

where H is the altitude and b is the wingspan. This "ground effect" increases the effective L/D, with the result of increasing the lift at a given angle of attack and decreasing the lift-induced drag.

These effects can be accurately accounted for in the take-off analysis. However, for the conceptual design they can be neglected, which will give a conservative (longer) estimate of the take-off distance.

In integrating Eq. [8.4] to determine s_G, we note that a is a complex function of velocity and thrust. The maximum thrust is also a function of velocity. Given this, there are two possible approaches that can be used in evaluating the integral:

1. Assume that T/W is constant, then

$$a = f_1 + f_2 V^2, \tag{8.12a}$$

where

$$f_1 = g \left(\frac{T}{W} - \mu \right) \tag{8.12b}$$

and

$$f_2 = \frac{g\rho}{2(W/S)} \left(\mu C_{L_G} - C_{D_0} - k C_{L_G} - \Delta C_{D_{0_{\text{flap}}}} - \Delta C_{D_{0_{LG}}} \right). \tag{8.12c}$$

Then

$$s_G = \int_0^{V_{TO}} \frac{dV^2}{f_1 + f_2 V^2} = \frac{1}{2 f_2} \ln \left[\frac{f_1 + f_2 V_{TO}^2}{f_1} \right]. \tag{8.13}$$

In following this approach, a good estimate for T/W is

$$\frac{T}{W} = \left(\frac{T}{W} \right)_{\text{max}} \quad \text{at } 0.7 V_{TO}. \tag{8.14}$$

2. Integrate numerically taking small velocity time steps, over which T/W is constant and equal to the maximum thrust at each upper time-step value of velocity.

The first approach will be used in the analysis with the spreadsheet developed in this chapter.

8.1.2 Rotation

In the rotation part of the take-off, the aircraft angle of attack is increased until $C_L = 0.8 C_{L_{\text{max}}}$. As a convention, this is assumed to take three seconds. During this maneuver, the aircraft velocity is V_{TO}, so that the ground distance is

$$S_R = 3 V_{TO}, \tag{8.15}$$

where V_{TO} has units of feet per second in order to obtain s_R in feet.

8.1.3 Transition

In the transition portion of the take-off, the aircraft leaves the ground and flies at a constant velocity along a circular arc of radius R_{TR}. During this maneuver, the load factor, n_{TR}, is

$$n_{TR} = \left(\frac{L}{W}\right)_{TR} = 1 + \frac{V_{TO}^2}{R_{TR}g}. \tag{8.16}$$

Now, the lift-to-weight ratio is

$$\left(\frac{L}{W}\right)_{TR} = \frac{C_{L_{TO}}\rho V_{TO}^2}{2(W/S)} = \frac{0.8C_{L_{max}}\rho(1.2V_s)^2}{2(W/S)}. \tag{8.17}$$

Upon substituting for V_s, all of the variables cancel, leaving

$$\left(\frac{L}{W}\right)_{TR} = 1.15 \tag{8.18}$$

Substituting this result into Eq. [8.16], we obtain

$$R_{TR} = \frac{V_{TO}^2}{0.15g}. \tag{8.19}$$

Following Eqs. [3.8 & 3.9], the rate of climb is

$$\frac{dH}{dt} = V_{TO}\sin\gamma. \tag{8.20}$$

At a constant climb velocity,

$$V_{TO}\sin\gamma = \frac{V_{TO}(T - D)}{W}. \tag{8.21}$$

The transition portion ends when $\gamma = \gamma_{CL}$. At this point, the radial vector has turned by an angle equal to γ_{CL}. Thus, from geometry, the horizontal (ground) distance subtended by the arc is

$$s_{TR} = R_{TR}\sin\gamma_{CL}. \tag{8.22}$$

and the altitude at the end of the transition portion is

$$H_{TR} = R_{TR}(1 - \cos\gamma_{CL}). \tag{8.23}$$

8.1.4 Climb

The climb portion starts at the end of transition and ends when the aircraft reaches a prescribed altitude, $H_{obstacle}$. Throughout this portion, the climb angle is γ_{CL}. Therefore, from geometry, the horizontal (ground) distance is

$$s_{CL} = \frac{H_{obstacle} - H_{TR}}{\tan\gamma_{CL}}. \tag{8.24}$$

The prescribed obstacle height is 50 feet for military and small civil aircraft and 35 feet for commercial aircraft. In aircraft designs with low wing loading at take-off, it is possible that the $H_{obstacle}$ altitude is reached during the transition portion. In this case, $s_{CL} = 0$.

8.1.5   Balanced Field Length

The previous subsections provided a breakdown of the take-off distance. The total take-off distance is the sum of these parts. This then specifies the minimum airfield length that is required for the aircraft to take-off.

There is, however, another important airfield length at take-off that is called the "balanced field length." This corresponds to the total airfield length, which is required for safety in the event that one engine of a multiengine aircraft fails. In this event, there is a reference velocity called the "decision speed." If the engine fails when the take-off velocity is below the decision speed, the field length needs to be sufficient to allow the aircraft to break to a stop. If it fails above the decision speed, the field length has to be long enough to allow the aircraft to take off and clear the necessary obstacle height. A schematic illustration of the balanced field length is shown in Figure 8.3.

The balanced field length can be estimated from the following fairly detailed empirical formula:

$$s_{\text{BFL}} = \frac{0.863}{1 + 2.3G}\left[\frac{W/S}{\rho g\left(0.8C_{L_{\max}}\right)} + H_{\text{obstacle}}\right]\left[\frac{1}{\frac{T_{av}}{W} - U} + 2.7\right] + \frac{655}{\sqrt{\rho/\rho_{\text{SL}}}}, \quad (8.25a)$$

$$\text{Jet Engine: } T_{av} = 0.75 T_{\text{static}}\left[\frac{5 + B}{4 + B}\right], \quad (8.25b)$$

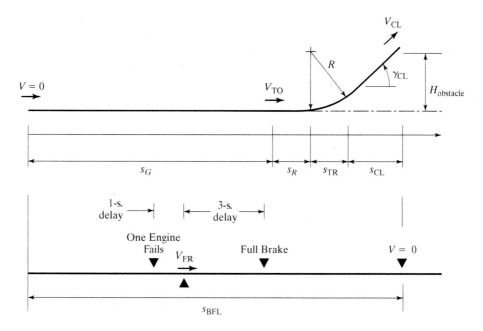

FIGURE 8.3: Schematic illustration of the balanced field length for an emergency aborted take-off, compared to standard take-off.

and

$$\text{Propeller Engine: } T_{av} = 5.75P \left[\frac{(\rho/\rho_{\text{SL}})\,N_e D_p^2}{P} \right]^{1/3}. \qquad (8.25c)$$

In Eq. [8.25],

$$G = \gamma_{\text{CL}} - \gamma_{\text{min}}, \qquad (8.26)$$

where $\gamma_{\text{min}} = 0.024$ for a two-engine aircraft, 0.027 for a three-engine aircraft, and 0.030 for a four-engine aircraft. U represents the drag produced by extended flaps and is given by

$$U = 0.01C_{L_{\text{max}}} + 0.02. \qquad (8.27)$$

B is the jet engine bypass ratio. P is the shaft horsepower for the reciprocating engine. N_e is the number of engines, and D_p is the propeller diameter. s_{BFL} has units of feet.

8.2   LANDING

The landing flight phase consists of an approach starting at an altitude of 50 feet, a touch-down at a velocity of V_{TD}, a free-roll at constant velocity, and a braking deceleration until the aircraft comes to rest. The distance required to accomplish this is the landing distance, s_L. A schematic representation of the landing phase is shown in Figure 8.4. Table 8.4 lists various minimum landing parameters for three classes of aircraft.

     In the first portion of the landing phase, the aircraft starts at an elevation of 50 feet and a velocity of

$$V_{50} = 1.3V_s. \qquad (8.28)$$

For military aircraft, $V_{50} = 1.2V_s$ (Table 8.4). The aircraft descends at a fixed angle, γ_{approach}, and velocity, V_{50}, until it reaches a transition height, H_{TR}. The horizontal distance for this portion of the landing is designated s_A.

     In the second portion of the landing, the aircraft flies along a constant radius circular arc. At the bottom of the arc, the aircraft touches down at the velocity

$$V_{\text{TD}} = 1.15V_s. \qquad (8.29)$$

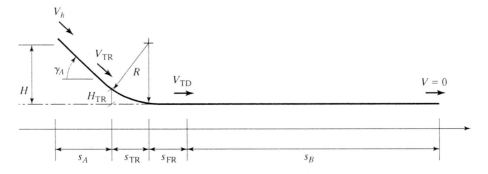

FIGURE 8.4: Schematic illustration of the landing flight phase, which is made up of four parts.

TABLE 8.4: Minimum take-off specifications.

	MIL-C5011A Military	FAR Part 23 Civil	FAR Part 25 Commercial
Velocities:			
	$V_{TO} \geq 1.2V_s$ $V_{CL} \geq 1.1V_s$	$V_{TO} \geq 1.3V_s$ $V_{CL} \geq 1.15V_s$	$V_{TO} \geq 1.3V_s$ $V_{CL} \geq 1.15V_s$
Breaking Friction Coefficient:			
	0.30	—	—
Field Length Definition:			
	Distance needed starting above 50-f obstacle	Distance needed starting above 50-f obstacle	160% of distance starting above 35-f obstacle

Just prior to touchdown, the aircraft will flare to produce a positive angle of attack such that

$$CL_{L_{\text{touchdown}}} = 0.8C_{L_{\max}},\qquad(8.30)$$

where $C_{L_{\max}}$ is based on an enhanced lift (flap) configuration. The horizontal distance for this portion of the landing is designated s_{TR}.

In the third portion of the landing, the aircraft rolls freely at a constant velocity of V_{TD}. The distance for this portion of the landing is s_{FR}.

In the final part of the landing, the aircraft brakes to a stop. This distance is designated s_B.

The total landing distance is the sum of distances in the four parts:

$$s_L = s_A + s_{TR} + s_{FR} + s_B.\qquad(8.31)$$

For landing, a worst case scenario is assumed in which the weight of the aircraft at landing, W_L, is taken to be

$$W_L = W_{TO} - 0.5W_{\text{fuel}}.\qquad(8.32)$$

That is, it is assumed to be carrying 50 percent of its take-off fuel weight.

Each of the portions of the landing phase will now be discussed in detail.

8.2.1 Approach

The approach is a constant velocity decent from a height of 50 feet to a height of H_{TR}. The decent angle can be determined from Eq. [8.21], which for landing is

$$V_{50} \sin \gamma_{\text{approach}} = \frac{V_{50}(T - D)}{W}.\qquad(8.33)$$

Most typically, the engines will be at idle so that $T = 0$, then

$$\gamma_{\text{approach}} = \sin^{-1} \frac{-D}{W}.\qquad(8.34)$$

The drag will be determined from Eq. [8.7] for full flaps and extended landing gear. As a reference, a transport aircraft decent angle would be no larger than 3°.

The approach distance can be found from Eq. [8.24], which for landing is

$$s_A = \frac{H_{TR} - 50}{\tan \gamma_{approach}}. \tag{8.35}$$

H_{TR} is obtained from the transition portion of the landing.

8.2.2 Transition

In the transition portion of the landing, the aircraft flies along a circular arc with radius, R_{TR}. The velocity decelerates slightly from V_{50} to V_{TD}, which is from 1.3 to $1.15V_s$. Using an average velocity of $V_{TR} = 1.23V_s$, which we assume is constant, we apply Eq. [8.17] to obtain

$$n_{TR} = \frac{L}{W}\bigg|_{TR} = 1.19. \tag{8.36}$$

R_{TR} can be computed from Eq. [8.19], which for landing is

$$R_{TR} = \frac{(1.23V_s)^2}{0.19g}. \tag{8.37}$$

H_{TR} can be found from Eq. [8.23], which for landing is

$$H_{TR} = R_{TR}\left(1 - \cos \gamma_{approach}\right). \tag{8.38}$$

From Eq. [8.22], the horizontal (ground) distance is

$$s_{TR} = -R_{TR} \sin \gamma_{approach}. \tag{8.39}$$

8.2.3 Free-Roll

During the free-roll portion of the landing, the aircraft maintains a constant velocity of V_{TD}. As a convention, this is assumed to last for three seconds. The distance covered is, therefore,

$$s_{FR} = 3V_{TD}. \tag{8.40}$$

where V_{TD} has units of feet per second to obtain s_{FR} in feet.

8.2.4 Braking

The braking portion is governed by the same equations that were used to analyze the ground-roll portion of take-off (Section 8.1.1). The deceleration in this case is given by Eq. [8.5] with the option that

1. the engines will be at idle so that $T = 0$ or
2. the engines are equipped with thrust reversers, in which case $-0.5T_{Max} \leq T \leq -0.4T_{Max}$.

TABLE 8.5: Rolling friction coefficients with brakes applied, for different runway surfaces.

Surface Type	μ
Concrete (wet or dry)	0.4–0.6
Hard Turf	0.4
Firm & Dry Dirt	0.3
Soft Turf	0.0
Wet Grass	0.3
Snow or Ice-Covered	0.07–0.10

The drag includes all of the quantities given in Eq. [8.7]. For the rolling friction, the friction coefficients are larger because of braking. These are given in Table 8.5.

Assuming that any reverse thrust is not a function of velocity, the braking distance is given by Eq. [8.13], which for landing is

$$s_B = \int_{V_{TD}}^{0} \frac{dV^2}{f_1 + f_2 V^2} = \frac{1}{2 f_2} \ln \left[\frac{f_1}{f_1 + f_2 V_{TD}^2} \right]. \tag{8.41}$$

The terms f_1 and f_2 are given by Eqs. [8.12(b & c)], with the only modification being that $T = 0$ or $T = -0.4 T_{Max}$ to $-0.5 T_{Max}$.

The total landing distance is the sum of the distances in each portion. In addition, the FAR-25 requires that the quoted landing distance for commercial aircraft be 1.6 times longer than calculated in order to account for pilot differences. Therefore, the total distance is

$$s_L = 1.6 [s_A + s_{TR} + s_{FR} + s_B]. \tag{8.42}$$

8.3 SPREADSHEET APPROACH FOR TAKE-OFF AND LANDING ANALYSIS

A spreadsheet named **to_l.xls** was developed to estimate take-off and landing distances for conceptual designs. This incorporates the different equations that were presented in this chapter. A sample with values corresponding to the conceptual SSBJ is shown in Figures 8.5 and 8.6.

The spreadsheet is divided into two parts corresponding to take-off or landing analysis. These are denoted by the bold titles. In each of these, the input parameters are grouped at the top.

Most of the output parameters are intermediate calculations that are needed to determine each portion of the take-off or landing distances. The distances associated with each portion are labeled with bold characters. These are summed up to give the total distances, whose value is highlighted in bold numerics.

In the case of take-off, depending on the input parameters (especially climb angle), it is possible that the altitude at the end of "transition" may reach or exceed the obstacle height. As a result, the distance for the "climb" portion will be negative. Therefore, in summing up the distances in each portion of take-off, the sign of s_{CL} is checked. If it is negative, s_{CL} was not added to the total take-off distance. The equation for this can be seen by selecting the spreadsheet cell next to s_{TO}.

Take-Off						
CD_0	0.04		mu_TO	0.05		
A	2		T_max (lb)	49026		
H (f)	1,000		f_LG	30.73		
CL_G	2		A_LG (f^2)	2.5		
W_TO (lb)	90,523		deltCD_0_flap	0.05		
S (f^2)	600		gamma_CL (deg)	3		
			H_obstacle (f)	35		
k	0.2		T/W	0.54		
rho (lbm/f^3)	0.07		f1 (f/s^2)	15.83		
W/S (lb/f^2)	150.87		deltCD_0_LG	0.13		
S (f^2)	600		f2 (f^−1)	−126.1E−6		
V_T-O (f/s)	306.49		R_TR (f)	19447.91		
q_T_O (lb/f^2)	108.63		H_TR (f)	26.65		
			S_G (f)	5470.32		
			S_R (f)	919.46		
			S_TR (f)	1017.82		
			S_CL (f)	159.28		
			S_T-O (f)	7566.88		
Landing						
W_L (lb)	23384		D_50 (lb)	19933.81		
W/S (lb/f^2)	38.97		gamma_A (deg)	−58.48		
V_50 (f/s)	168.75		gamma_A _act	−3		
V_TD (f/s)	149.28		R_TR (f)	4166.95		
q_50 (lb/f^2)	32.93		H_TR (f)	5.71		
q_TD (lb/f^2)	25.77		f1 (f/s^2)	−19.32		
mu _L	0.6		f2 (f^−1)	562.8E−6		
T_L (lb)	0					
			S_A (f)	845.1		
			S_TR (f)	218.08		
			S_FR (f)	447.85		
			S_B (f)	930.61		
			S_L (f)	2441.63		
			1.6(S_L) (f)	3906.61		

FIGURE 8.5: Spreadsheet (TO_L) for take-off and landing distance calculations (Part 1).

The height at the end of transition, H_{TR}, is displayed in a cell of the spreadsheet. This can be compared to the obstacle height in judging the climb angle.

In the landing analysis, the approach angle γ_A is calculated based on Eq. [8.34]. This value can be relatively large. Therefore, the spreadsheet allows an input denoted as $\gamma_{A_{act}}$, which will be the *actual* value used in determining the approach distance, s_A.

Different rolling friction coefficients, μ_{TO} or μ_L, are used based on take-off or landing conditions. The coefficients in the latter are usually larger because of braking. Both sets of values were listed in Tables 8.2 and 8.5.

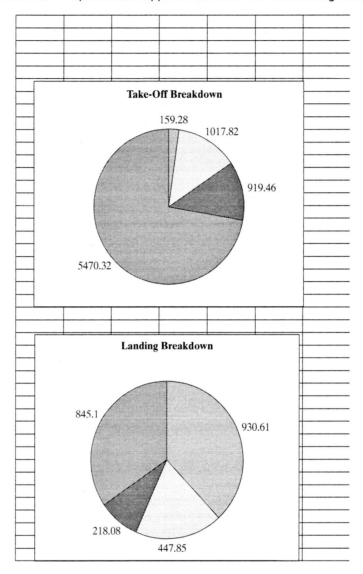

FIGURE 8.6: Spreadsheet (TO_L) for take-off and landing distance calculations (Part 2).

As usual, special care is used in noting the units in all inputs and calculated values. For example, the quantities f_1 and f_2 have specific units, which differ from each other. Therefore, the units of the input variables have to be followed meticulously.

8.3.1   Case Study: Take-Off and Landing

The input values in the spreadsheet correspond to the conceptual SSBJ that was proposed in Chapter 1. In the list of input parameters, C_{D_0}, A, W_{TO}, S, and T_{max} were specified or derived in previous spreadsheets. The wing area was increased by 15 percent to 600 f^2

in order to reflect the extension of slotted flaps. Other parameters that are specific to take-off and landing included the rolling friction coefficient, μ; projected area of the landing gear, A_{LG}; the added drag due to the flaps, $\Delta C_{D_{0_{flap}}}$; the climb and approach angles, γ_{CL} and γ_A; and the obstacle height at take-off, $H_{obstacle}$.

The values of the rolling friction coefficients corresponded to those of a concrete surface, which we associate with a built-up air field. The area of the landing gear is used in the calculation of the added drag due to the landing gear, $\Delta C_{D_{0_{LG}}}$. A_{LG} can be estimated from photographs of comparable aircraft. $\Delta C_{D_{0_{LG}}}$ is determined from Eq. [8.8], which involves the correlation factor, f_{LG}, given by Eq. [8.9]. Although f_{LG} is shown as an input parameter, it is actually computed based on the take-off weight.

The value of $\Delta C_{D_{0_{flap}}}$ was taken from Table 8.3. This corresponds to a slotted flap, which is deflected at a 50° angle.

The climb angle was chosen to be a rather modest 3°. This was well under the 10° angle that is possible for the wing loading, based on the analysis in Chapter 3. The obstacle height is 35 feet, which is the required value for commercial and civil aircraft.

The intermediate calculations include the take-off velocity, which is equal to $1.2V_s$; dynamic pressure, q_{TO}; the thrust-to-weight ratio, T/W; the quantities f_1 and f_2; and the transition radius, R_{TR}. The ground distance, s_G, is given by Eq. [8.13] and involves f_1 and f_2. f_2 is given by Eq. [8.12c]. We note that it is possible (and likely) that f_2 is negative. This makes it possible for the quantity inside the *log* in Eq. [8.13], namely,

$$\left[\frac{f_1 + f_2 V_{TO}^2}{f_1} \right]. \tag{8.43}$$

to be negative and, therefore, to make the *log* of this quantity undefined. As a result, care needs to be taken in the selection of the input values that lead to the values of f_1 and f_2.

For the input conditions shown in the spreadsheet, the ground distance portion of the take-off was 5470 feet, which accounted for 73 percent of the total take-off distance. The distance associated with the rotation, s_R, was based on Eq. [8.15]. This corresponded to 919 feet or 12 percent of the total distance.

The distance associated with the transition portion, s_{TR}, made up 13 percent of the total take-off distance. The smallest portion corresponded to climb, s_{CL}, which contributed only 2 percent to the total take-off distance.

When summed up, the total distance required for take-off was 7566 feet. This compared to 5238 feet, which came from the preliminary estimate using the take-off parameter based on historical data.

For the landing analysis, there are a few additional input conditions, which need to be specified. These are the weight at landing, W_L, and the approach angle, $\gamma_{A_{actual}}$. As mentioned, the maximum approach angle is determined in the spreadsheet. This uses Eq. [8.34], which involves the drag. The drag is determined from Eq. [8.7]. The velocity is the approach velocity at the 50 ft altitude, $V_{50} = 1.3V_s$. The values of the drag and lift coefficients were the same as those used at take-off.

In the case study, the maximum approach angle was found to be −58°. This value is much too large for a commercial aircraft. Therefore, a more representative value of −3° was used. The approach distance, s_A, was determined from Eq. [8.35]. With the values in the spreadsheet, this distance was found to be 845 feet. This represented 35 percent of the total landing distance.

The transition distance was found from Eq. [8.39]. This involved the transition radius, R_{TR}, which was found from Eq. [8.37]. These gave a transition distance of 218 feet, which was 9 percent of the total.

The free-roll distance, s_{FR}, was based on Eq. [8.40]. This used the touchdown velocity, $V_{TD} = 1.15V_s$. These gave a free-roll distance of 447 feet, which was 18 percent of the total.

Finally, the braking distance was computed from Eq. [8.41]. This utilized expressions for f_1 and f_2, which were found using parameters representative of the conditions at landing. For f_1, the engines were assumed to be idling, so that $T/W = 0$. Thrust reversing was not determined to be necessary since the breaking distance was ultimately small. The distance for breaking was found to be 930 feet, which was only 38 percent of the total landing distance.

When summed up, the total distance required for landing was 2441 feet. This value was multiplied by 1.6 according to FAR-25 regulations to give the final take-off distance of 3906 feet. This is nearly half of the value of 7922 feet, which came from the preliminary estimate in Chapter 3 using the landing parameter based on historical data.

This analysis required specifying the maximum lift coefficient. In this case, a value of $C_{L_{max}} = 2.0$ was used. The effect of the choice of the maximum lift coefficient on take-off and landing distances is shown in Figure 8.7. This indicates a greater sensitivity

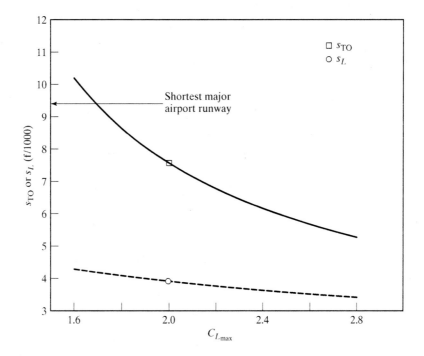

FIGURE 8.7: Effect of $C_{L_{max}}$ on take-off and landing distances for the conditions of the conceptual SSBJ. Symbols are selected values.

on the take-off distance. For reference, the arrow marks the shortest runway for a major airport in the United States. (See Table 1.4.) The distances corresponding to the value of $C_{L_{max}}$ used for the SSBJ are denoted by the symbols. These are well within the minimum distance for all the major airports.

The next chapter deals with the enhanced lift devices such as trailing-edge flaps and leading-edge slats. The design of these will have the goal of meeting the maximum lift coefficient that was specified for take-off and landing in this chapter.

8.4 PROBLEMS

8.1. Using the spreadsheet, determine how the take-off and landing distances change with
   1. the airport altitude;
   2. the wing loading;
   3. the thrust-to-weight ratio;
   4. the maximum lift coefficient.

   Keeping the first two fixed, which is better for shortening the distances, large T/W or $C_{L_{max}}$?

8.2. For the conditions of the case study aircraft, how does the take-off distance vary with the climb angle? What will be the largest possible climb angle? What parameter can be used to increase the climb angle?

8.3. For the conditions of the case study aircraft, how does the landing distance vary with the approach angle? What will be the largest possible approach angle? What parameter can be used to increase the approach angle?

8.4. Using this spreadsheet, determine the take-off distances for a range of take-off parameters, given by Eq. [3.3], then compare the take-off distances to those given by Eq. [3.5]. What do you conclude?

8.5. Using this spreadsheet, determine the landing distances for a range of landing parameters, given by Eq. [3.6], then compare the landing distances to those given by Eq. [3.7]. What do you conclude?

8.6. For the case study aircraft, what is the $C_{L_{max}}$ that is required to have total take-off and landing distances of under 1000 feet? Do you think that it is possible?

8.7. Describe the properties of a STOL aircraft. Use examples from the spreadsheet to substantiate your answer.

CHAPTER 9

Enhanced Lift Design

Photograph of the Lockheed–Martin C5-A Galaxy heavy-cargo transport in an approach to land with leading- and trailing-edge flaps fully extended. It has a maximum take-off weight of 840,000 pounds, and take-off and landing distances of 12,200 feet and 4,900 feet, respectively. (Courtesy of Lockheed Martin)

In the previous chapter, values of $C_{L_{\max}}$ were used to determine the take-off and landing distances. These assumed the use of some type of lift-enhancing devices. This chapter deals with such devices and provides a methodology for determining the 3-D lift and drag characteristics in a high-lift configuration. The goal is to achieve the necessary $C_{L_{\max}}$

values in order to satisfy the maximum lift requirements that are imposed by such flight phases as take-off and landing, and combat (maximum maneuverability).

In a 2-D wing, the lift at a given angle of attack can be increased by increasing camber (shifting α_{0_L}). However, the maximum achievable lift is limited to the amount of camber, or the angle of attack, at which the flow can no longer follow the curvature of the airfoil and separates (lower photograph at the start of Chapter 4). In most cases when this occurs, the flow separates first from the leading edge.

Lift-enhancing devices fall in two categories: passive and active. The passive devices also fall in two categories: trailing-edge devices, which primarily act to increase the camber of the airfoil section, and leading-edge devices, which primarily act to prevent leading-edge separation.

Passive lift enhancement is relevant to most of the aircraft designs, except STOL and ultra-STOL. Such designs must make use of active lift enhancement. This generally consists of using air streams that are directed over the upper surface of wing in order to energize the boundary layer and prevent flow separation.

9.1 PASSIVE LIFT ENHANCEMENT

The most common types of trailing-edge lift enhancing devices are plane flaps, split flaps, slotted flaps, and Fowler flaps. Slotted flaps include single, double, and triple segments. Examples of these are illustrated in Figure 9.1.

The most common types of leading-edge devices are a fixed slot, leading-edge flap, Krueger flap, and plain slats (slotted leading-edge flap). Examples of these are illustrated in Figure 9.2.

The effectiveness of any of these devices depends on their deflection angle, δ_f; the wing thickness-to-chord ratio; the ratio of the flap chord to the wing chord, c/c_f; and the wing sweep angle and aspect ratio. In most cases, $c/c_f \simeq 0.3$, and the maximum lift occurs at $\delta_f \simeq 40°$.

The plain flap is simply a deflection of the trailing edge of the airfoil section. This type is most widely used on smaller aircraft. The split flap is similar to the plane flap except that only the bottom half of the airfoil section deflects. The lift produced by a split flap is virtually the same as a plain flap, but the drag is larger. As a result they are rarely used now, but were popular on aircraft built during World War II.

A slotted flap is essentially a plane flap with the addition of a slot at the hinge point to allow high-pressure air from the lower side of the airfoil to pass over the upper surface of the flap. This has the effect of adding momentum to the boundary layer on the upper flap surface to allow larger flap deflections before the flow separates. A Fowler flap is a slotted flap that translates rearward, away from the wing. This has the benefit of increasing the slot width and increasing the effective wing area. An example of this can be seen in the photograph of the C5-A at the start of the chapter.

Further refinements of the slotted (Fowler) flap design involves using two or three flap segments. These designs can lead to extremely high lift coefficients ($C_{L_{\max}} \geq 3.0$), but are complex and require extra volume inside the wing to be stored during cruise. Single slotted flaps are most common on mid-size aircraft. Most larger commercial and transport aircraft use a multiple slotted flap arrangement.

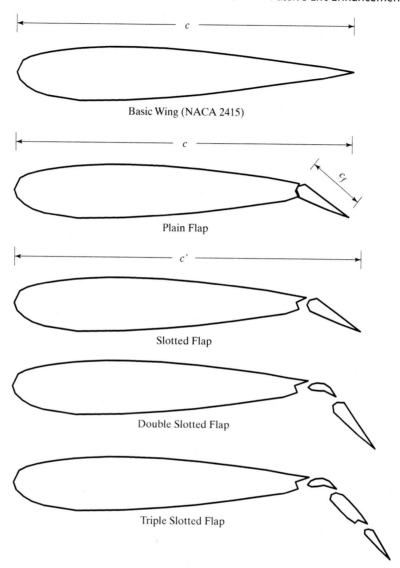

FIGURE 9.1: Schematic illustration of different trailing-edge flap configurations.

Although trailing-edge flaps increase lift at a given angle of attack, they do not increase the angle of stall, α_s, but actually causes it to decrease. This is the result of changes in the location of the stagnation line and local pressure gradient near the leading edge, which leads to a leading-edge flow separation. Sharper leading edges are more sensitive to this.

One solution for leading-edge separation is to increase the leading-edge radius. This is the principle effect of a leading-edge flap, which generally consists of a hinged portion of the leading edge, which deflects downward to effectively increase the leading-edge curvature. A variation on this is a Krueger flap, which consists of a hinged flap on the

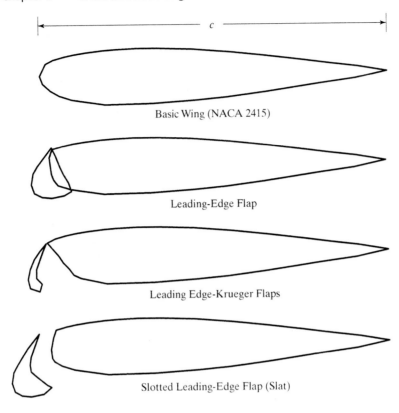

FIGURE 9.2: Schematic illustration of different leading-edge flap configurations.

lower side of the wing leading edge, which extends out into the flow. This approach is lighter in weight and, as a result, popular on large aircraft with large wing spans.

A leading-edge slot works the same way as a slotted flap by allowing air from the high-pressure lower surface to flow to the upper surface to add momentum to the boundary layer and prevent flow separation. A slotted leading-edge flap (slat) is the leading-edge equivalent of the trailing-edge slotted flap. In this, the leading edge is extended forward and downward to open the slot and simultaneously increase the wing section camber and area. As a result of the change in camber, there is also a small change in α_{0_L}. This is the arrangement used on the C5-A.

Of these devices, leading-edge flaps are more effective than slotted flaps on highly swept wings. They are also usually located over the outboard half-span of the wing in order to reduce the potential for wing-tip stall. The optimum leading-edge flap deflection is approximately 30–40°.

9.1.1 Lift Determination

Trailing-Edge Flaps. This section describes the approach to construct the 3-D lift coefficient versus angle of attack for the main wing with flaps and slats. This will be a modification of the C_L versus α construction that was presented in Chapter 4.

It involves determining the changes in $C_{L\max}$, α_{0_L}, and α_s that result from these passive lift devices. Additionally, the added drag due to the enhanced lift configuration is calculated so that it can be compared to the empirical value used in the take-off and landing analysis in the previous chapter.

Starting with the original 3-D lift coefficient versus angle of attack that was previously developed, the first step is to find the change in the angle of attack at zero lift, $\Delta\alpha_{0_L}$, produced by the addition of a trailing-edge flap. The formula to determine $\Delta\alpha_{0_L}$ depends on the type of trailing-edge flap. These are as follows.

For *plane flaps*,

$$\Delta\alpha_{0_L} = -\frac{dC_l}{d\delta_f}\frac{K'}{C_{l_\alpha}}\delta_f, \tag{9.1}$$

where C_{l_α} is the 2-D section lift coefficient, which from linear theory should be 2π (rad^{-1}), and K' is a correction for nonlinear effects, which is found from Figure 9.3.

The term $dC_l/d\delta_f$ is the change in the 2-D section lift coefficient with flap deflection. This is found from Figure 9.4 and is a function of c_f/c and t/c.

For *single slotted and Fowler flaps*,

$$\Delta\alpha_{0_L} = -\frac{d\alpha}{d\delta_f}\delta_f, \tag{9.2}$$

where $d\alpha/d\delta_f$ is found from Figure 9.5 as a function of c_f/c.

For *split flaps*,

$$\Delta\alpha_{0_L} = -\frac{k}{C_{l_\alpha}}(\Delta C_l)_{\frac{c_f}{c}=0.2}, \tag{9.3}$$

where k and $(\Delta C_l)_{\frac{c_f}{c}=0.2}$ are found from Figures 9.6 and 9.7, respectively. Note that as the subscript denotes, Figure 9.7 is for $\frac{c_f}{c} = 0.2$.

As a result of the flap deflection, the C_L-versus-α curve for the wing is translated along the α-axis by the amount $\Delta\alpha_{0_L}$. To complete the construction of the lift curve requires determining $C_{L\max}$ with flaps. To do this first requires finding $C_{L\max}$ for the **basic (unflapped) 3-D wing**, which was not done in Chapter 4.

Finding $C_{L\max}$ requires determining whether the main wing in the design has a "low" or "high" aspect ratio. This distinction is made because with high aspect-ratio wings, $C_{L\max}$ is primarily determined by the airfoil section shape, whereas, for a low aspect-ratio wing, $C_{L\max}$ is primarily determined by the wing planform shape.

A wing is considered to have a *high* aspect ratio when

$$A > \frac{4}{(C_1 + 1)\cos(\gamma_{\mathrm{LE}})}. \tag{9.4}$$

The coefficient, C_1, is a function of the taper ratio, λ, and can be found from Figure 9.8.

For *high* aspect-ratio wings, $C_{L\max}$ and α_s are found from

$$C_{L\max} = \left[\frac{C_{L\max}}{c_{l\max}}\right] C_{l\max} \tag{9.5}$$

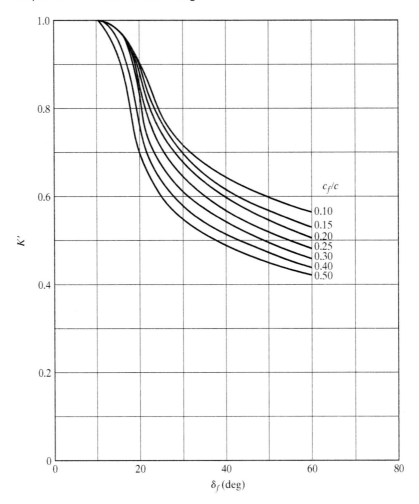

FIGURE 9.3: Nonlinear correction for plane flap $\Delta\alpha_{0_L}$ calculations. (From Ellison, 1969.)

and

$$\alpha_s = \frac{C_{L_{\max}}}{C_{L_\alpha}} + \alpha_{0_L} + \Delta\alpha_{C_{L_{\max}}}. \qquad (9.6)$$

Here $C_{L_\alpha} = dC_L/d\alpha$ is for the basic 3-D wing, which was determined in the construction in Chapter 4, and $C_{l_{\max}}$ is the maximum 2-D section lift coefficient, which is available from experiments and given in such reference books as "Theory of Wing Sections" for various section types.

The quantity $C_{L_{\max}}/C_{l_{\max}}$ is found from Figure 9.9. This is a function of the leading-edge sharpness parameter, Δy. The sharpness parameter can be found for different NACA airfoils in Figure 9.10.

The stall angle, α_s, is found from Eq. [9.6]. The only parameter not already determined is $\Delta\alpha_{C_{L_{\max}}}$. This can be obtained from Figure 9.11. It is a function of the leading-edge sharpness parameter, Δy.

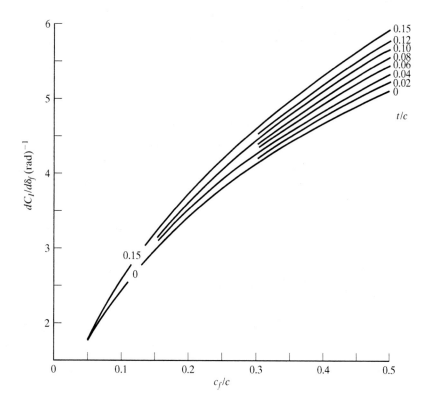

FIGURE 9.4: Change in 2-D lift coefficient with flap angle for plane flaps of different chord ratios. (From Ellison, 1969.)

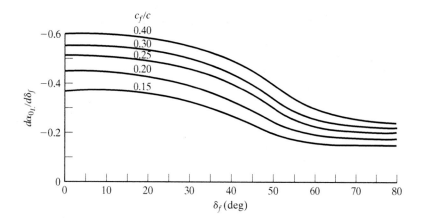

FIGURE 9.5: Wing section lift-effectiveness parameter for single slotted and Fowler flaps. (From Ellison, 1969.)

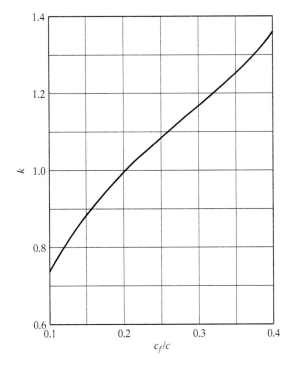

FIGURE 9.6: Coefficient, k for determining $\Delta\alpha_{0_L}$ for a split flap. (From Ellison, 1969.)

For *low* aspect-ratio wings, $C_{L_{\max}}$ and α_s are found from

$$C_{L_{\max}} = \left(C_{L_{\max}}\right)_{\text{base}} + \Delta C_{L_{\max}} \tag{9.7}$$

and

$$\alpha_s = \left(\alpha_{C_{L_{\max}}}\right)_{\text{base}} + \Delta\alpha_{C_{L_{\max}}}. \tag{9.8}$$

For Eq. [9.7], $\left(C_{L_{\max}}\right)_{\text{base}}$ is obtained from Figure 9.12, where it is a function of the leading-edge sweep angle; aspect ratio, A; Mach number parameter, β; leading-edge sharpness parameter; and coefficient, C_1, which was found from Figure 9.8.

$\Delta C_{L_{\max}}$ is found from Figure 9.13 where it is a function of the sweep angle, aspect ratio, Mach number, and the coefficient C_2, which can be found from Figure 9.14.

For Eq. [9.8], $\left(\alpha_{C_{L_{\max}}}\right)_{\text{base}}$ is obtained from Figure 9.15. $\Delta\alpha_{C_{L_{\max}}}$ is found from Figure 9.16. If $(C_2 + 1)A\tan(\Lambda_{\text{LE}})$ is greater than approximately 4.5, the result is a function of Mach number. Otherwise $\Delta\alpha_{C_{L_{\max}}}$ is a function of $A\cos(\Lambda_{\text{LE}})\left[1 + (2\lambda)^2\right]$.

At take-off and landing, the Mach number will be of the order of 0.2 or less. For combat, where high lift devices are used to enhance maneuverability, the Mach number will be approximately 0.8.

The construction of the C_L versus α curve for the basic (unflapped) wing can now be completed as shown for the $\delta = 0$ curve in Figure 9.17. This involves adding a smooth faired curve to the straight line originating at α_{0_L} and having the slope, $dC_L/d\alpha$, which was found in the wing analysis in Chapter 4.

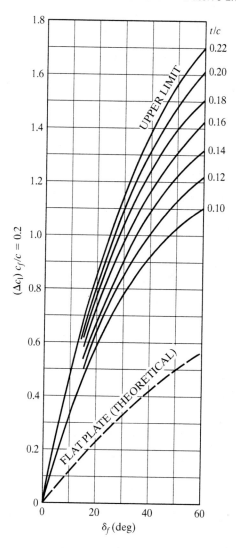

FIGURE 9.7: Coefficient, $(\Delta C_l)_{\frac{c_f}{c}=0.2}$ for determining $\Delta\alpha_{0_L}$ for a split flap. (From Ellison, 1969.)

To complete the construction of the C_L versus α curve for the *flapped wing*, the increment in $C_{L_{max}}$ due to the flaps needs to be determined. This is given as

$$\Delta C_{L_{max}} = \Delta C_{l_{max}} \frac{S_{WF}}{S_W} K_\Delta. \tag{9.9}$$

In this equation, $\Delta C_{l_{max}}$ is the increment in the maximum lift for a flapped 2-D wing. It is found graphically by first estimating the change in the stall angle for the flapped 2-D wing, $\Delta\alpha_s$.

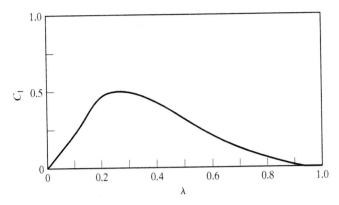

FIGURE 9.8: Coefficient C_1 used in the determination of high or low aspect-ratio classification. (From Ellison, 1969.)

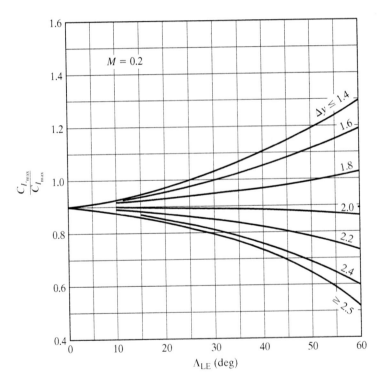

FIGURE 9.9: 3-D to 2-D maximum lift coefficient ratio for high aspect-ratio wings. (From Ellison, 1969.)

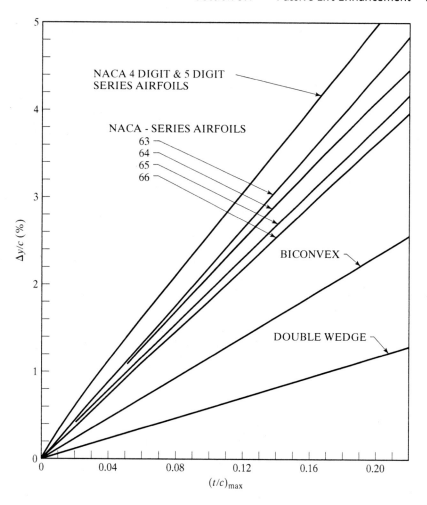

FIGURE 9.10: Leading-edge sharpness parameter as a function of the thickness-to-chord ratio for families of NACA airfoil sections. (From Ellison, 1969.)

Figure 9.18 is used for this purpose. It shows $\Delta\alpha_s$ as a function of trailing-edge flap deflection, based on airfoil data from Abbott and Von Doenhoff (1945).

$C_{l_{max}}$ for the 2-D flapped wing then approximately corresponds to the lift coefficient at the stall angle with flaps,

$$\alpha_{s_{flapped}} = \alpha_{s_{basic}} + \Delta\alpha_s. \tag{9.10}$$

Based on this,

$$\Delta C_{l_{max}} = \left(C_{l_{max}}\right)_{flapped} - \left(C_{l_{max}}\right)_{basic}. \tag{9.11}$$

S_{WF}/S_W in Eq. [9.9] is the ratio of the planform area of the wing having the same span as the flaps, to the total wing planform area. This is illustrated in Figure 9.19.

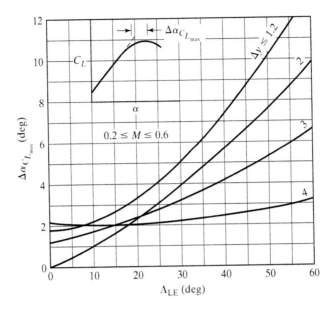

FIGURE 9.11: Angle-of-attack increment for the maximum lift coefficient of high aspect-ratio wings. (From Ellison, 1969.)

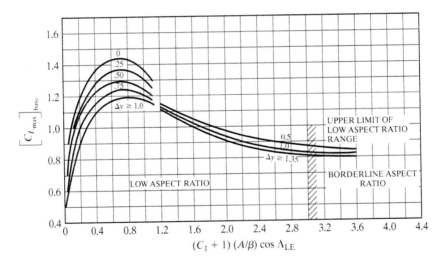

FIGURE 9.12: Subsonic maximum lift coefficient for basic low aspect-ratio wings. (From Ellison, 1969.)

Finally, K_Δ in Eq. [9.9] is an empirical correction that accounts for wing sweep, $\Lambda_{c/4}$. This is given as

$$K_\Delta = \left[1 - 0.08\cos^2(\Lambda_{c/4})\right]\cos^{3/4}(\Lambda_{c/4}). \tag{9.12}$$

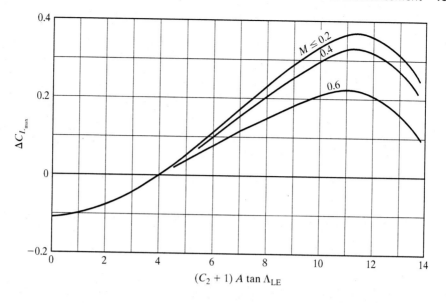

FIGURE 9.13: Subsonic maximum lift coefficient increment for basic low aspect-ratio wings. (From Ellison, 1969.)

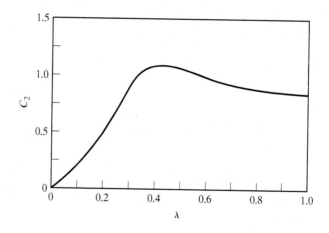

FIGURE 9.14: Taper ratio correction for coefficient C_2 used for determining maximum lift coefficient for basic low aspect-ratio wings. (From Ellison, 1969.)

In this expression, $\Lambda_{c/4}$ is the sweep angle of the quarter-chord line on the wing. This is usually close to the the sweep angle of the maximum thickness line of the wing.

With the calculation of $\Delta C_{L_{\max}}$, the maximum lift for the flapped 3-D wing is then

$$\left(C_{L_{\max}}\right)_{\text{flapped}} = \left(C_{L_{\max}}\right)_{\text{basic}} + \Delta C_{L_{\max}}. \tag{9.13}$$

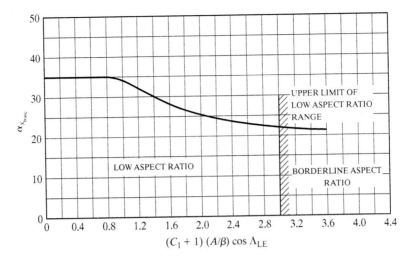

FIGURE 9.15: Angle-of-attack corresponding to subsonic maximum lift coefficient for basic low aspect-ratio wings. (From Ellison, 1969.)

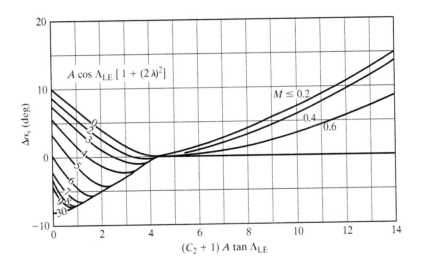

FIGURE 9.16: Angle-of-attack increment at subsonic maximum lift coefficient for basic low aspect-ratio wings. (From Ellison, 1969.)

This is illustrated in Figure 9.17 as the curve marked $\delta_f > 0$. Note that if the nonlinear (curved) portion of the lift curve stays the same with or without the flaps, $(\alpha_s)_{\text{flapped}} < (\alpha_s)_{\text{unflapped}}$.

Leading-Edge Flaps. Estimating the increment in $C_{L_{\max}}$ due to *leading-edge* flaps is not nearly as precise as it was for the trailing-edge flaps. An estimate can come

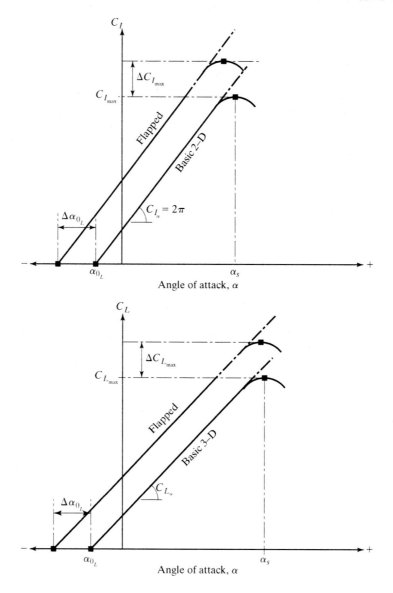

FIGURE 9.17: Schematic of the construction of lift versus angle-of-attack curves for flapped 2-D wing (top) and flapped 3-D wing (bottom).

from the following equation, which is similar to Eq. [9.9] and uses the same definition for S_{WF}/S_W as before:

$$\Delta C_{L_{max}} = \Delta C_{l_{max}} \frac{S_{WF}}{S_W} \cos(\Lambda_{LE}). \tag{9.14}$$

The 2-D maximum lift coefficient increment, $\Delta C_{l_{max}}$ for various types of leading-edge flaps is given in Table 9.1. The value of $\Delta C_{L_{max}}$ with leading-edge flaps from

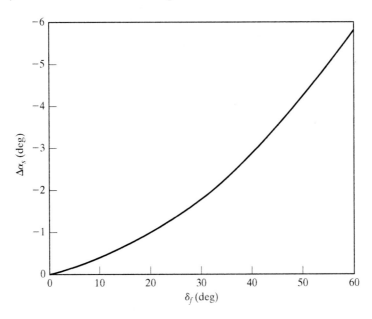

FIGURE 9.18: Change in stall angle due to flap deflection for a basic 2-D wing.

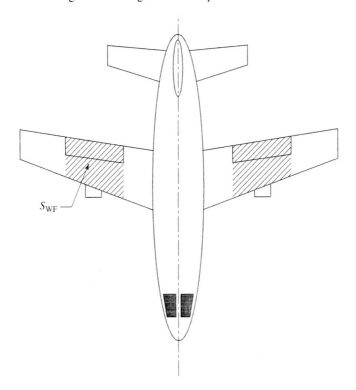

FIGURE 9.19: Illustration defining the flapped wing planform area, S_{WF}.

TABLE 9.1: 2-D lift coefficient increment for different leading-edge flap designs.

Type	$\Delta C_{l_{max}}$
Fixed Slot	0.2
Leading-Edge Flap	0.3
Kruger Flap	0.3
Slat	0.4

Eq. [9.14] is then added to the value with trailing-edge flaps, which was obtained from Eq. [9.13], to get the overall lift coefficient with passive lift devices such as these.

9.1.2 Drag Determination

When the trailing-edge flaps are deflected, they produce an increase in the base (zero lift) drag of the wing. This is expressed as an increase in the based drag coefficient given as

$$\Delta C_{D_0} = k_1 k_2 \frac{S_{WF}}{S_W}. \tag{9.15}$$

The coefficient k_1 is a function of the ratio of the flap to wing chords and is found from Figure 9.20 for the different types of flaps. The coefficient k_2 is a function of the flap deflection and can be found from Figure 9.21.

The value of ΔC_{D_0} found from Eq. [9.15] should be added to the base drag coefficient, C_{D_0}, for the main wing in order to calculate the overall drag at take-off and landing.

9.2 ACTIVE LIFT ENHANCEMENT

If one of the principle design drivers is short take-off and landing (STOL), it is likely that passive lift enhancement approaches will not provide a sufficient $C_{L_{max}}$. As an example of the requirements for take-off, we consider the take-off distance based on the simple take-off parameter, TOP, that was given in Eq. [3.3]. The relation for the take-off distance in this case was given in Eq. [3.5]. Incorporating the TOP into that equation gives

$$s_{TO} = 20.9 \frac{W}{S} \frac{W}{T} \frac{1}{C_{L_{max}}} + 87 \sqrt{\frac{W}{S} \frac{1}{C_{L_{max}}}}. \tag{9.16}$$

Here we have assumed that the altitude at take-off is sea level, so that $\sigma = \rho_{TO}/\rho_{SL} = 1$.

For STOL aircraft, $s_{TO} < 1000$ feet. For medium-size aircraft that would be designed to carry passengers or cargo, efficient cruise would dictate $W/S \simeq 40$ lb/f^2 and $T/W \simeq 0.2$.

Using these values, the conditions on $C_{L_{max}}$ for STOL are

$$\frac{4180}{C_{L_{max}}} + \frac{550}{\sqrt{C_{L_{max}}}} < 1000. \tag{9.17}$$

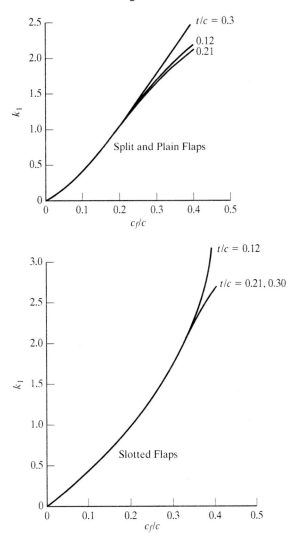

FIGURE 9.20: Chord factor, k_1, for drag increment due to different trailing-edge flaps. (Young, 1953.)

In order to satisfy Eq. [9.17], $C_{L_{max}} > 5.47$. The largest $C_{L_{max}}$ that is attainable by passive approaches is approximately 4.0. Therefore, other (active) approaches are needed if aircraft of this type are to be able to achieve such short take-off distances.

Figure 9.22 shows some of the common approaches used for active lift enhancement. These generally fall in three categories: upper surface blowing (USB); blown flaps where air is supplied either externally (EBF) or internally (IBF); and vectored thrust.

With USB, a high velocity air stream is directed over the upper surface of the main wing. This requires placing the engines above and forward of the wing.

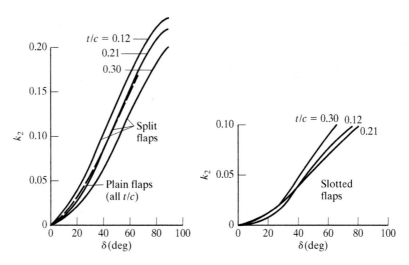

FIGURE 9.21: Flap deflection factor, k_2, for drag increment due to different trailing-edge flaps. (Young, 1953.)

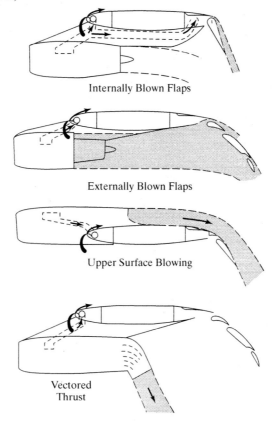

FIGURE 9.22: Examples of active lift enhancement approaches.

With blown flaps, high-velocity air is directed specifically at the trailing-edge flaps. For externally blown flaps (EBF), the air is supplied by the engine exhaust, and the engine is located below the wing. The flaps are slotted in this case so that high-momentum air can reach the upper surface and energize the boundary layer over the flaps. A portion of the air in this arrangement is also deflected downward. The YC-15, which was pictured in Figure 1.4, used this arrangement.

Internally blown flaps (IBF) duct a portion of the engine exhaust air only to the upper side of the trailing-edge flaps.

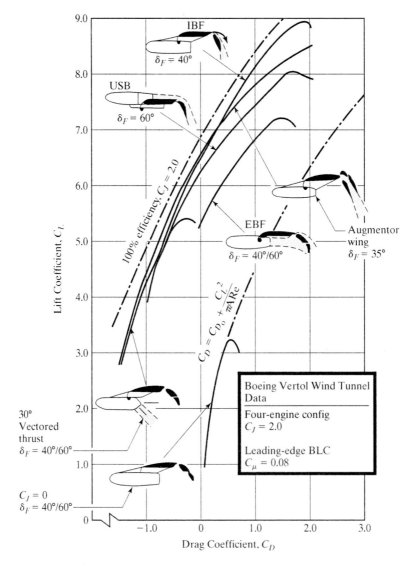

FIGURE 9.23: Drag polars for active lift enhancement approaches. (From the Boeing Co.)

In addition to the enhanced aerodynamic lift that these three approaches provide, they also generate a component of downward thrust. This results because of the Coanda effect, which is the ability of an air stream to follow a curved surface. When properly designed, the air stream on the upper surface leaves the trailing edge at the angle of the flaps.

Vectored thrust uses an articulated exit nozzle to direct the jet exhaust air downward. This gives a downward component of thrust, which is independent of any aerodynamic lift enhancement on the wing.

The effectiveness of these active approaches is summarized in Figure 9.23 in terms of the drag polar, C_L versus C_D. Any of these are capable of providing lift coefficients in excess of 7.0. The vectored thrust, USB, and IBF have lower drag coefficients than EBF. Also the effectiveness of the USB is a function of the jet coefficient defined as

$$C_j = \frac{\text{Thrust}}{q\,S_W}. \tag{9.18}$$

The value for C_j in Figure 9.23 is 2.0.

In all these approaches, there are additional factors that affect the selection of one over another. The IBF requires internal ducting that can be heavy and result in internal momentum losses. The USB blows hot exhaust air over the wing surface. This generally requires that portion of the wing to be covered with a heat-resistant material (stainless steel), which adds weight.

The EBF approach only directs the hot exhaust over the flaps, so that the area of heat-resistant material that has to be covered is less than with the USB. This makes the weight penalty less. This, and its relative simplicity, may be the reason that the USB approach appears to be the most popular means for active lift enhancement used by aircraft manufacturers.

9.3 SPREADSHEET FOR ENHANCED LIFT CALCULATIONS

The equations for estimating the 3-D lift coefficient for different trailing-edge and leading-edge flap designs have been incorporated into a spreadsheet called **flaps.xls**. A sample, which uses the conditions for the conceptual SSBJ proposed in Chapter 1, is shown in Figures 9.24 to 9.26.

Part of the spreadsheet input requires values from the various figures in this chapter. The relevant figures are noted next to the cells whose values are to be read from the respective figure.

The characteristics of the main wing are entered in the section titled "Wing Data" located at the top of the spreadsheet. These include

1. the airfoil type, such as NACA 64x;
2. the leading-edge sweep angle, Λ_{LE};
3. the taper ratio, λ;
4. the maximum thickness-to-chord ratio, $(t/c)_{max}$;
5. the Mach number parameter, β;
6. the aspect ratio, A;
7. the sweep angle of the maximum thickness line, $\Lambda_{t/c}$;

Wing Data:								
Airfoil	NACA	64x						
Λ_{LE}	62	deg						
λ	0.00							
t/c	0.04							
T–O Mach No.	0.1							
β	0.99							
A	2							
$\Lambda_{t/c}$	47.2	deg						
$C_{l\alpha}$ (no flap)	0.11	1/deg						
$C_{L\alpha}$ (no flap)	0.04	1/deg						
α_{0L}	−1	deg						
C_{lmax}	1.4							
α_s	12	deg						
Trailing-edge Flap Design:								
Flap type	slot	slot, plane or split						
S_f/S_w	0.60							
δ_f	40	deg						
c_f/c	0.25							
Delta $\alpha 0L$:								
	Plane Flap							
K'	0.58	Fig. 9.3						
$dC_l/d\delta_f$	0.5	Fig. 9.4						
$\Delta\alpha_{0l}$	−18.06	deg						
	Single Slotted & Fowler Flap							
$d\alpha/d\delta_f$	−0.4	Fig. 9.5						
$\Delta\alpha_{0l}$	−20	deg						
	Split Flaps							
k	1.1	Fig. 9.6						
ΔCl	0.8	Fig. 9.7						
$\Delta\alpha_{0l}$	−8	deg						
Aspect Ratio Criterion:								
C_1	0	Fig. 9.8						
High A criteria	8.52	Low						
Basic Wing-High Aspect Ratio:	.							
Δy	0.8	%	Fig. 9.10					
C_{Lmax}/C_{lmax}	1.3		Fig. 9.9					
C_{Lmax}	1.82							
$\Delta\alpha_{CLmax}$	12.5	deg	Fig. 9.11					
α_s	52.86	deg						

FIGURE 9.24: Spreadsheet (FLAPS) for 3-D flapped wing lift and drag calculations (Part 1).

Basic Wing-Low Aspect Ratio:

$(C1 + 1)$?	0.94	
$(C_{Lmax})_{base}$	1.2	Fig. 9.12
C_2	0	Fig. 9.14
$(C2 + 1)$?	3.76	
ΔC_{Lmax}	-0.02	Fig. 9.13
C_{Lmax}	1.19	
$(\alpha_{CLmax})_{base}$	34 deg	Fig. 9.15
$A \cos($?	0.94	
$\Delta \alpha_{Lmax}$	1 deg	Fig. 9.16
α_s	35 deg	

Effect of Trailing-edge Flap:

Flap type	slot	slot, plane or split
α_{01}	-21	
Basic 3–D α_s	52.86 deg	
Basic 3–D C_L	1.82	
2–D $\Delta \alpha_s$	-2.5 deg	Fig. 9.18
2–D $\Delta \alpha_{s\,flapped}$	9.5 deg	
$(C_{lmax})_{flapped}$	3.36	
ΔC_{lmax}	1.96	
$K\Delta$	0.72	
ΔC_{Lmax}	0.85	
C_{Lmax}	2.03	
3–D $\alpha_{s\,flapped}$	50.36 deg	

Leading-edge flap CL Max:

ΔC_{lmax}	0.3	Table 9.1
ΔC_{Lmax}	0.08	
C_{Lmax}	2.11	

Trailing-edge flap Added drag:

k_1	1.4	Fig. 9.20
k_2	0.08	Fig. 9.21
ΔC_{D0}	0.06	

Lift Curve Plotting:

	2–D (no flaps)		2–D (flaps)		3–D (no flaps)		3–D (flaps)	
	α	C_l	α	C_l	α	C_L	α	C_L
	-1	0	-21	0	-1	0	-21	0
	9.18	1.12	3.4	2.68	32.09	1.46	15.92	1.62
	12	1.4	9.5	3.36	52.86	1.82	50.36	2.03
	14.82	1.12	15.6	2.68	73.64	1.46	84.81	1.62

FIGURE 9.25: Spreadsheet (FLAPS) for 3-D flapped wing lift and drag calculations (Part 2).

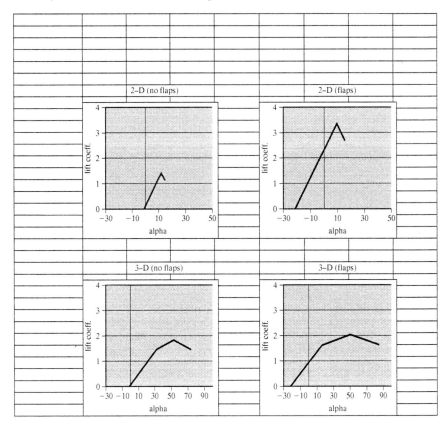

FIGURE 9.26: Spreadsheet (FLAPS) for 3-D flapped wing lift and drag calculations (Part 3).

8. the slope, $C_{l_\alpha} = dC_l/d\alpha$ for the 2-D wing;

9. the slope, $C_{L_\alpha} = dC_L/d\alpha$ for the 3-D wing;

10. the zero-lift angle of attack, α_{0_L};

11. the maximum lift coefficient for the 2-D airfoil, $C_{l_{max}}$;

12. the angle of attack corresponding to the maximum 2-D lift coefficient, α_s.

As noted before, the characteristics of the 2-D airfoil section can be obtained from sources such as "Theory of Wing Sections." The other properties of the wing were determined in the analysis and accompanying spreadsheet (wing.xls) in Chapter 4.

The input for the design of the trailing-edge flaps follows next in the section titled "Trailing-edge Flap Design." The input includes

1. the flap type: slotted (slot), plain, or split;

2. the ratio of the planform area of the portion of the wing, which has flaps, to the total wing area, S_{WF}/S_W, as was illustrated in Figure 9.19;

3. the flap deflection, δ_f;

4. and the ratio of the flap chord to wing chord, c_f/c.

The first step in the calculations is to determine the change in the zero-lift angle of attack, $\Delta\alpha_{0_L}$, produced by the flaps. This is done in the section entitled "Delta α_{0_L}." Here all three of the flap types are considered, so that the different designs can be compared. However, the value of $\Delta\alpha_{0_L}$ that is used in other calculations is the flap type designated in the flap design input (slot, plain, or split).

The next section of the spreadsheet titled "Aspect Ratio Criterion" applies the condition given in Eq. [9.4] to determine whether the wing has a "low" or "high" aspect ratio. The right-hand side of Eq. [9.4] is evaluated and displayed in the cell next to "High A criteria." This value is then compared to the wing geometric aspect ratio, A, and based on the condition is termed to be low or high. The result is displayed in the cell to the right of the numeric value of the criterion.

The designation of the aspect ratio criterion acts as a "switch" in subsequent calculations whereby only the appropriate formulas and results are used to determine the final values of $C_{L\max}$ and α_s with flaps. Therefore, for the next section of the spreadsheet "Basic Wing," only the appropriate part "High Aspect Ratio" or "Low Aspect Ratio" needs to be completed.

The "Basic Wing" section of the spreadsheet determines $C_{L\max}$ and α_s for the *basic 3-D wing* without flaps. For high aspect-ratio wings, this is based on Eqs. [9.5 & 9.6] and requires input from Figures 9.9, 9.10, and 9.11.

The calculations for a low aspect-ratio wing are a bit more extensive. These are based on Eqs. [9.7 & 9.8] and require input from Figures 9.12 through 9.16. Figure 9.12 requires the calculation of the parameter $(C_1 + 1)(A/\beta)\cos(\Lambda_{LE})$. This is denoted in the spreadsheet as $(C_1 + 1)\cdots$.

Figure 9.13 requires the calculation of the parameter $(C_2 + 1)A\tan(\Lambda_{LE})$. This is denoted as $(C_2+1)\cdots$. Also, Figure 9.16 requires knowing the parameter $A\cos(\Lambda_{LE})[1+(2\lambda)^2]$. This is denoted in the spreadsheet as $A\cos(\cdots$.

Having fully determined the characteristics of the basic 3-D wing, the effect of the trailing-edge flaps are next found. This is done in the section titled "Effect of Trailing-Edge Flap." The top of this section lists the characteristics of the basic 3-D wing, which have been automatically copied down from the appropriate parts of the spreadsheet. In this section, $\Delta C_{L\max}$ is found from Eq. [9.9]. This requires finding $\Delta C_{l\max}$ for the flapped 2-D airfoil section, which comes from Eq. [9.11].

To find this, we first determine $C_{l\max}$ for the flapped 2-D wing. This comes from calculating the angle of attack at $C_{l\max}$, which is 2-D$\alpha_{s_{\text{flapped}}}$ and given by

$$2\text{-D}\alpha_{s_{\text{flapped}}} = \alpha_s + 2\text{-D}\Delta\alpha_s. \tag{9.19}$$

2-D$\Delta\alpha_s$ is found from Figure 9.18.

From this, the $C_{l\max}$ for the 2-D flapped wing section comes from assuming a linear lift versus angle of attack up to stall, or

$$C_{l\max} = C_{l_\alpha}(\alpha_s - \alpha_{0_L}). \tag{9.20}$$

In the spreadsheet, this is denoted as $(C_{l\max})_{\text{flapped}}$.

Then $\Delta C_{L_{\max}}$ is found as

$$\Delta C_{L_{\max}} = (C_{l_{\max}})_{\text{flapped}} - C_{l_{\max}}. \tag{9.21}$$

Following Eq. [9.9], $\Delta C_{L_{\max}}$ is then calculated. The maximum lift coefficient for the 3-D flapped wing is then found as

$$C_{L_{\max}} = (C_{L_{\max}})_{\text{basic}} + \Delta C_{L_{\max}}. \tag{9.22}$$

The angle of attack at $C_{L_{\max}}$ is found by solving Eq. [9.10]. In the spreadsheet, the result is denoted as 3-D$\alpha_{s_{\text{flapped}}}$.

If leading-edge flaps are used, the augmented lift coefficient, $\Delta C_{L_{\max}}$ is found based on Eq. [9.14]. This is calculated in the section titled "Leading-edge flap CL Max." It uses a value of $\Delta C_{l_{\max}}$, which is taken from Table 9.1. The total $C_{L_{\max}}$ in this section is then the *sum* of the $C_{L_{\max}}$ values for the leading- and trailing-edge flaps.

Finally, the added drag due to trailing edge flaps, ΔC_{D_0}, is calculated in the section titled "Trailing-edge flap Added Drag." This is based on Eq. [9.15] and takes input from Figures 9.20 and 9.21. This drag coefficient uses the planform area of the main wing as the reference area to calculate the added drag force.

The bottom part of the spreadsheet plots the lift coefficient versus angle of attack for four wing configurations consisting of

1. the unflapped 2-D wing;
2. the flapped 2-D wing;
3. the unflapped 3-D wing;
4. the flapped 3-D wing.

These do not include the augmented lift due to leading-edge flaps. The table preceding the plots shows the values used in their construction. These are determined automatically based on the values calculated in the upper portions of the spreadsheet.

9.3.1   Case Study: Enhanced Lift Design

The parameters used in the spreadsheet shown in Figure 9.24 correspond to the conceptual supersonic business jet, which was proposed in Chapter 1. The general 2-D wing design was developed in the "wing" spreadsheet discussed in Chapter 4 and shown in Figures 4.14 and 4.15. Most of the values used as input at the top of the "flap" spreadsheet are taken from this previous 2-D wing design.

The 2-D section maximum lift coefficient, $C_{l_{\max}}$, and corresponding angle of attack, α_s, were taken from experimental data. The take-off Mach number was chosen to be 0.1, which then gave a value for β of approximately 0.9.

For the trailing-edge flap design, slotted flaps were selected. The slotted area of the wing would correspond to 60 percent of the total wing area. The maximum flap deflection was chosen to be 40°, and the maximum chord length of the flaps would be 20 percent of the average wing chord. The flap areas have been added to the drawing of the SSBJ. This is shown in Figure 9.27.

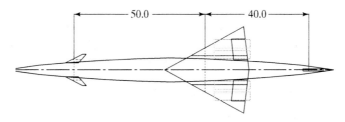

FIGURE 9.27: Top-view drawing showing the locations of the main wing flaps on the conceptual supersonic business jet.

Although slotted flaps were selected, representative values were input to all the trailing-edge flap categories to see how they affected $\Delta\alpha_{0_L}$. We observe that the largest value occurs for the slotted flaps, closely followed by the plane flap. The split flaps resulted in a significantly lower change in the zero-lift angle of attack, which could indicate that they are less effective for this wing design.

The aspect-ratio criterion categorized the main wing as low aspect ratio. Therefore, only the low aspect-ratio calculations of the basic wing are relevant. Numbers were input into the high aspect-ratio section only or reference.

Comparing the different combinations, the trailing-edge flaps increased the maximum lift coefficient by approximately 2.4 times, from 1.4 to approximately 3.4.

When the 3-D effects were included, however, the increase in the maximum lift coefficient with flaps decreased from 2.4 to 1.7 times. This gave an overall maximum lift coefficient of 2.0.

This decrease in the flap effectiveness for the 3-D wing is primarily due to the small aspect ratio and, most importantly, to the high sweep angle used in this design.

Although the 3-D effects lowered the maximum lift coefficient, they substantially increased the stall angle. Comparing the unflapped cases, the stall angle increased from 12 to 35°.

The trailing-edge flap will always have the tendency to decrease the stall angle. For both the 2-D and 3-D wings, the decrease was approximately 2.5°. However, for the 3-D wing, this change was a significantly smaller percentage of the unflapped value.

The leading-edge flaps increased the value of $C_{L_{max}}$ from 2.03 to 2.114, or approximately 5 percent. This relatively modest improvement was primarily limited by the large leading-edge sweep angle. Recall that $C_{L_{max}} = 2$ was used in the take-off and landing analysis in the previous chapter.

The increase in the drag coefficient caused by the trailing-edge flaps, $\Delta C_{D_{flaps}}$, was found to be 0.06 compared to the value of 0.05 that was taken from Table 8.3 for the

TABLE 9.2: Summary of results of 3-D flapped wing for conceptual SSBJ.

$C_{L_{max}}$	α_s	$\Delta C_{D_{flaps}}$	s_{TO}	s_L
2.11	50.36°	0.063	7695 f	3887 f

take-off and landing analysis in the previous chapter. When the new value is used, it increases s_{TO} from 7566 to 7695 feet and decreases s_L from 3906 to 3887 feet.

Overall the design met the requirements for the maximum lift coefficient needed at take-off and landing. A summary is given in Table 9.2.

9.4 PROBLEMS

9.1. Consider an aircraft with the main wing design that follows.
For this, determine the maximum lift coefficient, $C_{L_{max}}$; the angle of attack at the maximum lift, α_s; and the zero-lift angle of attack, α_{0_L} for a flap design consisting of plane flaps, $\delta_f = 30°$, $S_{WF}/S_W = 0.6$, and $c_f/c = 0.25$. Make a plot of C_L versus α. What is the added drag coefficient for this flap design?
- Airfoil: NACA 63x
- $t/c = 0.12$
- $\lambda = 0.30$
- $A = 9$
- $\Lambda_{LE} = 30°$
- $\Lambda_{t/c} = 25.5°$
- $\beta = 0.98$ (TO and L)
- $\alpha_{0_L} = -2.8°$
- $C_{l_{max}} = 1.2$
- $\alpha_s = 12.3°$
- $C_{l_\alpha} = 0.117$
- $C_{L_\alpha} = 0.107$

9.2. For the wing in Problem [9.1], what is the maximum possible lift coefficient if leading-edge flaps were also used? How does the lift enhancement with leading-edge flaps in this case, compare to that for the case study supersonic business jet (originally given in the spreadsheet)? What are the main differences in the two designs that account for the difference in effectiveness of the leading-edge flaps?

9.3. Repeat Problem [9.1], changing the flap design from a plane flap to a *slotted* flap.

9.4. For the properties of the main wing in Problem [9.1], generate a table that shows the change in $C_{L_{max}}$ as a function of the flap deflection angle, δ_f. Use $S_{WF}/S_W = 0.6$ and $c_f/c = 0.25$. Plot the results.

9.5. For the properties of the main wing in Problem [9.1], generate a table that shows the change in $C_{L_{max}}$ as a function of the flap chord length, c_f/c. Use a plane flap with $S_{WF}/S_W = 0.6$ and $\delta_f = 30°$. Plot the results.

9.6. For the properties of the main wing in Problem [9.1], generate a table that shows the change in $C_{L_{max}}$ as a function of the wing t/c. Use a plane flap with $S_{WF}/S_W = 0.6$, $\delta_f = 30°$, and $c_f/c = 0.4$. Plot the results.

9.7. With the main wing in Problem [9.1], for the same taper ratio, how large of a sweep angle is possible before it is considered to have a low aspect ratio? How does this change if the taper ratio, λ, is 1? Explain these answers in terms of 3-D lift effects.

9.8. Consider the conditions of the case study supersonic business jet, which were originally given in the spreadsheet. Determine the change in $C_{L_{max}}$ for a range of flap deflections.

9.9. Consider the conditions of the case study supersonic business jet, which were originally given in the spreadsheet. Determine the change in $C_{L_{max}}$ for a range of flap chord lengths, c_f/c.

9.10. A NACA $64_1 - 212$ airfoil section has a $C_{l_{max}}$ of 1.3. If this section is used on a main wing with $A = 6$, $\lambda = 0.5$, and $\Lambda_{t/c} = 35°$, is it possible to achieve a $C_{L_{max}} = 1.65$

with trailing-edge flaps? List all assumptions, and show all work in answering this question.

9.11. A NACA $64_1 - A112$ airfoil section has a $C_{l_{max}}$ of 0.95. A historic aircraft that used this section for its main wing recorded a modest $C_{L_{max}} = 1.15$ with *split* flaps. The main wing had $A = 3.5$, $\lambda = 0.5$, and $\Lambda_{t/c} = 45°$. Determine the conditions of the flap design that would give this $C_{L_{max}}$. List all assumptions, and show all work. How might you improve on the flap performance?

9.12. Design a wing (ie., select the wing section type, planform shape, and flap design) that will give the highest possible maximum lift coefficient for a passive lift enhancement. Show all your work.

CHAPTER 10

Structural Design
and Material Selection

10.1 STRUCTURAL LOADS
10.2 INTERNAL STRUCTURE DESIGN
10.3 MATERIAL SELECTION
10.4 SPREADSHEET FOR STRUCTURE DESIGN
10.5 CASE STUDY: STRUCTURAL ANALYSIS
10.6 PROBLEMS

Photograph of an Apex wing section undergoing loading tests. The Apex is designed to fly at high altitudes of up to 100,000 feet. It is designed for flight loads of 5 Gs. The wing section is a high strength-to-weight boron/graphite composite construction. (NASA Dryden Research Center Photo Collection.)

At this point in the design, the wings, fuselage, horizontal and vertical tail sections, and engines have been designed. These make up a majority of the external configuration of the aircraft. This chapter deals with the general design of the *internal* structure of an aircraft in order to withstand the loads that occur during the different phases of a flight plan.

The design of the structure is based on a load limit, which is the largest expected load. For aerodynamic forces, this is related to the aerodynamic load factor, n. Load factors were designated in Chapter 3 for some of the flight phases, such as intercept, and with maximum and sustained turn rates. These will be considered here in the design of the structure.

In addition to the loads that occur at different flight phases, the following are also considered:

1. the loads produced when flying at the highest possible angle of attack, without stalling;
2. the loads that occur at a dive speed equal to the $1.5V_{\text{cruise}}$; and
3. the loads produced by wind gusts, such as those that can occur in thunder storms or from clear-air turbulence.

The *largest* load factor from any of those in this group will be considered to be the "design load factor," which will be the basis for the design of the internal structure.

This chapter divides the structural design into three categories: the determination of structural loads, the general design of the internal structure, and the selection of materials.

The design of the internal structure and the material selection clearly go hand-in-hand. The use of higher strength materials can reduce the size or number of structural elements. However, the structure weight is an important factor that also needs to be considered. Therefore, the structure design and material selection should be done together.

10.1   STRUCTURAL LOADS

In determining the structural loads, the main wing, horizontal and vertical stabilizers, and fuselage are considered separately. The loads on these are a due to a combination of static and dynamic weights, and aerodynamic forces.

10.1.1   Load Factors

The load factor, n, was first introduced in Chapter 3 in the selection of the wing loading. It is defined as the ratio of the lift to the weight, $n = L/W$. In level flight, the lift produced by the wings equals the weight, so that $n = 1$. However, during maneuvers such as climb to altitude, acceleration to high speed, or sustained or instantaneous turns associated with combat, significantly larger load factors can occur. Since these set the limit on the internal structure, it is important that the maximum load factor be determined.

Intercept.   For an intercept flight phase, the load factor explicitly appeared in the expression for wing loading in Eq. [3.20]. This was the condition that minimized the drag-to-weight ratio and thereby maximized excess power. Rearranging that equation to solve for n, we obtain

$$n = \frac{q}{W/S}\sqrt{\frac{C_{D_0}}{k}}, \tag{10.1}$$

where again, q is the dynamic pressure, C_{D_0} is the base drag coefficient for the wing, and $k = 1/\pi Ae$ with $e \simeq 0.8$.

Instantaneous Turn Rate.   With instantaneous turn rate, the load factor was given in Eq. [3.24]. This is reproduced in

$$n = \sqrt{\left(\frac{\dot{\psi}_{inst} V}{g}\right)^2 + 1}. \tag{10.2}$$

Here the turn rate, $\dot{\psi}_{inst}$, is the instantaneous turn rate, which has units of radians per second.

Sustained Turn Rate.   Recall that in a sustained turn, the speed and altitude are maintained so that the thrust equals the drag, and the load factor is constant.

An expression for the maximum sustained load factor as a function of the wing loading, which is needed to achieve a specified sustained turn rate, was given in Eq. [3.32]. This is reproduced in

$$n = \left[\frac{q\pi Ae}{W/S}\left[\left(\frac{T}{W}\right)_{max} - \frac{qC_{D_0}}{W/S}\right]\right]^{1/2}. \tag{10.3}$$

In terms of the maximum sustained turn rate, the load factor is

$$n = \sqrt{\left(\frac{\dot{\psi}_{sust} V}{g}\right)^2 + 1}, \tag{10.4}$$

where $\dot{\psi}_{sust}$ is the maximum sustained turn rate with units of radians per second.

Climb.   In the analysis of wing loading effect on climb, it was assumed that $n = 1$. However, an expression can be derived, which relates the climb rate to the load factor.

By definition,

$$n = \frac{L}{W} = \frac{C_L q S}{W}. \tag{10.5}$$

The climb gradient is given as

$$G = \sin \gamma = \frac{(T - D)}{W}. \tag{10.6}$$

Eq. [3.12] expressed the total drag as the sum of the base drag, with drag coefficient, C_{D_0}, and lift-induced drag. Using Eq. [10.5], the D/W can be put in terms of the load factor as

$$\frac{D}{W} = n\frac{C_{D_0}}{C_L} + \frac{1}{n}\frac{C_L}{\pi Ae}. \tag{10.7}$$

Substituting for D/W in Eq. [10.6] and solving for n, we obtain

$$n = \frac{(T/W - G) \pm \left[(T/W - G)^2 - (4C_{D_0}/\pi Ae)\right]^{0.5}}{2C_{D_0}/C_L} \tag{10.8a}$$

with the condition that

$$\frac{T}{W} \geq G + 2\sqrt{\frac{C_{D_0}}{\pi Ae}}. \tag{10.8b}$$

As an example, with $C_{D_0} = 0.007$, $C_L = 1.2$, $A = 2$, and a climb angle of 11 degrees, the minimum thrust-to-weight ratio that satisfies Eq. [10.8b] is 0.27, and the load factor given by Eq. [10.8a] is 6.4.

Take-Off Transition.   The transition phase of take-off is a climb at constant radius. Analysis determined that the load factor was constant and equal to $n = 1.15$. Although this is not likely to be the largest load factor, for consistency it needs to be considered.

High Angle of Attack.   A high load factor can result from an instantaneous change in the angle of attack during level flight. The load factor in terms of the dynamic pressure and lift coefficient is

$$n = \frac{q C_L}{W/S}. \tag{10.9}$$

This would be applied for example, at cruise conditions, to determine the extreme load factor condition. To illustrate this, we take the conditions at the start of cruise for the conceptual SSBJ, where $q = 531$ lb/f^2 and $W/S = 157$ lb/f^2. The maximum lift coefficient, without flaps, is approximately 1.0. Applying this to Eq. [10.9], the load factor is 3.4.

Dive Condition.   The maximum dynamic pressure is produced in a dive. As a standard, the dive velocity it taken as $V_{\text{dive}} = 1.5 V_{\text{cruise}}$. Therefore, the dynamic pressure increases by the factor of 1.5^2 or 2.25.

We can again illustrate this by using the conditions of the conceptual SSBJ. With $C_{L_{\text{cruise}}} = 0.2$, under a dive condition, the load factor would be $n = 1.52$.

10.1.2   *V–n* Diagrams

A $V-n$ diagram shows the flight load factors that are used for the structural design as a function of the air speed. These represent the maximum expected loads that the aircraft will experience. These load factors are referred to as the "limit" load factors. Load factor standards for aircraft are covered by FAR-25 for transports, and FAR-23 for normal, utility, acrobatic, and commuter aircraft. The military specification is covered in MIL-A-8661A. A summary of load limits is given in Table 10.1.

TABLE 10.1: Maximum load factors for various aircraft based on FAR-25 and 23.

Aircraft Type	Load Factor
General Aviation (normal)	$-1.25 \leq n \leq 3.1$
General Aviation (utility)	$-1.8 \leq n \leq 4.4$
General Aviation (acrobatic)	$-3.0 \leq n \leq 6.0$
Homebuilt	$-2 \leq n \leq 5$
Commercial Transport	$-1.5 \leq n \leq 3.5$
Fighter	$-4.5 \leq n \leq 7.75$

TABLE 10.2: Load factors for transport aircraft based on FAR-25.

W_{TO}(lbs)	n_{max}
≤ 4100	3.8
$4100 < W_{TO} \leq 50,000$	$2.1 + (24,000/(W_{TO} + 10,000))$
$>50,000$	2.5

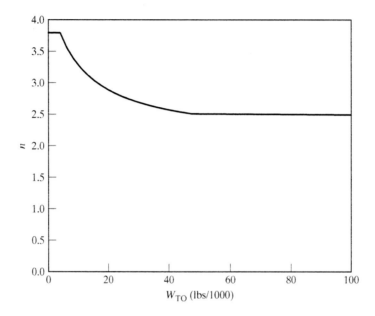

FIGURE 10.1: Positive load limit for transport aircraft based on FAR-25.

In addition to Table 10.1, FAR-25 defines a maneuver load factor for transport aircraft, which varies with the design gross take-off weight. This is listed in Table 10.2 and plotted in Figure 10.1.

An example of a $V-n$ diagram for maximum maneuver load factors is shown in Figure 10.2. The curve from $n = 0$ to point "A" represents the maximum normal-component load produced by high angle-of-attack flight. The equation for this curve is given by Eq. [10.9]. The maximum value at point "A" is determined by the FAR standard for the particular type of aircraft (Table 10.1). This limit corresponds to the horizontal line "A–D."

Point "D" occurs at the highest flight velocity, which is the "dive" velocity. Recall that $V_{dive} = 1.5 V_{cruise}$. The point V_c in Figure 10.2 represents the cruise velocity. At cruise, $n = 1$, which is shown as the dashed horizontal line. The intersection of that line with the "0–A" curve corresponds to the stall velocity, V_s, which is the minimum speed at which the aircraft can maintain level flight.

The line "H–F" represents the largest negative load factor for a particular type of aircraft (Table 10.1). These are generally less than the maximum positive load factor.

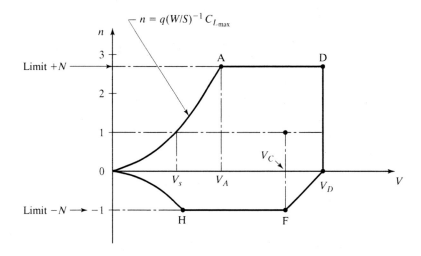

FIGURE 10.2: $V-n$ diagram showing the envelope of load factors for a maneuvering aircraft.

The curve "0–H" is also given by Eq. [10.9], but does not extend as far because of the lower negative load limit.

Point "F" corresponds to the intersection of the negative load limit and the maximum design cruise velocity, V_C. The negative load factor envelope is then closed by the line at point "F" to the $n = 0$ point at the dive velocity, V_D.

Gust Loads.   Gust loads are unsteady aerodynamic loads that are produced by atmospheric turbulence. They represent a load factor that is added to the aerodynamic loads, which were presented in the previous sections.

The effect of a turbulent gust is to produce a short-time change in the effective angle of attack. This change can be either positive or negative, thereby producing an increase or decrease in the wing lift and a change in the load factor, $\Delta n = \pm \Delta L / W$.

Figure 10.3 shows a model for the effect of a gust on an aircraft in level flight. The aircraft has a forward velocity, V. The turbulent gust produces small velocity components, v and u. At that instant, the velocity component in the aircraft flight direction is $V + v$. In level flight, the mean velocity component normal to the flight direction is $U = 0$. Therefore, the total normal velocity is u.

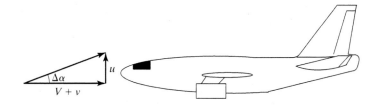

FIGURE 10.3: Model for gust load effect on a aircraft in level flight.

In most cases, u and v are much less than the flight speed, V. Therefore, $V + v \simeq V$. Based on this assumption, the effective angle of attack is

$$\Delta \alpha = \tan^{-1} \frac{u}{V}. \tag{10.10}$$

Because u is small compared to V,

$$\Delta \alpha \simeq \frac{u}{V}. \tag{10.11}$$

The incremental lift produced by the small change in the angle of attack is

$$\Delta L = \frac{1}{2} \rho V^2 S C_{L_\alpha} \Delta \alpha. \tag{10.12}$$

Substituting Eq. [10.11] then gives

$$\Delta L = \frac{1}{2} \rho V S C_{L_\alpha} u. \tag{10.13}$$

The incremental load factor is then

$$\Delta n = \frac{\rho u V C_{L_\alpha}}{2W/S}. \tag{10.14}$$

The peak load factor is then the sum of the mean load factor at cruise ($n = 1$) and the fluctuation load factor, namely,

$$n_{\text{peak}} = n + \Delta n. \tag{10.15}$$

The gusts that result from atmospheric turbulence occur in a fairly large band of frequencies. Therefore, their effect on an aircraft depends on factors that affect its frequency response. In particular, the frequency response is governed by an equivalent mass ratio, μ, defined as

$$\mu = \frac{2W/S}{\rho g \bar{c} C_{L_\alpha}}, \tag{10.16}$$

where $\bar{c}$ is the mean chord of the main wing and g is the gravitational constant. Note that μ is dimensionless, so that in British units, $g_c = 32.2 f - lb_m/lb_f - s^2$ is required in the numerator.

The mass ratio, μ, is a parameter in a response coefficient, K, which is defined differently for subsonic and supersonic aircraft, namely,

$$K = \frac{0.88\mu}{5.3 + \mu} \quad (subsonic) \tag{10.17a}$$

or

$$K = \frac{\mu^{1.03}}{6.95 + \mu^{1.03}} \quad (supersonic). \tag{10.17b}$$

TABLE 10.3: Statistical gust velocity values.

Flight Condition	Altitude Range (f)	$\hat{u}$ (f/s)
High Angle of Attack	0–20,000	66
Level Flight	0–20,000	50
Dive Condition	0–20,000	25
High Angle of Attack	50,000	38
Level Flight	50,000	25
Dive Condition	50,000	12.5

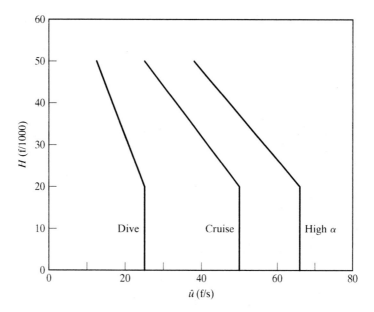

FIGURE 10.4: Variation of gust velocity, $\hat{u}$, with altitude for different flight conditions.

The normal component of the gust velocity, u, is the product of the statistical average of values taken from flight data, $\hat{u}$, and the response coefficient, or

$$u = K\hat{u}. \tag{10.18}$$

Table 10.3 gives values of $\hat{u}$. The variation with altitude is presented in Figure 10.4

Considering Eq. [10.17], we observe that turbulent gusts have a greater effect on aircraft with a lower wing loading. Therefore, a higher wing loading is better to produce a "smoother" flight, as well as in lowering the incremental structural loads.

$V-n$ Diagram Gust Envelope.   The effect of the additive gust loads can be seen in the $V-n$ diagram shown in Figure 10.5. This is shown in blue to contrast it with the load factor envelope for maneuvers alone.

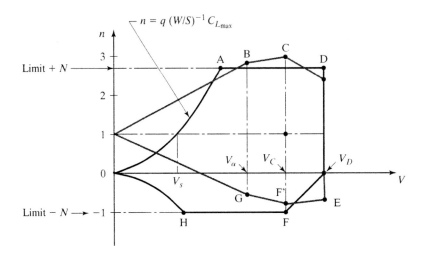

FIGURE 10.5: $V-n$ diagram showing the envelope of load factors including gusts (blue) on a maneuvering aircraft.

Point "B" corresponds to the maximum lift at the highest angle of attack plus the load factor for a gust with $\hat{u} = 66$ f/s.

Point "C" refers to the load factor at the design cruise velocity, V_c plus that for a gust with $\hat{u} = 50$ f/s.

Point "D" corresponds to the load factor at the dive velocity, V_D plus that for a gust with $\hat{u} = 25$ f/s.

Points "E", "F", and "G" correspond to the addition of loads from negative gusts at the velocities corresponding to dive, V_D; cruise, V_c; and maximum lift, V_α, respectively.

Plots like Figure 10.5, which superpose the maneuver loads with the gust loads, are important for determining the conditions that produce the highest load factors. The largest values are the ones used in the structural design.

10.1.3 Design Load Factor

The "limit load factor" denoted in Figures 10.2 and 10.5, is the highest of all the maneuvering load factors plus the incremental load due to turbulent gusts,

$$n_{\text{limit}} = n_{\text{max}} + \Delta n. \tag{10.19}$$

In order to provide a margin of safety to the structural design, the limit load factor is multiplied by a "safety factor," SF. The standard safety factor used in the aircraft industry is 1.5. This value was originally defined in 1930 because it corresponds to the ratio of the tensile ultimate load strength to yield strength of 24ST aluminum alloy, a material commonly used on aircraft. Over the years since it was first designated, this safety factor has proved to be reliable.

The "design load factor" is then defined as the product of the limit load factor and the safety factor,

$$n_{\text{design}} = 1.5 n_{\text{limit}}. \tag{10.20}$$

This factor represents the ultimate load that the internal structure is designed to withstand. In the selection of materials used in the design of the structure, the material ultimate stress will be divided by the n_{limit} to guarantee that the material will not fail up to the design load limit.

10.1.4   Wing Load Distribution

The loads on the wing are made up of aerodynamic lift and drag forces, as well the concentrated or distributed weight of wing-mounted engines, stored fuel, weapons, structural elements, etc. This section will consider these as the first step in designing the internal structure for the wing.

Spanwise Lift Distribution.   As a result of the finite aspect ratio of the wing, the lift distribution varies along the span, from a maximum lift at the root, to a minimum lift at the tip. The spanwise lift distribution should be proportional to the shape of the wing planform. It can readily be calculated using a vortex panel method. However, if the wing planform is elliptic in shape, with a local chord distribution, $c(y)$ given as

$$c(y) = \frac{4S}{\pi b}\sqrt{1 - \left(\frac{2y}{b}\right)^2}, \tag{10.21}$$

an analytic spanwise lift distribution exists. This is given as

$$L^E(y) = \frac{4L}{\pi b}\sqrt{1 - \left(\frac{2y}{b}\right)^2}, \tag{10.22}$$

where L^E is the total lift generated by the wing with an elliptic planform. In both these expressions, y is the spanwise coordinate of the wing, with $y = 0$ corresponding to the wing root, and $y = \pm b/2$ corresponding to the wing tips. A schematic is shown in Figure 10.6.

The analysis of the elliptic planform wing shows that it results in an elliptic lift distribution in the spanwise direction. This is the basis for a semi-empirical method for estimating the spanwise lift distribution on untwisted wings with general trapezoidal planform shapes. The method is attributed to Schrenk (1940) and assumes that the spanwise lift distribution of a general untwisted wing has a shape that is the *average* between the actual planform chord distribution, $c(y)$, and that of an elliptic wing. In this approach, the area under the spanwise lift distribution for the elliptic or general planform, must equal the total required lift.

For the trapezoidal wing, the local chord length, $c(y)$, varies along the span as

$$c(y) = c_r\left[1 - \frac{2y}{b}(1 - \lambda)\right], \tag{10.23}$$

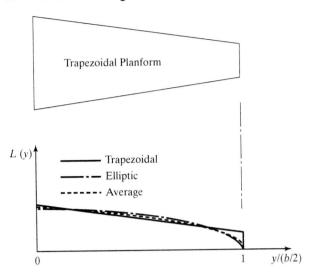

FIGURE 10.6: Schematic representation of two spanwise lift distributions for an elliptic and trapezoidal planform shape, and the average of the two lift distributions using Schrenk's (1940) approximation.

where again, c_r is the root chord length and λ is the taper ratio. Following the elliptic wing, we can take the spanwise lift distribution to vary like the spanwise chord variation. Therefore,

$$L^T(y) = L_r \left[1 - \frac{2y}{b}(1 - \lambda) \right], \tag{10.24}$$

where L_r is the local lift value at the location of the wing root ($y = 0$).

Now the total lift must equal the value found by integrating the lift distribution in the spanwise direction. Therefore,

$$L = \int_{-b/2}^{b/2} L(y)dy = 2L_r \int_0^{b/2} \left[1 - \frac{2y}{b}(1 - \lambda) \right] dy. \tag{10.25}$$

Evaluating the integral, we obtain

$$L = \frac{L_r b(1 + \lambda)}{2}. \tag{10.26}$$

With this, we have an expression for L_r, which gives the necessary total lift for the trapezoidal lift distribution, namely,

$$L_r = \frac{2L}{b(1 + \lambda)} \tag{10.27}$$

and, therefore,

$$L^T(y) = \frac{2L}{b(1+\lambda)}\left[1 - \frac{2y}{b}(1-\lambda)\right]. \qquad (10.28)$$

As a check, for a planar wing ($\lambda = 1$), $L^T(y) = L/b$, which is the correct lift per span.

To use Schrenk's method, it is necessary to graph the spanwise lift distributions given in Eq. [10.19] for the elliptic planform and Eq. [10.28] for the trapezoidal planform. In each case, L is the required total lift. The approximated spanwise lift distribution is then the local average of the two distributions, namely,

$$\bar{L}(y) = \frac{1}{2}\left[L^T(y) + L^E(y)\right]. \qquad (10.29)$$

An example of this corresponds to the dotted curve in Figure 10.6.

It should be pointed out that Schrenk's method does not provide a suitable estimate of the spanwise lift distribution for highly swept wings. In that instance, a panel method approach or other computational method is necessary.

Added Flap Loads.   Leading-edge and trailing-edge flaps enhance the lift over the spanwise extent where they are placed. The lift force is assumed to be uniform in the region of the flaps and to add to the local spanwise lift distribution that is derived for the unflapped wing.

The determination of the added lift force produced by the flaps requires specifying a velocity. For this, the velocity is taken to be twice the stall value, $2V_s$, with flaps down.

Spanwise Drag Distribution.   The drag force on the wing varies along the span, with a particular concentration occurring near the wing tips. An approximation that is suitable for the conceptual design is to assume that

1. the drag force is constant from the wing root to 80 percent of the wing span and equal to 95 percent of the total drag on the wing;
2. the drag on the outward 20 percent of the wing is constant and equal to 120 percent of the wing total drag.

In most cases, the wing structure is inherently strong (stiff) in the drag component direction because the relevant length for the bending moment of inertia is the wing chord, which is large compared to the wing thickness. Therefore, the principle bending of the wing occurs in the lift component direction. The design of the internal structure of the wing is then primarily driven by the need to counter the wing-thickness bending moments.

Concentrated and Distributed Wing Weights.   Other loads on the wing, besides the aerodynamic loads, are due to concentrated weights, such as wing-mounted engines, weapons, fuel tanks, etc., and due to distributed loads such as the wing structure.

TABLE 10.4: Statistical weights of major components on various types of aircraft.

Component	Multiplier			Factor*
	Combat	Transport/ Bomber	General Aviation	
Main Wing	9.0	10.0	2.5	S_W
Horizontal Tail	4.0	5.5	2.0	S_W
Vertical Tail	5.3	5.5	2.0	S_W
Installed Engine	1.3	1.3	1.4	Uninstalled W_{engine}
Landing Gear	0.033 (Navy, 0.045)	0.043	0.057	W_{TO}
	(Navy, 0.045)			W_{TO}
Fuselage	4.8	5.0	1.4	$S_{fuse-wetted}$

*__Note:__ Areas have units of f^2 and weights have units of pounds.

Since the structure is being designed at this step, it is difficult to know precisely what the final weight will be. Therefore, historic weight trends for different aircraft are used to make estimates at this stage of the design. A refined weight analysis will be done later as the initial step in determining the static stability coefficients for the aircraft.

Table 10.4 gives historic weights for the major components of a range of different aircraft. These include the main wing, horizontal and vertical tails, fuselage, installed engine, and landing gear. The weights of these components are determined from the table as

$$W(\text{lb}_f) = \text{Multiplier} \times \text{Factor}, \qquad (10.30)$$

where the "multiplier" is a number that corresponds to a general type of aircraft and the "Factor" is a reference portion of the aircraft, such as the wing planform area, S_W, or the fuselage wetted area, $S_{fuse-wetted}$.

Engines and landing gear mounted on the wing can be treated as concentrated loads. The wing structure will be considered as a distributed load. It is reasonable to consider that the weight of a spanwise section of the wing would scale with the wing chord length, so that with a linear tapered wing, the distributed weight would decrease in proportion to the local chord from the root to the tip.

10.1.5 Shear and Bending Moment Analysis

The wing structure can be considered to be a cantilever beam, which is rigidly supported at the wing root. The critical loads that need to be determined are the shear forces and bending moments along the span of the wing. These take into account the loads on the wing produced by the aerodynamic forces and component weights, which were discussed in the previous section. A generic load arrangement is listed in Table 10.5 and illustrated in Figure 10.7.

To determine the shear force and bending moments along the span, it is useful to divide the wing into spanwise segments of width Δy. A schematic of such an element is shown in Figure 10.8.

TABLE 10.5: Load summary for generic wing which is illustrated in Figure 10.7.

Load Type	Magnitude (lb_f)	$y/(b/2)_{start}$	$y/(b/2)_{end}$
Lift (unflapped)	10,000	0	1
Lift (flapped)	5000	0	0.4
Fuel	3000	0	0.4
Engine	3000	0.3	0.3
Structure	4000	0	1

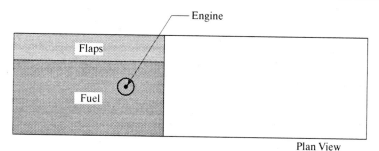

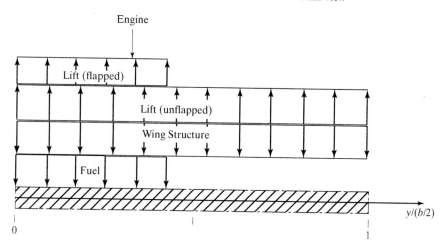

FIGURE 10.7: Schematic representation of spanwise load distribution on a generic trapezoidal wing ($\lambda = 1$).

As an example, the element shows a distributed load, $W(y)$. The resultant load acting on the element is then $W(y)\Delta y$. In the limit as Δy goes to zero, Δy approaches the differential length, dy, and the resultant load is $W(y)dy$.

The element shear force, V, is related to the resultant load as

$$W = \frac{dV}{dy}.$$

(10.31)

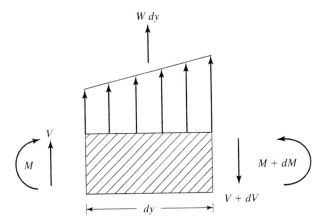

FIGURE 10.8: Schematic representation of shear loads and bending moments on a spanwise element of the wing.

The bending moment, M, acting on the element is related to the shear force by

$$V = \frac{dM}{dy}. \tag{10.32}$$

In integral form,

$$V = \int W \, dy \tag{10.33}$$

and

$$M = \int V \, dy. \tag{10.34}$$

These integrals can be approximated by sums, namely,

$$V = \sum_{i}^{N} W_i \Delta y \tag{10.35}$$

and

$$M = \sum_{i}^{N} V_i \Delta y, \tag{10.36}$$

where N is the number of elements over which the wing span is divided. Of course, the sums approximate the integrals better as the number of elements becomes large; however, a reasonably good estimate for the conceptual design can be obtained with approximately 20 elements over the half-span of the wing.

In order to make these definite integrals, the integration (summation) needs to be started where the shear and moment are known. With the wing, this location is at the wing tip ($y = b/2$), where $V(b/2) = M(b/2) = 0$.

Note that in this case, the resultant load on an element is $W_i = W(y)\Delta y$ which is the quantity inside the sum in Eq. [10.35]. If the index, i, in Eq. [10.35] indicates the elements along the wing span, with $i = 1$ signifying the one at the wing tip, then

$$V_1 = 0;$$

$$V_2 = W_1 + W_2;$$

$$V_3 = W_1 + W_2 + W_3 = V_2 + W_3;$$

$$V_4 = V_3 + W_4;$$

$$\vdots$$

$$V_N = V_{N-1} + W_N. \tag{10.37}$$

Note that the shear on element N must equal the sum of the resultant loads on the wing. In reality, there might be a small discrepancy due to the finite number of elements in which the wing span is subdivided. However, with a large enough number of elements (for example 20), the difference should be small.

The bending moment on the wing is given by Eq. [10.36]. For the moments along the wing span, one should also start at the wing tip where the moment on that element is zero. Then following the format in Eq. [10.37],

$$M_1 = 0;$$

$$M_2 = V_1 + \Delta y V_2;$$

$$M_3 = V_1 + \Delta y V_2 + \Delta y V_3 = M_2 + \Delta y V_3;$$

$$M_4 = M_3 + \Delta y V_4;$$

$$\vdots$$

$$M_N = M_{N-1} + \Delta y V_N. \tag{10.38}$$

These formulas provide a good approximation of the distribution of the shear and moment along the span of the wing. An example of their use is given in the spreadsheet that accompanies this chapter.

An example of the application of these equations is shown in Figure 10.9. The loads correspond to those listed in Table 10.5 and illustrated in Figure 10.7.

The top plot in Figure 10.9 illustrates Schrenk's approximation of the spanwise lift distribution for the finite span wing. The solid curve corresponds to the lift distribution for the trapezoidal wing. This is constant along the span because the taper ratio (λ) in this example is 1. (See Eq. [10.28].) The long-dashed curve corresponds to the spanwise lift distribution for an equivalent elliptic planform airfoil given by Eq. [10.19]. The short-dashed curve is the average of the trapezoidal and elliptic distributions given by Eq. [10.29]. This is the lift distribution that is used in evaluating the shear and moment distribution for the wing.

The total spanwise load distribution, W, shown in Figure 10.9 for a generic wing includes all of the weight and lift components. For this, the wing was divided into 20

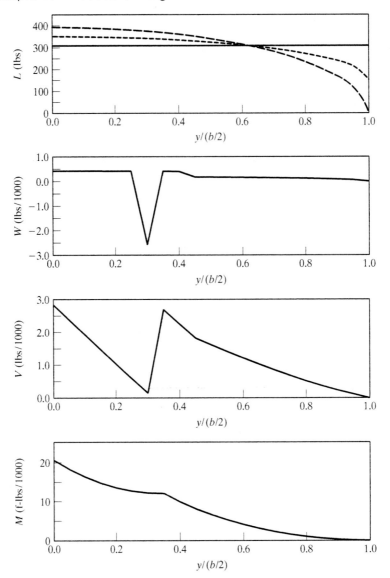

FIGURE 10.9: Spanwise distributions of lift force, L; weight, W; shear loads, V; and bending moment, M, for the load distribution listed in Table 10.5 and illustrated in Figure 10.7.

spanwise elements. The sharp negative spike in the load distribution marks the location of the engine. The more gradual dip in the loads near $y/(b/2) = 0.4$ corresponds to the outboard edge of the flaps.

The spanwise distribution of the shear load, V, comes from Eq. [10.37]. This shows that the largest shear is at the wing root, with the second largest shear being at the location of the engine.

The moment distribution, M, in Figure 10.9 is based on Eq. [10.38]. It reflects the wing cantilever structure, whereby the largest moment is at the wing root. The small peak in the moment distribution near $y/(b/2)$ is due to the engine.

10.1.6   Fuselage Load Distribution

The fuselage can be considered to be supported at the location of the center of lift of the main wing. The loads on the fuselage structure are then due to the shear force and bending moment about that point.

The loads come from a variety of components, for example, the weights of payload, fuel, wing structure, tail structure, engines, fuselage structure, and tail control lift force. Figure 10.9 illustrates a typical load distribution. Note that the coordinate along the fuselage is denoted as x and the length of the fuselage is L. Table 10.6 gives an

TABLE 10.6: Load summary for fuselage that is illustrated in Figure 10.10.

Load Type	Magnitude (lbs)	x/L_{start}	x/L_{end}	$x/L_{Resultant}$	$M@x_{CL}$ (f-lbs)
Fuel	10,000	0.3	0.5	0.4	-1000
Payload	5000	0.1	0.6	0.35	-750
Fuse. Struct.	9000	0	1	0.5	0
Engine	3000	0.7	0.7	0.7	600
Wing Struct.	2000	0.3	0.6	0.45	-100
Tail Struct.	3000	0.8	1	0.9	1200
Tail Lift	-125	0.9	0.9	0.9	-50

$x_{CL} = 0.5$

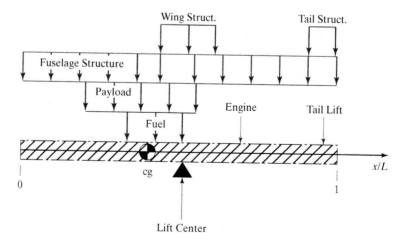

FIGURE 10.10: Schematic representation of forces acting on a generic fuselage.

example of a load breakdown and the resulting moments about the location of the wing center of lift.

For static stability in the pitching direction, the balance of the loads about the center of lift should result in a nose-down moment that has to be offset by the horizontal stabilizer downward lift force. (This would be upward lift in the case of a canard.) This is determined by finding the moments produced by the product of the resultants of the respective loads, and the distance from the location where they act to the location of the center of lift. For the conditions in Table 10.6, the net moment needed to balance the fuselage is −50 f-lbs. This moment is produced by a downward lift force of 125 lbs at the location of the center of lift of the horizontal tail.

There is obviously some leeway in setting the locations of many of the items in Table 10.6. Their placement affects the static stability as well as the performance of the aircraft. Placing the center of gravity too far forward of the center of lift can make an aircraft too stable and require too large of a control force from the horizontal tail to maintain a level pitching attitude. For an aft tail, this situation leads to larger lift-induced drag on the horizontal tail and main wing.

The static stability analysis will be discussed in detail in the next chapter. However, a simplified measure of the pitching stability that can be used for locating some of the heavier components on the fuselage is called the static margin (SM).

The static margin is defined as

$$SM = \frac{x_{np} - x_{cg}}{\bar{c}}. \tag{10.39}$$

where x_{np} is the location of the neutral lift point of the main wing and $\bar{c}$ is the mean aerodynamic chord (m.a.c). As a first approximation, we can neglect the lift-induced moment of the main wing so that x_{np} corresponds to the center of lift, x_{CL}.

For static stability in the pitching direction, the static margin is positive (SM > 0). Nominal values for the static margin for a large spectrum of aircraft gives a range of $3 \leq SM \leq 10$.

A consideration in the placement of the fuel is how the location of the center of mass will shift as the fuel weight is reduced over a flight plan. In the case of a long-range aircraft, the static margin can change significantly from the start of cruise to the end of cruise. The placement of the fuel should be such that the static margin always remains positive.

10.1.7 Shear and Bending Moment Analysis

The fuselage structure can be considered to be a beam that is simply supported and balancing at x_{CL}. As with the wing, we wish to determine the resulting shear forces and bending moments along the length of the fuselage. The procedure for this is the same as for the wing, namely, to divide the fuselage into discrete elements along its length. It is useful if one of the elements is at the x-location of x_{CL}.

The elemental shear forces and bending moments follow the formulas given in Eqs. [10.37 & 10.38], with the exception that Δy in the case of the wing is replaced by Δx for the fuselage. The equations for determining shear force and bending moment for the fuselage are given in Eqs. [10.40 & 10.41]:

$$V_1 = W_1;$$

$$V_2 = W_1 + W_2;$$

$$V_3 = W_1 + W_2 + W_3 = V_2 + W_3;$$

$$V_4 = V_3 + W_4;$$

$$\vdots$$

$$V_N = V_{N-1} + W_N \tag{10.40}$$

and

$$M_1 = V_1;$$

$$M_2 = V_1 + \Delta x V_2;$$

$$M_3 = V_1 + \Delta x V_2 + \Delta x V_3 = M_2 + \Delta x V_3;$$

$$M_4 = M_3 + \Delta x V_4;$$

$$\vdots$$

$$M_N = M_{N-1} + \Delta x V_N. \tag{10.41}$$

The summation starts at one end of the fuselage ($x = 0$ or $x = L$). In contrast to the wing, the shear force in the first element (at one end) is considered to be the load on that element (W_1), and the moment is considered to be the shear on that element (V_1). Starting at the selected end, the summation then continues across each element to the other end, as with the wing.

In this process, the shear force and moment can be found for the summation of all the loads, or separately for the individual loads, with the total shear and moment being the sum of the individual shear and moment distributions. In either approach, it is important to include the concentrated *reaction load* that occurs at the point of support, $x = x_{CL}$.

The inclusion of the resultant force was not necessary for the wing, because only half of the wing span was considered and the point of support was at one end (root). For the fuselage, if the resultant load is properly included, the force on that element minus the shear at x_{CL} should equal the sum of the total load across all of the elements.

An example of the application of these equations is shown in Figure 10.11. The loads correspond to those listed in Table 10.6, and illustrated in Figure 10.10. The shear, V, shows a reversal of sign at $x = x_{CL}$ as a result of the resultant force that acts at the point of support at the wing lift center. As a result, the shear is zero at the leading and trailing points of the fuselage.

The moment, M, in Figure 10.11 is also a maximum at the point of support of the fuselage, which corresponds to the wing lift center. These values set the maximum stress condition for the structural design and dictate the internal structural layout of the fuselage.

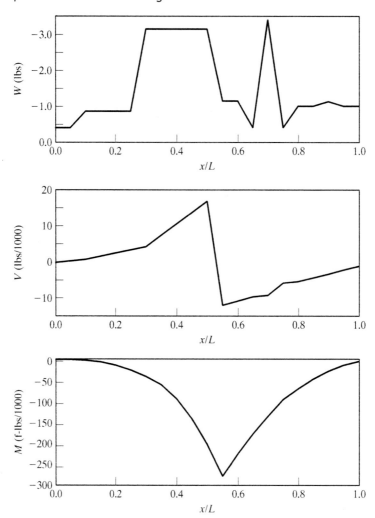

FIGURE 10.11: Lengthwise distributions of the weight, W; shear loads, V; and bending moment, M, for the load distribution listed in Table 10.6 and illustrated in Figure 10.10.

10.2 INTERNAL STRUCTURE DESIGN

The *conceptual* design is mainly concerned with the gross aspects of the structural design. The complete structural design is primarily completed later in the preliminary and detailed steps of the aircraft design. At this stage, the concern is on general structural aspects, which includes the selection of materials to best withstand the maximum loads, while also seeking a low structure weight.

The type of structure used on aircraft since the 1930s is called "semi-monocoque." The word "monocoque" is French for "shell only." The semi-monocoque design uses a thin sheet-metal skin (shell) to resist *tensile* loading and an internal frame of light-weight stiffeners to resist *compressive* loading.

FIGURE 10.12: Photograph of fuselage section for a Boeing 767 that illustrates longerons and bulkheads used in the semi-monocoque construction. Note that the section is inverted. (Courtesy of the Boeing Company.)

In the fuselage, the frame elements that run along its length are called "longerons." Those elements that run around the internal perimeter of the fuselage are called "bulkheads." The structural criterion for the cross-section size of longerons and the spacing of bulkheads is to resist compressive buckling. An example of the internal structural components in the fuselage can be seen in the photograph of a Boeing 767 fuselage section in Figure 10.12.

In the wing, the prevalent design consists of having a central internal beam that runs along its span. The beam is referred to a wing "spar" and is designed to withstand the shear and tensile stresses caused by the shear forces and bending moment. The beam cross-section can range from a hollow square or rectangular ("box") shape to an I-shape.

The wing profile is formed by "ribs," which are cut into the shape of the airfoil cross-section and attach to the central wing spar. A thin sheet-metal skin is attached to the ribs in order to build up the complete wing shape. As a structural element, the skin primarily adds torsional stiffness to the wing.

The horizontal and vertical tail surfaces can be constructed in the same way as the main wing. Alternately, because of their smaller size and because there is no use for their internal volume, they can be be fabricated of full-depth stabilizing material such as foam-plastic or honeycomb material. Honeycomb material is made by bonding very thin corrugated sheets together to form internal hexagonal cells (similar to a bee honeycomb), which run through the material. When the honeycomb material is sandwiched between two thin metal sheets, it forms an extremely rigid and light-weight structural element. This form of construction is excellent for other nonstructural elements such as flaps, fillets, and landing gear doors in order to reduce weight.

10.2.1 Structural Analysis

Tensile Loading. In the semi-monocoque design, the sheet metal covering is designed to withstand the tensile loads. With the fuselage, supported at the center of lift for a positive load factor, the tensile loads are on the top surface. For a negative load factor, they would be on the bottom surface. In most cases, the positive load factor is the larger of the two, so that it will dictate the structural design.

The tensile force acting on the fuselage skin is due to the moment at x_{CL}. The tensile stress is then

$$\sigma_T = \frac{M_{x_{\mathrm{CL}}} R}{I}, \tag{10.42}$$

where R is the half-height of the fuselage at x_{CL} and I is the bending moment of inertia. In the case of a *circular* cross-section fuselage, R is the radius of the fuselage; I is

$$I = \frac{\pi}{4}(R^4 - r^4), \tag{10.43}$$

where $r = R - t$; and t is the thickness of the sheet metal skin. Since $t \ll R$,

$$I \approx \frac{\pi}{2} R^3 t. \tag{10.44}$$

Therefore, from Eq. [10.42],

$$\sigma_T = \frac{2M_{x_{\mathrm{CL}}}}{\pi R^2 t}. \tag{10.45}$$

The stress in the skin must be less than the ultimate tensile stress for the material divided by the design load factor, or

$$\sigma_T \leq \sigma_{T_u}/n_{\mathrm{design}}. \tag{10.46}$$

From this, using Eq. [10.45], the *minimum* skin thickness is

$$t_{\mathrm{min}} = \frac{2M_{x_{\mathrm{CL}}} n_{\mathrm{design}}}{\pi \sigma_{T_u} R^2}. \tag{10.47}$$

In many cases, a more desirable cross-section shape for the fuselage is elliptic in order to give a higher ceiling height. For an *elliptic* cross-section fuselage, where the major axis is the vertical height of the fuselage, the bending moment of inertia about the minor axis (due to $M_{x_{\mathrm{CL}}}$) is

$$I = \frac{\pi}{4}(A^3 B - C^3 D), \tag{10.48}$$

where A is the major axis radius, B is the minor axis radius, and $C = A - t$ and $D = B - t$, where again t is the fuselage skin thickness.

For $t \ll R$,

$$I = \frac{\pi}{4}(A^2 B + A^3)t. \tag{10.49}$$

In this case, the *minimum* skin thickness is

$$t_{\mathrm{min}} = \frac{4M_{x_{\mathrm{CL}}} n_{\mathrm{design}}}{\pi \sigma_{T_u}(AB + A^2)}. \tag{10.50}$$

In either case of a circular or elliptic cross-section fuselage, Eq. [10.47 or 10.50] provides values for the minimum skin thickness needed to withstand the tensile load produced by the maximum bending moment. The required thickness depends on the material property, σ_{T_u}. Values for materials typically used are presented in the next section.

Compressive Loading.   In the semi-monocoque design, the longerons are designed to withstand the compressive loads. With the fuselage, supported at the center of lift
for a positive load factor, the compressive loads are on the lower side. For a negative
load factor, they are on the upper side. Again in most cases, the positive load factor is
the larger of the two and dictates the structural design.

Structural failure under compression for the longerons usually occurs due to buckling. Therefore, this will set the structural design limit. The criterion for buckling comes
from the Euler column formula, given as

$$F_E = \frac{C\pi^2 EI}{L^2},$$
(10.51)

where F is the critical column load to produce buckling, L is the unsupported length,
and C is a factor that depends on how the column is fixed at its ends. For pinned ends,
$C = 1$, whereas $C = 4$ for fixed ends. The longerons are often supported by compara-
tively flexible ribs or bulkheads, which are free to twist or bend. Thus, a value of $C = 1$
is appropriate. If the bulkheads are rigid enough to provide restraint to the longerons, a
value of $C = 1.5$ can be used.

Using Eq. [10.51], the critical stress is

$$\sigma_E = \frac{C\pi^2 E}{(L/\rho)^2},$$
(10.52)

where the radius of gyration is given as

$$\rho = \sqrt{\frac{I}{A}},$$
(10.53)

with I being the bending moment of inertia and A being the cross-section area of the
column.

In order to prevent a structural failure in the longerons, the actual compressive
stress must be less than the buckling stress divided by the design load factor, namely,

$$\sigma < \sigma_E / n_{\text{design}}.$$
(10.54)

The next step is then to determine the actual compressive stress in the longerons. This
requires setting the configuration of longerons around the fuselage.

An example arrangement of longerons around a circular cross-section fuselage of
radius, R, is shown in Figure 10.13. These consist of a series of circular cross-section
elements that are equally spaced in a symmetric pattern about the vertical centerline of
the fuselage. The angular distance between longerons is $90°/N$, where N is the number
of longerons in a $90°$ arc-segment of the fuselage. In the example drawn in Figure 10.13,
the angular distance is $22.5°$, corresponding to $N = 4$.

The longerons in the lower half-plane of the fuselage are under compression due
to the bending moment, M, for a positive load factor. For the design, the maximum
stress is $\sigma_{\text{max}} = M y_{\text{max}}/I$, where y_{max} corresponds to the largest vertical distance of a
longeron from the center of bending, which for this shape, is the horizontal centerline of
the fuselage.

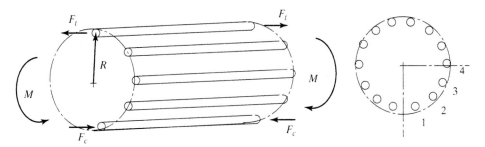

FIGURE 10.13: Sample cross-sectional view of longeron placement around fuselage perimeter.

By definition, the bending moment of inertia is

$$I = \int y^2 dA. \tag{10.55}$$

For discrete longerons of equal cross-section area, A_l,

$$I = A_l \Sigma(y^2). \tag{10.56}$$

For the general arrangement shown in Figure 10.7, for $N = 4$ and using symmetry,

$$I = 4A_l R^2 \left[\cos^2(90°/4) + \cos^2(2 \cdot 90°/4) + \cos^2(3 \cdot 90°/4) + \cos^2(4 \cdot 90°/4) \right]. \tag{10.57}$$

Also, $y_{max} = R\cos(90°/N) = R\cos(22.5°)$, so that the maximum stress in a longeron is

$$\sigma_{max} = \frac{M\cos(22.5°)}{4A_l R \left[\cos^2(22.5°) + \cos^2(45°) + \cos^2(67.5°) \right]}. \tag{10.58}$$

In order that the longeron not buckle due to this stress,

$$n_{design}\sigma_{max} < \frac{C\pi^2 E}{(L/\rho)^2}. \tag{10.59}$$

Substituting for σ_{max} from Eq. [10.58] after evaluating the cosine terms and grouping the pre-determined terms on the right-hand side, we obtain

$$\frac{L^2}{I_l R} < \frac{C\pi^2 E}{0.154 n_{design} M}, \tag{10.60}$$

where I_l is the bending moment of inertia for the longeron cross-section shape.

Equation [10.60] illustrates some of the design decisions that need to be made with regards to the longerons. Ultimately, the maximum unsupported length, L, needs to be decided. It depends on the material used for the longerons through the modulus of elasticity, E, and on the cross-section area and shape through, I_l. The cross-section shape should be one that has a large bending moment of inertia. Typical shapes are either hollow circular or rectangular tubes, or open S-shaped beams. The material should be one that has a large strength-to-weight ratio. Examples of these will be given in the next section.

The final value of L will become the spacing between bulkheads in the fuselage. If the spacing of bulkheads were determined due to other design decisions, that value of L would be used, and the cross-section properties of the longerons that was needed to prevent buckling would then be determined.

10.3   MATERIAL SELECTION

The criterion commonly used in the selection of structural materials for aircraft is that they have a minimum weight. Achieving this generally requires a synergy between the structural design and the material selection. Other factors that can also enter into the selection of materials are resistance to corrosion, fatigue behavior, creep characteristics, and machining and fabrication ability.

The two types of loading that were analyzed with regards to the fuselage were pure tension (for the skin) and pure compression (for the longerons). In addition, the main wing spar will experience pure bending. Expressions that relate applied loads, F, in these cases, to the induced stresses are

$$\sigma_t = \frac{F}{A}, \tag{10.61}$$

$$\sigma_c = \frac{\pi^2 EI}{AL^2}, \tag{10.62}$$

and

$$\sigma_b = \frac{My}{I} \tag{10.63}$$

for the tensile, compressive, and bending stresses, respectively.

The weight of a structural member can be expressed in terms of the density, ρ and volume, $V = Lbt$. Therefore, substituting for the areas in the stress equations and solving for the respective weights,

$$W_t = \frac{FL\rho}{\sigma_t}, \tag{10.64}$$

$$W_c = \frac{L^2 b\rho}{\pi} \left(\frac{12\sigma_c}{E}\right)^{1/2}, \tag{10.65}$$

and

$$W_b = L\rho \left(\frac{12Mb}{\sigma_b}\right)^{1/2}. \tag{10.66}$$

For the same applied loads, Eqs. [10.64–10.66] allow comparisons to be made between the weights of materials with different ultimate stress limits. For example, for tensile loading,

$$\left[\frac{W_1}{W_2}\right]_t = \frac{\rho_1}{\rho_2}\frac{\sigma_{u_{t_1}}}{\sigma_{u_{t_2}}}. \tag{10.67}$$

For compressive loading,

$$\left[\frac{W_1}{W_2}\right]_c = \frac{\rho_1}{\rho_2}\left(\frac{E_2}{E_1}\right)^{1/3}. \tag{10.68}$$

For bending,

$$\left[\frac{W_1}{W_2}\right]_b = \frac{\rho_1}{\rho_2}\left(\frac{\sigma_{u_{t_1}}}{\sigma_{u_{t_2}}}\right)^{1/2}. \tag{10.69}$$

Table 10.7 shows values of the weight ratios from Eqs. [10.67–10.69], for different materials using as a reference (material subscript 2) the properties of aluminum alloy 2024-T3. The rows of the table are listed in the order of highest to lowest ultimate stress.

Analyzing the materials in Table 10.7 on the basis of minimizing weight, for bending and compressive buckling, the lower density materials (bottom three) are better than the aluminum alloys. For tensile loading, there is not a significant difference between the materials listed.

Therefore, for structures such as the fuselage skin, which are designed on the basis of tensile loading, the selection of the material can be made to include other aspects such as corrosion resistance, finish, cost, etc.

For structural elements that are designed to resist buckling and bending, it would appear that a magnesium alloy would be preferable. However, again other considerations may be important, such as that magnesium has a greater tendency to corrode compared to an aluminum alloy. Although stainless steel has the highest ultimate stress and modulus of elasticity, in all aspects of loading, it has a significant weight penalty. However, in applications where elements are exposed to high temperatures, or where corrosion is a high concern, stainless steel may be the best material of choice.

Fiber-reinforced plastics consist of high-strength fibers that are bonded together with resins and built up in layers to form an integral structure. Early examples consisted of glass fibers bonded with polyester resin. Currently, high-modulus fibers such as boron, silicon carbide, graphite, and beryllium are used along with new resins such as cycloaliphatic epoxies and polymeric resins. The new fibers and resins can be combined in a unidirectional pre-impregnated form that allows the design of structures that are expressly tailored to specific load arrangements and applications. Through the use

TABLE 10.7: Strength-to-weight comparisons for different aircraft materials.

Material	σ_u (10^3 lb/in.^{2})	ρ (lb/in.^{2})	E (10^3 lb/in.^{2})	$\left[\frac{W_1}{W_2}\right]_t$ Tension	$\left[\frac{W_1}{W_2}\right]_c$ Comp./Buckling	$\left[\frac{W_1}{W_2}\right]_b$ Bending
Stainless Steel	185	0.286	26.0	1.23	2.12	1.72
2024-T3 Al	66	0.100	10.5	1	1	1
7075-T6 Al	77	0.101	10.4	0.87	1.01	0.93
Magnesium Alloy	40	0.065	0.5	1.07	0.77	0.83
Laminated Plastic	30	0.050	2.5	1.10	0.83	0.74
Spruce Wood	9.4	0.016	1.3	1.09	0.31	0.42

of composite materials such as these, the total weight of an aircraft can be reduced by more than 35 percent. The wing pictured at the start of this chapter is constructed from a boron/graphite composite.

Another means of reducing the structure weight is through the use of honeycomb sandwich elements. In this case, a thick light-weight core of honeycomb material is bonded between two thin facing layers of high tensile-strength material. In such an arrangement, the resisting bending moment is

$$M = \sigma_f t_f t_c, \tag{10.70}$$

where σ_f is the stress in the facing layers, with thickness t_f, and t_c is the thickness of the core material, with $t_f \ll t_c$. This is a particularly good method of construction for the vertical and horizontal stabilizers because the internal volume is not needed for fuel storage.

A minimum-weight analysis of such sandwich structures shows that with regards to bending moment, the minimum weight is obtained when the weight of the facing layers is the same as that of the core material. When the minimum-weight sandwich structure is compared to a solid-beam element made from the same material as the facing material, to resist the same bending moment, the weight of the solid beam is

$$\frac{W_{\text{sandwich}}}{W_{\text{solid}}} = 1.63 \sqrt{\frac{\rho_c}{2\rho_f}}, \tag{10.71}$$

where ρ_c and ρ_f are the densities of the core and facing materials, respectively.

To see the impact of the weight savings, if the facing material is 2024-T3 aluminum alloy from Table 10.7, and the density of the honeycomb material is approximately 10 times less than that, the weight fraction is

$$\frac{W_{\text{sandwich}}}{W_{\text{solid}}} = 0.37. \tag{10.72}$$

In this case, the honeycomb sandwich structure is only 37 percent of the weight of the equivalent solid structure. When it is noted that this ratio is smaller than that of any of the other materials subjected to bending in Table 10.7, the great benefit it has towards achieving light-weight structures is clearly evident.

10.3.1 Material Properties and Applications

Aluminum. Aluminum remains the most widely used metal for aircraft applications. The reasons are its low material cost, relatively high strength-to-weight ratio, ease of machining, and good corrosion resistance.

The most common aluminum alloy used on aircraft is 2024 (or 24ST), which is referred to as "duralumin" or hard aluminum. It can be heat treated to an ultimate tensile strength of 65 Ksi and has excellent fracture toughness and crack propagation resistance.

The 7xxx aluminum alloys provide the highest strength and are extensively used in high-stress applications. The most widely used of these is the 7075 alloy. However, 7050 and 7010 have better corrosion resistance. To further improve corrosion properties, alloys are clad in a thin layer of pure aluminum.

Steel.   Steel alloys are used in applications requiring high strength and fatigue resistance. Examples are wing attachment fittings and landing gears. Steel is also used in high-heat conditions such as for fire-walls and engine mounts. Stainless steel is used if corrosion resistance is important.

The cost of steel is approximately one-sixth that of aluminum. It is also relatively easy to machine.

The properties of steel alloys can be strongly modified by heat treating and hardening. The procedures for heat treating and the material properties that result can be found in material handbooks.

Titanium.   Titanium has a better strength-to-weight ratio than aluminum and has a working temperature limit almost as high as steel. It has excellent corrosion resistance, equaling or exceeding that of most steels and all aluminum alloys.

However, titanium is extremely difficult to machine and form. Most titanium alloys must be formed at temperatures over 1000°F and with very high forming stresses. It is also seriously affected by impurities. These impurities include most gases, such as hydrogen, oxygen, and nitrogen. Exposure to these gases causes embrittlement, which reduces its impact toughness. As a result, any hot-working, for example welding, has to be done under special conditions.

These problems keep the cost of the raw and finished material relatively high, to approximately five to ten times that of aluminum. However, in high-performance applications, the special properties of titanium outweigh the added cost. It is extensively used in jet engine parts and in high-stress airframe components such as in the landing gear and in the spindle elements of all-moving ("flying") horizontal tails. Titanium is also sometimes used as the substructure to graphite/epoxy composite skins because it does not cause galvanic corrosion. Figure 10.14 shows a photograph of the Lockheed SR-71, which makes extensive use of titanium to withstand the high temperatures of supersonic flight.

Magnesium.   As illustrated in Table 10.6, magnesium has a very good strength-to-weight ratio and is well suited for applications for strength in bending and resistance to buckling. It tolerates high temperatures and is easily formed by casting, forging, or machining. It is commonly used for engine mounts, wheels, control hinges, brackets, stiffeners, and fuel tanks.

One difficulty with magnesium is that it corrodes easily. As a result, it needs to be coated with a protective finish. In addition, it is flammable. As a result, Mil Specs advise against using magnesium except when significant weight reductions are necessary. It should be used only in areas that are easy to inspect for corrosion, and in applications where its protective coating is not subject to abrasion, such as occurs near leading edges or in hot engine exhaust.

Heat-Resistant Steels.   The most prominent of heat-resistant steels is Inconel. Inconel 600 has a lower strength than other steel alloys, but has a working temperature of up to 2000°F. It is weldable and best suited to low-stressed parts that are exposed to high temperatures. Inconel 718 retains the strength of the best steel alloys, but has better creep resistance up to 1300°F. It was used extensively on the X-15 aircraft.

FIGURE 10.14: Photograph of Lockheed SR-71 supersonic aircraft, which has an external structure that is largely made from titanium to withstand the high flight temperatures. (NASA Dryden Research Photo Collection.)

Composites.   The merit of composites towards minimum weight design were discussed in the previous section. The most commonly used advanced composite used today is graphite–epoxy. It is easily molded and has a high strength-to-weight ratio. It is however approximately 20 times more expensive than aluminum.

Boron–epoxy composites were initially used for complete part fabrication, such as the F-111 horizontal tail and F-4 rudder. It, however, costs four times more than graphite–epoxy. As a result, it is now most commonly used to fabricate smaller stiffening elements.

Aramid, which is marketed under the name "Kevlar," is used with an epoxy matrix to fabricate low-stress components. Its advantage is that it has a more gradual failure compared to other composites. For example, when added to graphite–epoxy, a graphite–aramid–epoxy composite provides better ductility. It has been used in the Boeing 757 for fairings and landing gear doors.

Composites using an epoxy resin matrix have a maximum temperature limit of 350°F, and are usually only used in applications where the temperature does not exceed

260°F. Composites for higher temperature applications use Polymide resin for the matrix material. Some of these are usable up to 600°F, although a more typical upper temperature limit is 350°F.

For even higher temperature applications, metal–matrix composites are used. These use aluminum or titanium as the matrix material, and boron, silicon carbide or Aramid as the fiber.

Despite the significant weight reductions offered by composite materials, there are a number of problems in using them in aircraft designs. The first of these is that elements

TABLE 10.8: Characteristics of metals used in aircraft.

Material	σ_{t_u} 10^3lb/in^2	σ_{t_y} 10^3lb/in^2	σ_{c_u} 10^3lb/in^2	E 10^6lb/in^2	ρ lb/in^3	Characteristics
Alloy Steels						
4130 Normalized	95	75	75	29	0.283	weldable
4130 (180HT)	180	163	173	29	0.283	weldable
4330 (220HT)	220	186	194	29	0.283	
300M (280HT)	280	230	247	29	0.283	
4340 (260HT)	260	215	240	29	0.283	
Stainless Steel						
301 (full hard)	185	140	98	26	0.286	weldable
15-5 PH	115	75	99	28.5	0.283	weldable
15-5 PH (190HT)	190	170	143	28.5	0.283	also other HTs
PH 15-7 Mo	190	170	179	29	0.277	
17-4 PH	190	170		28.5	0.282	weldable
Heat-Resistant Steels						
A286	140	95	95	29.1	0.287	up to 1300°F
Inconel 600	80	30	30	30	0.304	up to 2000°F
Inconel 718	170	145		29.6	0.297	up to 1300°F
Inconel X-750	155	100	100	31.0	0.300	up to 1500°F
Aluminum Alloy						
2024-T3 (clad)	58	39	42	10.5	0.100	common use
2219-T87	62	50	50	10.5	0.102	creep resistant
6061-T6	42	36	35	9.9	0.098	weldable
7075-T6	76	66	67	10.3	0.101	high strength
7178-T6	83	73	73	10.3	0.102	high strength
Magnesium						
HK 31A	34	24	22	6.5	0.0674	up to 700°F
HM 21A	30	21	17	6.5	0.0640	up to 800°F
Titanium Alloy						
6AL-4V	160	145	150	16.0	0.160	up to 750°F
13V-11Cr-3AL	170	160	162	15.5	0.174	up to 1000°F

TABLE 10.9: Characteristics of some composite materials used in aircraft.

Material*	σ_{t_u} 10^3lb/in.^{2}	σ_{c_u} 10^3lb/in.^{2}	E 10^6lb/in.^{2}	ρ lb/in.^{3}	Characteristics
Graphite–Epoxy	180	180	21	0.056	60% fiber vol.
Boron–Epoxy	195	353	30	0.073	50% fiber vol.
Graphite–Polyimide	204	111	20		
S-Fiberglass–Epoxy	219	73.9	7.70	0.074	
E-Fiberglass–Epoxy	105	69	4.23	0.071	45 % fiber vol.
Aramid–Epoxy	200	40	11	0.052	60% fiber vol.

*Note: In all cases, the fiber orientation is aligned in the principle strain direction.

made of composite materials do not accept concentrated loads well. Therefore, care must be taken to distribute loads and limit or re-enforce holes and cut-outs that concentrate stresses.

The second problem is that the strength of composites is affected by a number of factors during manufacturing including moisture content, cure cycle, matrix voids, and the exact ratio of fibers to resin. Over their lifetime, their strength is affected by temperature and ultraviolet ray exposure.

The final problem with composites is that internal failure is hard to detect and, when detected, difficult to repair.

Various properties of materials including metal alloys and composites are listed in Tables 10.8 and 10.9. These are intended as a summary of some of the commonly used materials for aircraft and not intended to replace more extensive tables contained in materials handbooks.

10.4   SPREADSHEET FOR STRUCTURE DESIGN

Three spreadsheets are contained in the file **loads.xls**. The spreadsheets separately calculate the maximum load factor, and the shear force and bending moment distributions on the wing and fuselage. Any of the three spreadsheets in the file are selectable by the tool bar at the bottom of the display. These are titled "load factor," "fuselage," and "wing." The three spreadsheets use some common information, which is referenced as "linked" variables. Most of these originate in the "wing" spreadsheet in this file.

10.4.1   Load Factors

The spreadsheet for determining load factors considers the different flight phases including intercept, and instantaneous and sustained turns, as well as limiting conditions set by the highest angle-of-attack flight and dive conditions. In addition, the added load factor resulting from wind gusts is calculated. The largest of the load factors is taken as the load limit. The load limit is then multiplied by the 1.5 safety factor to give the design load factor that should be used in the structural design. Figures 10.15 and 10.16 show the load factor spreadsheet for conditions that are relevant to the conceptual SSBJ proposed in Chapter 1.

Structural Design and Material Selection

Acceleration (Intercept)			Sustained Turn Rate		
H (f)	6,000		H (f)	55,000	
Mach No.	0.8		Mach No.	2.1	
W/S (lb/f^2)	170		ψ-dot	0.78	deg/s
V (f/s)	866.88		V (f/s)	1925.7	
rho (lbm/f^3)	0.06		rho (lbm/f^3)	0.01	
q (lbf/f^2)	752.36		q (lbf/f^2)	531.07	
n	0.83		n	1.29	

Instantaneous Turn Rate			Climb		
H (f)	55,000		CD_0	0.01	
Mach No.	2.1		C_L	1.2	
ψ-dot	0.94	deg/s	γ(deg)	11	
V (f/s)	1925.7		G (rad)	0.19	
rho (lbm/f^3)	0.01		T/W_min	0.27	
q (lbf/f^2)	531.07		T/W_actual	0.27	
n	1.4		T/W-G	0.07	
			n +	6.43	
			n −	6.37	

Highest α			Max q (dive)		
H (f)	55,000		Max q (dive)		
Mach No.	2.1		H (f)	55,000	
C_L	1.00		Mach No.	2.1	
WS (lb/f^2)	157		C_L	0.20	
			W/S (lb/f^2)	157	
V (f/s)	1925.7				
rho (lbm/f^3)	0.01		V_dive (f/s)	2888.55	
q (lbf/f^2)	531.07		rho (lbm/f^3)	0.01	
n	3.38		q_dive (lbf/f^2)	1194.91	
			n	1.52	

Gust Loads (cruise)		
W/S (lb/f^2)	157.00	
H (f)	55,000	
Mach No.	2.1	
V (f/s)	2888.55	
rho (lbm/f^3)	0.01	
$C_{L\alpha}$	2.5	1/rad
Uhat (f/s)	30	Table 10.2
μ	628.7613	
k	0.9910	
u (f/s)	29.7292	
Δn	0.20	
n_peak	1.20	

FIGURE 10.15: Load factor portion of the spreadsheet (LOADS) for maneuver and gust load factor calculations (Part 1).

Design Criteria					
F.S.	1.5				
load limit	3.38				
des. load	5.07				

FIGURE 10.16: Load factor portion of the spreadsheet (LOADS) for maneuver and gust load factor calculations (Part 2).

The load factor calculation for the acceleration flight phase are based on the solution of Eq. [10.1]. This takes as input altitude, Mach number, and wing loading. The dynamic pressure, q, is calculated from these input values. The aspect ratio, A, and C_{D_0} are global variables that are linked to values input in the "wing" spreadsheet in this file.

The load factor calculations for the instantaneous and sustained turn rates are based on the solution of Eq. [10.2 or 10.4], which are identical. These take as input the altitude, Mach number, and instantaneous or sustained turn rates in units of degrees per second. All of the other quantities are calculated based on these input values.

The load factor calculations for climb are based on Eq. [10.8]. This uses C_{D_0}, which is linked to the value in the "wing" spreadsheet in this file, the 3-D unflapped lift coefficient, and the angle of attack in degrees. The result is the solution of a quadratic equation so that there are two solutions. The quantity inside the radical must be positive for a real load factor. This puts a condition on the thrust-to-weight ratio. The minimum required value, T/W_{min}, is calculated. The actual thrust-to-weight needs to be input in the spreadsheet next to the cell marked T/W_{actual}. The load factors for the plus and minus operations in Eq. [10.8] are denoted $n+$ and $n-$. When $T/W_{actual} = T/W_{min}$, then $n+$ equals $n-$.

The load factor corresponding to high angle-of-attack flight is based on Eq. [10.9]. This takes as input the altitude, Mach number, maximum (unflapped) 3-D lift coefficient, and wing loading.

The load factor corresponding to maximum dynamic pressure of a dive condition is based on a velocity that is 1.5 times the cruise velocity. The input values in this case are altitude, cruise Mach number, lift coefficient, and wing loading.

The peak load factor due to wind gusts is calculated based on Eqs. [10.14 & 10.15]. The input values include the wing loading (cruise, dive, or other), the altitude, and Mach number. The value of C_{L_α} is linked to the variable that is input in the "wing" spreadsheet in this file. The gust velocity, u, is found from Eq. [10.18], which involves the statistical gust velocity, $\hat{u}$, and response coefficient, K. Values of $\hat{u}$ come from Table 10.3 or Figure 10.4. The value of K comes from Eq. [10.17a or 10.17b] based on whether the Mach number is subsonic or supersonic, respectively. These involve the mass ratio, μ, which is found from Eq. [10.16]. μ is a function of the mean aerodynamic chord (m.a.c.), which is linked to the value that is input in the "wing" spreadsheet portion of this file.

These lead to the incremental load factor, Δn. At cruise, the load factor is by definition $n = 1$. Therefore the the total load factor including gust loading is $n_{peak} = 1 + \Delta n$, which is denoted in the lowest cell in this portion of the spreadsheet.

The bottom of the spreadsheet lists the maximum of the calculated load factors. This is listed as the load limit. It is then multiplied by the 1.5 factor of safety (FS) to obtain the design load factor.

10.4.2 Wing Load Distribution

The spreadsheet for the analysis of the load distribution on the wing is referenced as "wing" in the **loads.xls** file. Figures 10.17 through 10.20 show the spreadsheet for the conditions of the conceptual SSBJ proposed in Chapter 1.

The input values are entered in the spreadsheet at the top, in the area labeled "wing data." These data include the Mach number (M), main wing area (S), aspect ratio (A), leading-edge sweep angle (Λ_{LE}), taper ratio (λ), wing base drag coefficient (C_{D_0}), and the change in 3-D lift coefficient with angle of attack (C_{L_α}). For a full design, this information comes from the 3-D wing design spreadsheet, **wing.xls**. The other quantities—wing span (b), wing leading-edge effective Mach number (M_{eff}), root and tip chord lengths (c_r and c_t), and mean aerodynamic chord length (m.a.c)—are calculated from the wing input data.

Other data that are required as input are the magnitudes and types of loads that are acting on the wing. These are input in the section labeled "Load Summary." This would include the total lift at cruise, L; the incremental lift produced by flaps, ΔL; the weight of fuel; the weights of any engines; and the total weight of the wing structure. In specifying the weights, the spanwise region over which they act also needs to be indicated. This is specified by the locations $y/(b/2)_{start}$ and $y/(b/2)_{end}$. For a concentrated load, the start and end locations will be the same. It is assumed that the wing span is divided into 20 elements. Therefore, the load per element, dw, is calculated in the load summary. For a concentrated load, dw equals the total load value, which acts on the element at the specified location of the concentrated load.

For the lift without flaps, the spanwise distribution of the lift is calculated in the spreadsheet. Therefore, no value of dw is calculated in the "Load Summary". For the incremental lift produced by flaps, ΔL, the spanwise lift distribution is assumed to be constant. Therefore, dw is ΔL divided by the number of elements encompassing the flapped portion of the wing span.

The portion of the spreadsheet that is labeled "Load Spanwise Distribution" calculates the spanwise distribution of the 3-D lift on the wing; sums that with the other forces acting on the wing; and calculates the shear force and bending moment distribution. The second column (B) of this portion of the spreadsheet defines the regions of elements into which the spanwise length of the wing is divided. Here only a symmetric half of the wing with length $b/2$ is considered.

The half span is divided into 20 elements. The second column lists the dimensionless locations, $0 \leq y/(b/2) \leq 1$. The first column lists the computed dimensional location, based on the wing half-span length ($b/2$).

The process for estimating the spanwise lift distribution is based on Schrenk's approximation in which the distribution is an average based on the actual planform shape and that of an elliptic planform wing. The spanwise distribution for an elliptic planform wing that will produce the same total lift is given in the column labeled "L(y)-ellip." This is based on Eq. [10.19]. The spanwise distribution based on the actual trapezoidal planform shape is given in the column labeled "L(y)-trap." This comes from Eq. [10.28]. Finally the average of these two lift distributions is contained in the column labeled "L-bar." This follows Eq. [10.29]. The sum of the spanwise lift distributions is shown at the bottom of each column. Note that the total is for the half wing span, so that twice

Wing Data						
M	2.10			b	32.2	ft
S	519	ft^2		b/2	16.11	ft
A	2.0			M$_{eff}$	0.99	
Λ_{LE}	62	deg		c$_r$	32.2	ft
λ	0.00			c$_t$	0.0	ft
C_D_0	0.007			m.a.c.	21.5	ft
C_L_α	0.044	1/deg				

Load Summary					
Load Type	Magnitude	y/(b/2)_start	y/(b/2)_end	dw	
L (unflap)	88817	0	1		
ΔL (flap)	61034	0	0.4	6781.56	
fuel	20000	0	0.4	2222.22	
engine (1)	3250	0.3	0.3	3250	
engine (2)	3250	0.1	0.1	3250	
structure	2335	0	1		

Load Spanwise Distribution

y(ft)	y/(b/2)	L(y)-ellip	L(y)-trap	L-bar	V-lift (lb)	M-lift (f-lb)	ΔL-flap	ΔV-flap
0.00	0.00	3510.00	5513.50	4511.75	57275.68	380304.43	6781.56	61034.00
0.81	0.05	3505.61	5237.83	4371.72	52763.93	334171.72	6781.56	54252.44
1.61	0.10	3492.41	4962.15	4227.28	48392.21	291673.00	6781.56	47470.89
2.42	0.15	3470.29	4686.48	4078.38	44164.93	252695.48	6781.56	40689.33
3.22	0.20	3439.09	4410.80	3924.94	40086.55	217122.83	6781.56	33907.78
4.03	0.25	3398.55	4135.13	3766.84	36161.61	184835.11	6781.56	27126.22
4.83	0.30	3348.33	3859.45	3603.89	32394.77	155708.74	6781.56	20344.67
5.64	0.35	3287.99	3583.78	3435.88	28790.88	129616.36	6781.56	13563.11
6.44	0.40	3216.97	3308.10	3262.54	25355.00	106426.74	6781.56	6781.56
7.25	0.45	3134.53	3032.43	3083.48	22092.46	86004.56	0.00	0.00
8.05	0.50	3039.75	2756.75	2898.25	19008.98	68210.18	0.00	0.00
8.86	0.55	2931.43	2481.08	2706.25	16110.73	52899.40	0.00	0.00
9.67	0.60	2808.00	2205.40	2506.70	13404.48	39923.01	0.00	0.00
10.47	0.65	2667.37	1929.73	2298.55	10897.77	29126.37	0.00	0.00
11.28	0.70	2506.64	1654.05	2080.35	8599.23	20348.75	0.00	0.00
12.08	0.75	2321.65	1378.38	1850.01	6518.88	13422.50	0.00	0.00
12.89	0.80	2106.00	1102.70	1604.35	4668.87	8171.87	0.00	0.00
13.69	0.85	1849.01	827.03	1338.02	3064.52	4411.33	0.00	0.00
14.50	0.90	1529.98	551.35	1040.66	1726.50	1943.02	0.00	0.00
15.30	0.95	1096.00	275.68	685.84	685.84	552.41	0.00	0.00
16.11	1.00	0.00	0.00	0.00	0.00	0.00	0.00	0.00
		56659.61	**57891.76**	**57275.68**			**61034.00**	

FIGURE 10.17: Wing structure shear and moment spreadsheet (Part 1).

the sum gives the total lift. Differences between the spanwise sum and the original total lift are due to the discretizing of the wing span into a finite number of elements. This difference would decrease as the number of elements increases.

A plot of the three lift distributions is shown below the columns. Series 1 refers to the elliptic distribution. In this case, the lift at the wing tip is always zero. Series 2 refers

ΔM-flap	W-fuel	V-fuel	M-fuel	W-structure	V-structure	M-structure	W-engine	V-engine
245799.23	−2222.22	**−20000.00**	**−80545.02**	−144.95	**−1521.97**	**−8989.75**	0	**−6500.00**
196639.39	−2222.22	−17777.78	−64436.01	−137.70	−1377.02	−7763.88	0	−6500.00
152941.74	−2222.22	−15555.56	−50116.90	−130.45	−1239.32	−6654.75	−3250	−6500.00
114706.31	−2222.22	−13333.33	−37587.68	−123.21	−1108.87	−5656.54	0	−3250.00
81933.08	−2222.22	−11111.11	−26848.34	−115.96	−985.66	−4763.40	0	−3250.00
54622.05	−2222.22	−8888.89	−17898.89	−108.71	−869.70	−3969.50	0	−3250.00
32773.23	−2222.22	−6666.67	−10739.34	−101.46	−760.99	−3269.00	−3250	−3250.00
16386.62	−2222.22	−4444.44	−5369.67	−94.22	−659.52	−2656.06	0	0
5462.21	−2222.22	−2222.22	−1789.89	−86.97	−565.30	−2124.85	0	0
0.00	0.00	0.00	0.00	−79.72	−478.33	−1669.53	0	0
0.00	0.00	0.00	0.00	−72.47	−398.61	−1284.25	0	0
0.00	0.00	0.00	0.00	−65.23	−326.14	−963.19	0	0
0.00	0.00	0.00	0.00	−57.98	−260.91	−700.50	0	0
0.00	0.00	0.00	0.00	−50.73	−202.93	−490.35	0	0
0.00	0.00	0.00	0.00	−43.48	−152.20	−326.90	0	0
0.00	0.00	0.00	0.00	−36.24	−108.71	−204.31	0	0
0.00	0.00	0.00	0.00	−28.99	−72.47	−116.75	0	0
0.00	0.00	0.00	0.00	−21.74	−43.48	−58.37	0	0
0.00	0.00	0.00	0.00	−14.49	−21.74	−23.35	0	0
0.00	0.00	0.00	0.00	−7.25	−7.25	−5.84	0	0
0.00	0.00	0.00	0.00	0.00	0.00	0.00	0	0
	−20000.00			**−1514.73**				

FIGURE 10.18: Wing structure shear and moment spreadsheet (Part 2).

to the spanwise distribution based on the trapezoidal planform. Series 3 is the average of the two, which then is the estimated 3-D spanwise lift distribution.

The next two columns in this section of the spreadsheet (F & G), compute the spanwise distribution of the shear force and bending moment due to the 3-D spanwise lift distribution. The method for determining these follows Eqs. [10.37 & 10.38]. In the

M-engine	Total V	Total M
−26177.13	**90287.71**	**510391.76**
−20941.70	81361.57	437669.51
−15706.28	72568.22	372136.82
−10470.85	67162.07	313686.73
−7853.14	58647.56	259591.03
−5235.43	50279.24	212353.34
−2617.71	42061.78	171855.92
0.00	37250.02	137977.25
0.00	29349.02	107974.21
0.00	21614.12	84335.03
0.00	18610.37	66925.93
0.00	15784.59	51936.21
0.00	13143.57	39222.51
0.00	10694.84	28636.02
0.00	8447.03	20021.85
0.00	6410.17	13218.19
0.00	4596.39	8055.12
0.00	3021.03	4352.96
0.00	1704.76	1919.67
0.00	678.59	546.57
0	0.00	0.00

FIGURE 10.19: Wing structure shear and moment spreadsheet (Part 3).

case of the wing, the maximum values should occur at the wing root. These have been shown in bold font to highlight them. As a check, the maximum shear force should equal the total spanwise lift.

The shear and moment distributions (columns F & G) based on the spanwise lift distributions alone are also plotted in the bottom part of the spreadsheet. Series 1 is

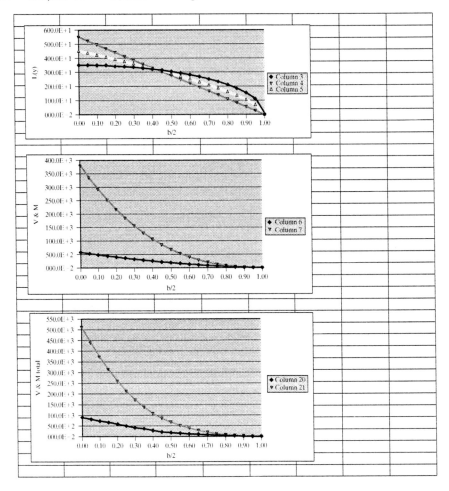

FIGURE 10.20: Wing structure shear and moment spreadsheet (Part 4).

the shear force, with units of pounds. Series 2 is the bending moment, with units of foot-pounds.

The next three columns (H, I & J) deal with the spanwise loads produced by the flapped portion of the wing. For these, the elemental force, dw, due to the incremental lift was placed into column H in the appropriate $y/(b/2)$ cells that correspond to the elements in the flapped portion of the wing. The incremental lift in the elements in the unflapped portion of the wing is zero. The spanwise distributions of the shear force and bending moments due to the incremental lift are then automatically calculated in columns I and J, respectively.

Columns K, L, and M refer to the loads produced by the weight of fuel stored in the wing. The fuel is assumed to produce a distributed load, with an elemental force, $-dw$, which is applied to the appropriate $y/(b/2)$ cells in column K. Note that the force vector is opposite to the lift vector so that the sign on the force is negative. The force distribution in column K is then integrated to obtain the shear force and moment distributions that are listed in columns L and M, respectively.

Columns N, O, and P next deal with the weight of the wing structure. The wing weight is assumed to be distributed in proportion to the planform shape, namely,

$$W_{\text{wing}}(y) = \frac{2L}{b(1+\lambda)}\left[1 - \frac{2y}{b}(1-\lambda)\right]. \tag{10.73}$$

The force distribution that is computed in column N is then integrated to obtain the shear force and moment distributions listed in columns O and P, respectively.

The force due to the weight of the engine is assumed to be concentrated in a single spanwise element. All of the other elements have zero force due to the weight of the engine. This appears in column Q in this portion of the spreadsheet. This force distribution is integrated to obtain the shear force and moment in columns R and S, respectively.

Finally, the total shear force and moment distributions are obtained as the sum of the shear and moment distributions from the respective force distributions. These are plotted at the bottom of the spreadsheet. Series 1 is the shear force, with units of pounds. Series 2 is the bending moment, with units of foot-pounds. The maximum values, at the wing root, are then used in the design of the internal structure.

10.4.3 Fuselage Load Distribution

The spreadsheet for the analysis of the load distribution on the fuselage is referenced as "fuselage" in the **loads.xls** file. Figures 10.21 though 10.24 show this spreadsheet for the conditions of the SSBJ proposed in Chapter 1. The layout of the spreadsheet is similar to the one for the load analysis on the wing. The required input information is entered at the top of the spreadsheet in the sections labeled "Fuselage Data," "Load Summary," and "Wing Center of Lift."

The only fuselage datum that is necessary is the length of the fuselage, L. The wing center of lift is the dimensionless location (x/L) where the center of lift acts on the fuselage. This is the point of balance of the fuselage about which the static margin is measured.

Other data that are required as input are the magnitudes and types of loads that are acting on the fuselage. These are input in the "Load Summary." This would include the weight of fuel stored in the fuselage, payload, engines, and the weight of the wing and tail structures (excluding engines). In specifying the weights, the spanwise region over which they act also needs to be indicated. This is specified by the locations x/L_{start} and x/L_{end}. For a concentrated load, the start and end locations will be the same. As with the wing, the fuselage is divided into 20 elements. This gives the load per element, dw, which is calculated in the load summary. For a concentrated load, dw equals the total load value, which acts on the element that corresponds to that location.

It is assumed that the fuselage does not generate lift. If this is not the case, a lift load distribution could be easily added to the spreadsheet.

Computed in the load summary is the location of the resultant of each of the distributed loads. The resultant force and the location where it acts are used to compute the moment about the center of lift location. These are listed in the column labeled "$M@C_{\text{lift}}$," with a positive moment corresponding to a nose-up (clockwise) motion.

			Fuselage Structure Analysis					

Fuselage Data

Fuse Length	126	ft						

Load Summary (fuselage)

Load Type	Magnitude (lbs)	x/L_start	x/L_end	resultant x/L	M @ C_lift f-lb (+ cw)	dw		
Fuel	40000	0.4	0.7	0.55	−2000	5714.29		
Payload	5000	0.1	0.6	0.35	−1250	454.55		
Structure	11000	0	1	0.5	−1100	523.81		
Engine(s)	13000	0.6	0.6	0.6	0	13000		
Wing Struct.	4600	0.4	0.8	0.6	0	511.11		
Tail Struct.	4800	0.8	1	0.9	1440	960		
				M Sum	**−2910**			
Tail Lift (req)	−9700	0.9	0.9	0.9	−2910	−9700		

Wing Center of Lift

L_ctr (x/L)	0.6							

x (ft)	x/L	W-fuel	V-fuel	M-fuel	W-payload	V-payload	M-payload	W-struct
0.00	**0.00**	0.00	0.00	2.52E+05	0.00	0.00	1.58E+05	−523.81
6.30	**0.05**	0.00	0.00	2.52E+05	0.00	0.00	1.58E+05	−523.81
12.60	**0.10**	0.00	0.00	2.52E+05	−454.55	0.00	1.58E+05	−523.81
18.90	**0.15**	0.00	0.00	2.52E+05	−454.55	454.55	1.58E+05	−523.81
25.20	**0.20**	0.00	0.00	2.52E+05	−454.55	909.09	1.55E+05	−523.81
31.50	**0.25**	0.00	0.00	2.52E+05	−454.55	1363.64	1.49E+05	−523.81
37.80	**0.30**	0.00	0.00	2.52E+05	−454.55	1818.18	1.40E+05	−523.81
44.10	**0.35**	0.00	0.00	2.52E+05	−454.55	2272.73	1.29E+05	−523.81
50.40	**0.40**	−5714.29	0.00	2.52E+05	−454.55	2727.27	1.15E+05	−523.81
56.70	**0.45**	−5714.29	5714.29	2.52E+05	−454.55	3181.82	9.74E+04	−523.81
63.00	**0.50**	−5714.29	11428.57	2.16E+05	−454.55	3636.36	7.73E+04	−523.81
69.30	**0.55**	−5714.29	17142.86	1.44E+05	−454.55	4090.91	5.44E+04	−523.81
75.60	**0.60**	−5714.29	**22857.14**	3.60E+04	−454.55	**4545.45**	2.86E+04	−523.81
81.90	**0.65**	−5714.29	−11428.57	−1.08E+05	0.00	0.00	0.00E+00	−523.81
88.20	**0.70**	−5714.29	−5714.29	−3.60E+04	0.00	0.00	0.00E+00	−523.81
94.50	**0.75**	0.00	0.00	0.00E+00	0.00	0.00	0.00E+00	−523.81
100.80	**0.80**	0.00	0.00	0.00E+00	0.00	0.00	0.00E+00	−523.81
107.10	**0.85**	0.00	0.00	0.00E+00	0.00	0.00	0.00E+00	−523.81
113.40	**0.90**	0.00	0.00	0.00E+00	0.00	0.00	0.00E+00	−523.81
119.70	**0.95**	0.00	0.00	0.00E+00	0.00	0.00	0.00E+00	−523.81
126.00	**1.00**	0.00	0.00	0.00E+00	0.00	0.00	0.00E+00	−523.81
		−40000.00			−5000.00			−11000.00

FIGURE 10.21: Fuselage structure shear and moment spreadsheet (Part 1).

As discussed in Section 10.1.4, static stability to pitch-up requires that there be a negative restoring moment. The *sum* of the moments produced by the forces on the fuselage are computed in the cell next to the label "M sum." For level flight, the horizontal tail has to offset the negative moment required for stability. Therefore, the downward lift force that is needed to produce the offset moment is computed in the row titled "Tail

Lift (req)." The tail lift force corresponds to the needed restoring moment divided by the distance from where the resultant force acts, to the location of the center of lift. This value appears in the magnitude column of the load summary. This force should always be negative for a statically stable aircraft with an aft tail design.

The static margin (SM) is computed in the spreadsheet to provide an indication of the static stability to pitch-up. This follows the definition given by Eq. [10.39]. For this, the center of gravity is computed by summing the moments about the leading point of the fuselage ($x = 0$) and dividing by the sum of the forces. The static margin is required to be positive for static stability. The stability condition is indicated as "stable" or "unstable" in the cell next to the static margin value.

Once the values and locations of forces have been resolved, the lower part of the spreadsheet determines the shear and moment distributions. The arrangement for these is similar to that used for the wing. The fuselage is divided into 20 elements. These are specified by the x/L locations listed in the second column (B) of this portion of the spreadsheet. The dimensional location of the elements are computed and listed in the first column (A).

Columns C, D, and E refer to the loads produced by the weight of fuel stored in the fuselage. The fuel is assumed to produce a distributed load, with an elemental force, $-dw$, which is applied to the appropriate x/L cells in column C. Note that the force vector is opposite to the wing lift vector so that the sign on the force is negative. The force distribution in column C is integrated to obtain the shear force following Eq. [10.40]. Note that the concentrated reaction load that occurs at the point of support was included in the formula for the shear in the cell corresponding to $x/L = x_{CL}/L$. This location is shown in a box in this portion of the spreadsheet. The moment distribution is found by integrating the shear force distribution. This follows the method given in Eq. [10.41].

The same procedure is used in calculating the shear force and bending moments for the other weights including the payload, fuselage structure, tail structure, and engines. The downward lift of the horizontal tail is modeled as a concentrated force, and the subsequent shear and moment distributions are calculated in columns V and W.

The total shear and moment distributions are presented in columns X and Y. These are the element sums of the shear and moment distributions for the respective weights and forces. These are plotted in the bottom part of the spreadsheet, where Series 1 refers to the shear force with units of pounds and Series 2 refers to the bending moment with units of foot-pounds. As expected, the maximum shear and bending occurs at the point of support at the location of the center of lift, $x/L = x_{CL}/L$.

10.5 CASE STUDY: STRUCTURAL ANALYSIS

10.5.1 Load Factor

We continue the case study corresponding to a conceptual supersonic business jet (SSBJ) that was proposed in Chapter 1. The spreadsheets that were shown in Figures 10.15 and 10.16 corresponded to the conditions for the conceptual SSBJ. These were used to determine the design load factor for the aircraft.

The conditions for evaluating the loads due to acceleration are representative of the acceleration to the cruise Mach number following takeoff. This gave a load factor of 1.71

Pitching Stability		
x_cg	70.92	f
x_cg/L	0.56	
S.M.	0.22	**stable**

V-struct	M-struct	W-engine	V-engine	M-engine	W-wing	V-wing	M-wing	W-tail
0.00	1.42E+05	0	0.00	0.00E+00	0	0.00	2.50E-11	0
523.81	1.42E+05	0	0.00	0.00E+00	0	0.00	2.50E-11	0
1047.62	1.39E+05	0	0.00	0.00E+00	0	0.00	2.50E-11	0
1571.43	1.32E+05	0	0.00	0.00E+00	0	0.00	2.50E-11	0
2095.24	1.22E+05	0	0.00	0.00E+00	0	0.00	2.50E-11	0
2619.05	1.09E+05	0	0.00	0.00E+00	0	0.00	2.50E-11	0
3142.86	9.24E+04	0	0.00	0.00E+00	0	0.00	2.50E-11	0
3666.67	7.26E+04	0	0.00	0.00E+00	0	0.00	2.50E-11	0
4190.48	4.95E+04	0	0.00	0.00E+00	−511.11	0.00	2.50E-11	0
4714.29	2.31E+04	0	0.00	0.00E+00	−511.11	511.11	2.50E-11	0
5238.10	−6.60E+03	0	0.00	0.00E+00	−511.11	1022.22	−3.22E+03	0
5761.90	−3.96E+04	0	0.00	0.00E+00	−511.11	1533.33	−9.66E+03	0
6285.71	−7.59E+04	−13000	0.00	0.00E+00	−511.11	**2044.44**	−1.93E+04	0
−4190.48	−1.16E+05	0	0.00	0.00E+00	−511.11	−2044.44	−3.22E+04	0
−3666.67	−8.91E+04	0	0.00	0.00E+00	−511.11	−1533.33	−1.93E+04	0
−3142.86	−6.60E+04	0	0.00	0.00E+00	−511.11	−1022.22	−9.66E+03	0
−2619.05	−4.62E+04	0	0.00	0.00E+00	−511.11	−511.11	−3.22E+03	−960
−2095.24	−2.97E+04	0	0.00	0.00E+00	0	0.00	0.00E+00	−960
−1571.43	−1.65E+04	0	0.00	0.00E+00	0	0.00	0.00E+00	−960
−1047.62	−6.60E+03	0	0.00	0.00E+00	0	0.00	0.00E+00	−960
−523.81	0.00E+00	0.00	0.00	000.0E−2	0	0	0	−960
		−13000			−4600			−4800

FIGURE 10.22: Fuselage structure shear and moment spreadsheet (Part 2).

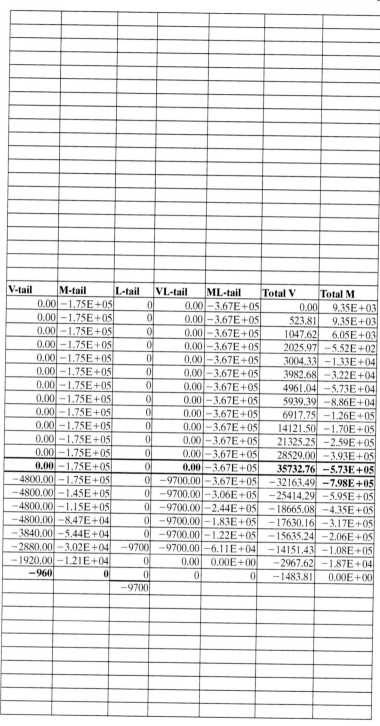

V-tail	M-tail	L-tail	VL-tail	ML-tail	Total V	Total M
0.00	−1.75E+05	0	0.00	−3.67E+05	0.00	9.35E+03
0.00	−1.75E+05	0	0.00	−3.67E+05	523.81	9.35E+03
0.00	−1.75E+05	0	0.00	−3.67E+05	1047.62	6.05E+03
0.00	−1.75E+05	0	0.00	−3.67E+05	2025.97	−5.52E+02
0.00	−1.75E+05	0	0.00	−3.67E+05	3004.33	−1.33E+04
0.00	−1.75E+05	0	0.00	−3.67E+05	3982.68	−3.22E+04
0.00	−1.75E+05	0	0.00	−3.67E+05	4961.04	−5.73E+04
0.00	−1.75E+05	0	0.00	−3.67E+05	5939.39	−8.86E+04
0.00	−1.75E+05	0	0.00	−3.67E+05	6917.75	−1.26E+05
0.00	−1.75E+05	0	0.00	−3.67E+05	14121.50	−1.70E+05
0.00	−1.75E+05	0	0.00	−3.67E+05	21325.25	−2.59E+05
0.00	−1.75E+05	0	0.00	−3.67E+05	28529.00	−3.93E+05
0.00	−1.75E+05	0	**0.00**	−3.67E+05	**35732.76**	**−5.73E+05**
−4800.00	−1.75E+05	0	−9700.00	−3.67E+05	−32163.49	**−7.98E+05**
−4800.00	−1.45E+05	0	−9700.00	−3.06E+05	−25414.29	−5.95E+05
−4800.00	−1.15E+05	0	−9700.00	−2.44E+05	−18665.08	−4.35E+05
−4800.00	−8.47E+04	0	−9700.00	−1.83E+05	−17630.16	−3.17E+05
−3840.00	−5.44E+04	0	−9700.00	−1.22E+05	−15635.24	−2.06E+05
−2880.00	−3.02E+04	−9700	−9700.00	−6.11E+04	−14151.43	−1.08E+05
−1920.00	−1.21E+04	0	0.00	0.00E+00	−2967.62	−1.87E+04
−960	**0**	0	0	0	−1483.81	0.00E+00
		−9700				

FIGURE 10.23: Fuselage structure shear and moment spreadsheet (Part 3).

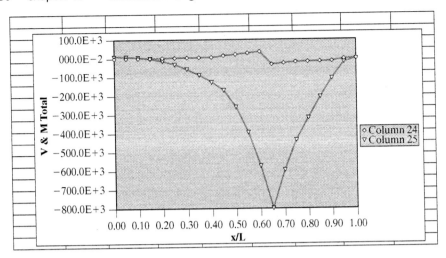

FIGURE 10.24: Fuselage structure shear and moment spreadsheet (Part 4).

The load factors corresponding to the sustained and instantaneous turn rates were based on conditions that might be representative of maneuvers after take-off or prior to landing. These are at Mach 0.8 and an altitude of 25,000 feet. The values of the turn rates were taken from the wing loading analysis, so that they were consistent with that part if the design. For these conditions, the sustained turn gave a load factor of 1.2. The instantaneous turn gave a slightly higher value of 1.22.

The conditions for maximum angle-of-attack flight were based on the same altitude as for the turn rate load factor. The lift coefficient was taken to be 80 percent of the maximum lift coefficient, without flaps. The Mach number was set to be 1, which corresponded to a velocity of 1019 f/s. These conditions gave a load factor of 3.04. As was illustrated in Figure 10.5, this high-α condition corresponds to the upper limit of positive load factors used in the design.

The conditions for the maximum dynamic pressure corresponded to a dive velocity of 1.5 times the cruise velocity. This was taken at the cruise altitude of 55,000 feet. The velocity in this case is 2888 ft/s. Based on Figure 10.5, the load factor for diving conditions would not exceed the high-α condition. Therefore, the lift coefficient was varied until the load factor was 3.04. The lift coefficient value that gave this load factor was 0.40.

The load factor due to gusts was determined for a number of conditions. The value shown in the spreadsheet corresponds to the cruise Mach number and altitude and the wing loading at the start of cruise. The wing C_{L_α} was based on the 3-D wing design performed in the **wing.xls** spreadsheet. The value of $\hat{U}$ was taken from Table 10.3 and Figure 10.4, for cruise at 50,000 foot altitude. The mass ratio, μ, was based on Eq. [10.16], and the response coefficient, K, came from Eq. [10.17b] for a supersonic Mach number. The choice of the equation in the spreadsheet is automatic based on Mach number being greater than or less than one. The mass ratio and response coefficient then combined according to Eq. [10.18] to give a fluctuation velocity of 24.7 f/s. This resulted in a dynamic load factor of $\Delta n = 0.11$ and a peak load factor at cruise of $n_{\text{peak}} = 1.11$. This is a relatively low value as a result of the relatively high wing loading.

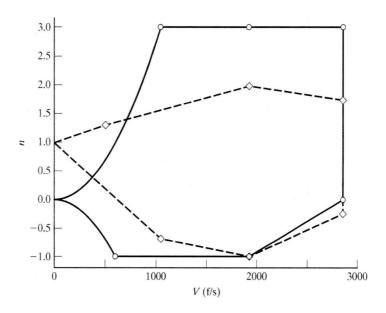

Solid: Maneuver Loads
Dashed: Gust Loads

FIGURE 10.25: Maneuvering and gust load factor envelope for conceptual SSBJ.

A plot of the load factor envelope for the conceptual SSBJ for maneuvering flight and gusts is shown in Figure 10.25. Based on Table 10.1, for an aircraft of this type, the load factor should fall in the range $-1.25 \le n \le 3.1$. The range for this design is $-1.0 \le n \le 3.04$. Using a load limit of 3.04, and the standard factor of safety of 1.5, the design load factor is 4.57.

10.5.2 Wing Load Distribution

The spreadsheets that were shown in Figures 10.17 through 10.20 correspond to the wing loading conditions of the conceptual SSBJ. The wing data used in determining the spanwise load distribution primarily come from the 3-D wing design spreadsheet **wing.xls**. For this, cruise conditions were used. This corresponds to $n = 1$, so that the resulting maximum loads obtained in this analysis can simply be multiplied by the design load factor to give the maximum design loads for the structure.

The total lift (unflapped) is based on the weight at the start of cruise (88,817 pounds). The incremental lift due to full flaps is based on 80 percent of the maximum lift coefficient of the 3-D flapped wing. The flapped portion of the wing corresponds to 60 percent of the wing area ($S_f/S_w = 0.6$). These values came from the spreadsheet for the flap design, **flaps.xls**. The total weight of fuel at take-off is approximately 40,000 lbs. It is estimated that approximately half of that would be stored in each half span of the main wing.

From the engine design spreadsheet, **engine.xls**, the scaled weight of an uninstalled engine was found to be approximately 2500 lbs. The installed weight was estimated using the factor of 1.3 from Table 10.4. This gave an installed engine weight of 3250 pounds. Two engines will be mounted under each wing.

The weight of the wing structure was estimated using Eq. [10.30] and Table 10.4. Because of the Mach number similarities, the factor for the wing weight was chosen to be that for a combat aircraft. This gave a total wing weight estimate of 4671 lbs. This value was divided by two to give the weight of one-half of the wing.

The spanwise lift distribution is found using Schrenk's approximation of an average between an elliptic distribution and one which varies in span according to the planform trapezoidal shape. Strictly speaking, this approximation is not accurate enough for wings that are as highly swept as those in this design. As such, it probably overestimates the total lift, which is at least conservative from the standpoint of the structure load estimates.

Given this, the plot of the spanwise lift distribution in Figure 10.20 shows a smooth decrease in the lift from the maximum value at the root, to zero-lift at the tip. The latter is due to having a zero taper ratio for the main wing planform. The total lift that is summed over each of the elements exceeds half of the total required lift for the half-wing area by about 40 percent. This is due to the discretization of the wing into constant lift elements. If more elements than the 20 were used, the difference would be smaller. As such, it gives a more conservative (larger) estimate of the loads, which is sufficient for a conceptual design.

The spanwise distributions of the shear force and moments accounting for all the forces on the wing showed a relatively smooth decay from a maximum at the root, to zero at the wing tip. The maximum shear force on the wing is estimated to be 90,280 pounds. The maximum bending moment is estimated at approximately 510,391 f-lbs. These values will be used in the design of the main wing spar that is the primary structural element in the wing.

10.5.3 Fuselage Load Distribution

The spreadsheets that were shown in Figures 10.21 through 10.24 correspond to the loading conditions on the fuselage of the conceptual SSBJ. The fuselage length is 126 feet, and the point of support (center of lift) is estimated to be at $x/L = 0.6$.

The nonexpendable payload is 5000 pounds, all of which is carried inside the fuselage. The structure weight of the fuselage is estimated to be approximately 11,000 pounds based on Table 10.4 for a fighter aircraft. The fuselage wetted area is 2284 f^2, which came from the fuselage design spreadsheet, **fuse.xls**.

The weight of the tail structure is also estimated from Table 10.4. This was taken to consist of the *sum* of the weights of the vertical tail (2750 pounds) and the horizontal tail (2076 pounds), although in one of the concepts, the horizontal tail is a forward canard and, therefore, will not be placed at the same location as the vertical tail.

In the design, the engines are mounted on the wing. Their weight and that of the fuel stored in the wing, and the weight of the wing structure, are exerted on the fuselage along the point of the wing attachment. The respective weights of these components are then included in the load summary. From the point of determining the static margin, their magnitudes and locations are listed separately.

In viewing the component weights, the largest of these correspond to the fuel and engines. Therefore, care is needed to locate these so that the aircraft is statically stable (positive static margin) and that the downward lift force on the horizontal tail, which is needed to offset the pitch-down moment, is not excessively large. For example, the engines were located so that they were at the location of the center of lift. Having too large of a downward lift on the horizontal tail requires greater lift from the main wing and, therefore, results in higher lift-induced drag. In the arrangement in the spreadsheet, the static margin is relatively low, but positive. The force exerted by the tail is 9700 pounds. As a check, as the fuel weight is decreased (even to zero), the static margin remains positive.

Although an aft tail is shown in the spreadsheet, this type of aircraft shows the utility of a forward (canard) horizontal tail. The forward position moves the weight of the horizontal tail forward of the center of gravity. In addition, the lift needed to offset the pitch-down moment is then positive and, therefore, augments the lift of the main wing. As a result, the lift-induced drag is reduced.

The fuselage was divided into 20 elements for determining the shear force and moment distributions. The center of lift was estimated to be at $x/L = 0.6$. This location is boxed in the columns of load distributions. The point and distributed load values were then input into the respective columns in the x/L elements in which the nonzero weights or forces acted. It was also important to include the reaction force in the formula for the shear force in the cell corresponding to the center of lift. For accuracy, the reaction force is the sum of the forces in all the cells, which is indicated at the bottom of each weight column.

The shear and moment distributions for all of the weights and forces were summed in the last two columns. These are plotted at the bottom of the spreadsheet. As expected, the maximum shear and moment occur at the point of support, which is the center of lift. The maximum shear force is 35,732 pounds. The maximum bending moment is −798,000 f-lbs. These values will be used in the design of the fuselage skin, longerons, and bulkheads, which will be the primary structural elements in the fuselage.

Using the maximum bending moment, Eq. [10.47] is used to estimate the minimum thickness of the fuselage skin. One of the possible selected material is 2024-T3 clad aluminum, with $\sigma_{t_u} = 58 \times 10^3$ lbs/in.^{2} (Table 10.7). The diameter of the fuselage at the location of the maximum moment is 9.0 feet. The design load factor is 4.57 based on the previous analysis. Substituting these into Eq. [10.47] gives $t_{min} = 0.163$ in.

We also can examine how the skin thickness changes with other materials. Table 10.10 lists values of t_{min} based on the 2024-T3 aluminum, a different aluminum alloy, and a stainless steel alloy. The 7178-T6 aluminum alloy has a higher ultimate tensile stress (Table 10.8); therefore, t_{min} is reduced to 0.114 in. The third column of

TABLE 10.10: Effect of different materials on the minimum fuselage skin thickness for the case study aircraft.

Material	t_{min} (in.)	W/W_{2024}
2024-T3	0.163	1
7178-T6	0.114	0.70
301-Stainless	0.051	0.90

Table 10.10 lists the weight fraction using the 2024-T3 aluminum as a reference. The two aluminum alloys have the same density, so that by virtue of the smaller thickness required, the weight is reduced, $W/W_{2024} = 0.70$. The 301-stainless steel alloy has a significantly higher tensile stress than either of the aluminum alloys. As a result, the skin thickness is substantially reduced, $t_{min} = 0.051$ in. However, the density is larger than that of aluminum. When compared to the 2024 aluminum, the reduced thickness still gives a 10 percent weight reduction. However, compared to the 7178 aluminum, 2024 aluminum is about 20 percent heavier. Therefore, based on weight, 7178 aluminum is a better choice.

The maximum spacing between bulkheads can be determined from Eq. [10.60]. Recall that this was based on having longerons azimuthally spaced 22.5° apart around the fuselage. If we assume that the longerons are made from thin-walled tubing, with radius R_l and thickness t_l, then $I_l = (\pi/2)R_l^3 t_l$. Substituting this into Eq. [10.60] gives

$$\frac{2L^2}{\pi R R_l^3 t_l E} < \frac{C\pi^2}{0.154 n_{design} M}. \tag{10.74}$$

Solving for the maximum bulkhead spacing, L_{max}, this gives

$$L_{max} = \sqrt{\frac{C\pi^3 R R_l^3 t_l E}{0.308 n_{design} M}}. \tag{10.75}$$

A relatively standard seamless hollow tubing, which comes in a large variety of diameters and wall thicknesses, is made from 6061-T6 aluminum alloy. Using a standard 2-inch diameter, with a wall thickness of 0.035 inches, we find, that $L_{max} = 0.52$ feet. This value is based on an end coefficient of $C = 1$. Increasing the tubing wall thickness to 0.083, increases L_{max} to 0.80 feet. This then corresponds to the maximum spacing between bulkheads that is allowable based on the buckling failure criterion.

The next chapter deals with a more detailed analysis of the static stability, including the roll and yaw directions. These will lead to criteria for sizing the control surfaces, including the vertical tail rudder.

10.6   PROBLEMS

10.1. Using the conditions of the conceptual supersonic business jet that are given in the spreadsheets in Figures 10.8 through 10.10, determine the largest spacing of bulkheads if the cross-section shape of the longerons is U-channel, with a material thickness of 0.083 in. and a channel height of 2 in. Use 6061-T6 aluminum alloy as the material. Other than structural, can you think of any advantages of using a U-channel over a thin-walled tubing for the longerons?

10.2. On the basis of minimum weight, how does titanium 6AL-4V compare to aluminum alloys for tension, compression, and bending loads? For what components of an aircraft might it be well suited?

10.3. List all the characteristics of an aircraft that affect its response to turbulent gust loads. What type of a design would minimize this effect, and how might this change other performance characteristics, such as range, maneuverability, take-off and landing, etc.?

10.4. How would wing sweep affect an aircraft's response to turbulent gust loads? Does larger sweep make it better or worse? Explain.

10.5. How might the minimum thickness for a fuselage skin vary for different materials such as titanium 6AL-4V, magnesium HK 31A, and S-fiberglass–epoxy? What might be the advantages and disadvantages of each?

10.6. How might the maximum distance between bulkheads needed to prevent buckling in longerons around a fuselage vary for different materials such as titanium 6AL-4V, magnesium HK 31A, S-fiberglass–epoxy, and boron–epoxy? What might be the advantages and disadvantages of each?

10.7. Design a minimum weight I-beam wing spar to withstand the maximum bending moment of 500,000 f-lbs in the conceptual supersonic business jet. The wing t/c_{max} is 0.04 and the root chord length is $c_r = 40.8$ feet. Include all the beam dimensions and the material specifications.

10.8. Find the weight ratio of a honeycomb sandwich beam to that of a solid beam whose material is the same as that of the sandwich beam face. The sandwich core density is 0.015 lb/in.^{3}. Find the weight ratio for the different face materials listed;

   1. 2024-T3 aluminum;
   2. 6AL-4V titanium;
   3. Inconel 600;
   4. S-fiberglass;
   5. boron–epoxy;
   6. graphite–epoxy.

10.9. For the case study aircraft, plot how the static margin changes with the fuel weight ranging from 40,000 pounds to 0 pounds. Repeat this for the conditions given in Table 10.5. Discuss the differences between the two cases.

10.10. Using the spreadsheet, generate tables and plots that show how the maximum bending moment on the wing changes with aspect ratio for three different taper ratios of 0, 0.5, and 1. From this, how do you think the wing weight would vary with aspect ratio in these three cases?

CHAPTER 11

Static Stability and Control

Photograph of a Northrop YB-49 Flying Wing. This aircraft was designed on a true flying wing principle, with no tail surfaces. It used "elevon" controls and was the first aircraft to incorporate artificial stability augmentation. (U.S. Air Force Museum Archives)

In the previous chapter, it was necessary to consider the pitching stability in order to estimate the force on the fuselage due to the horizontal tail lift, which counters the pitch-down moment. For this, rough estimates were made of the major weight components.

This chapter examines a more complete analysis of the static stability, which includes all three directions of motion consisting of

1. pitch;
2. roll;
3. yaw.

The application of these will lead to a method for sizing control surfaces. In addition, a more extensive set of empirical relations are presented to obtain improved estimates of the component weights. The locations of these components will provide a final estimate of the location of the center of gravity. In addition, the sum of these weights will provide a final check on the original structure weight that was presumed in the initial estimate of the weight at take-off.

11.1 REFINED WEIGHT ESTIMATE

The refined weight estimates are based on formulas that relate different characteristics of aircraft to their component weights. These formulas involve coefficients found by a minimum error fit to a large set of aircraft. In order to improve the result of the fits, the total set has been subdivided into smaller sets based on general mission requirements. These correspond to categories of aircraft consisting of combat/fighter, long-range/transport, and general aviation.

The formulas and coefficients are generally trade secrets of aircraft manufacturers because of their value in the conceptual design. However, useful estimates can be obtained based on the formulas published by Staton (1968, 1969) and Jackson (1971). The following are based on these.

The formulas are presented in a general format that involves coefficients whose values are given in subsequent tables. The different rows in the tables correspond to the different categories of aircraft. This approach is a convenient way for utilizing automated spreadsheets for evaluating the formulas for all of the types of aircraft.

11.1.1 Wing Weight

The general formula for the wing weight is

$$W_{\text{wing}} = C_1 C_2 C_3 W_{\text{dg}}^{C_4} n^{C_5} S_w^{C_6} A^{C_7} (t/c)^{C_8} (C_9 + \lambda)^{C_{10}} (\cos \Lambda)^{C_{11}} S_f^{C_{12}} q^{C_{13}} W_{\text{fw}}^{C_{14}} \quad (11.1)$$

with coefficients given in Table 11.1.
The parameters used in Eq. [11.1] correspond to

A being the aspect ratio of the wing;

K_{dw} being a coefficient, which is 0.768 for a delta wing and 1.0 otherwise;

K_{vs} being a coefficient, which is 1.19 for a variable-sweep wing and 1.0 otherwise;

n being the design load factor;

q being the dynamic pressure at cruise conditions (lbs/f^2);

S_w being the planform area of the main wing (f^2);

S_f being the planform area of the flapped portion of the main wing (f^2);

TABLE 11.1: Coefficients for Eq. [11.1].

	C_1	C_2	C_3	C_4	C_5	C_6	C_7	C_8	C_9
Fighter	0.0103	K_{dw}	K_{vs}	0.500	0.500	0.622	0.785	−0.4	1.0
Transport	0.0051	1	1	0.557	0.577	0.649	0.500	−0.4	1.0
Gen. Av.	0.0090	1	1	0.490	0.490	0.758	0.600	−0.3	0

	C_{10}	C_{11}	C_{12}	C_{13}	C_{14}
Fighter	0.050	−1.0	0.04	0	0
Transport	0.100	−1.0	0.10	0	0
Gen. Av.	0.004	−0.9	0	0.006	0.0035

t/c being the maximum thickness-to-chord ratio of the wing;

W_{dg} being the design gross weight (lbs);

W_{fw} being the weight of fuel stored in the wing (lbs);

Λ being the sweep angle of the maximum thickness line (°);

λ being the taper ratio.

11.1.2   Horizontal Tail Weight

The general formula for the weight of the horizontal tail is

$$W_{h-tail} = C_1 \left(1 + \frac{F_w}{b_{ht}}\right)^{C_2} W_{dg}^{C_3} n^{C_4} S_{ht}^{C_5} L_{ht}^{C_6} K_y^{C_7} (\cos \Lambda_{ht})^{C_8} A_{ht}^{C_9} (t/c)^{C_{10}} \lambda_{ht}^{C_{11}} q^{C_{12}},$$

(11.2)

with coefficients given in Table 11.2.
The parameters used in Eq. [11.2] correspond to

A_{ht} being the aspect ratio of the horizontal stabilizer;

b_{ht} being the horizontal stabilizer span dimension (f);

F_w being the fuselage width at the location of the horizontal stabilizer (f);

TABLE 11.2: Coefficients for Eq. [11.2].

	C_1	C_2	C_3	C_4	C_5	C_6	C_7	C_8	C_9
Fighter	0.5503	−2.00	0.260	0.260	0.806	0	0	0	0
Transport	0.0379	−0.25	0.639	0.100	0.750	−1	0.704	−1	0.116
Gen. Av.	0.0092	0	0.414	0.414	0.896	0	0	0.034	0.043

	C_{10}	C_{11}	C_{12}
Fighter	0	0	0
Transport	0	0	0
Gen. Av.	−0.120	−0.020	0.168

K_y being the pitching radius of gyration ($\simeq 0.3 L_t$) (f);

L_{ht} being the distance between the 1/4-m.a.c locations of the main wing and horizontal tail (f);

n being the design load factor;

q being the dynamic pressure at cruise conditions (lbs/f^2);

S_{ht} being the planform area of the horizontal tail (f^2);

t/c being the maximum thickness-to-chord ratio of the horizontal tail;

W_{dg} being the design gross weight (lbs);

Λ_{ht} being the sweep angle of the maximum thickness line (°);

λ_{ht} being the taper ratio.

11.1.3 Vertical Tail Weight

The general formula for the weight of the vertical tail is

$$W_{v-tail} = C_1 K_{rht} \left(1 + C_2 \frac{H_{ht}}{H_{vt}}\right)^{C_3} W_{dg}^{C_4} n^{C_5} S_{vt}^{C_6} M^{C_7} L_{vt}^{C_8} \left(1 + \frac{S_r}{S_{vt}}\right)^{C_9}$$
$$\cdot A_{vt}^{C_{10}} (C_{11} + \lambda_{vt})^{C_{12}} (\cos \Lambda_{vt})^{C_{13}} (t/c)^{C_{14}} K_z^{C_{15}} q^{C_{16}}, \tag{11.3}$$

with coefficients given in Table 11.3.

The parameters used in Eq. [11.3] correspond to

A_{vt} being the aspect ratio of the vertical stabilizer;

H_{ht} being the height of the horizontal stabilizer above the fuselage centerline (0 for a conventional tail) (f);

H_{vt} being the height of the vertical stabilizer above the fuselage centerline (same as vertical tail span) (f);

K_z being the yawing radius of gyration ($\simeq L_{vt}$) (f);

K_{rht} being a coefficient, which is 1.047 for a rolling ("flying") tail and 1.0 otherwise;

L_{vt} being the distance between the 1/4-m.a.c locations of the main wing and vertical tail (f);

TABLE 11.3: Coefficients for Eq. [11.3].

	C_1	C_2	C_3	C_4	C_5	C_6	C_7	C_8	C_9
Fighter	0.4520	1	0.500	0.488	0.488	0.718	0.341	−1	0.348
Transport	0.0026	1	0.225	0.556	0.536	0.500	0	−0.5	0
Gen. Av.	0.0076	0.2	1	0.376	0.376	0.873	0	0	0

	C_{10}	C_{11}	C_{12}	C_{13}	C_{14}	C_{15}	C_{16}
Fighter	0.223	1	0.250	−0.323	0	0	0
Transport	0.350	0	0	−1	−0.50	0.875	0
Gen. Av.	0.357	0	0.039	−0.224	−0.49	0	0.122

M being cruise Mach number;

n being the design load factor;

q being the dynamic pressure at cruise conditions (lbs/f^2);

S_r being the planform area of the vertical tail rudder (f^2);

S_{vt} being the planform area of the vertical tail (f^2);

t/c being the maximum thickness-to-chord ratio of the vertical tail;

W_{dg} being the design gross weight (lbs);

Λ_{vt} being the sweep angle of the maximum thickness line (°);

λ_{vt} being the taper ratio.

11.1.4 Fuselage Weight

The general formula for the weight of the fuselage is

$$W_{fuse} = C_1 C_2 C_3 W_{dg}^{C_4} n^{C_5} L^{C_6} L_t^{C_7} D^{C_8} S_f^{C_9} W^{C_{10}} (1 + K_{ws})^{C_{11}} q^{C_{12}} + C_{13}, \qquad (11.4)$$

with coefficients given in Table 11.4.
The parameters used in Eq. [11.4] correspond to

b_w being the main wing span (f);

D being the structural depth of the fuselage (f);

K_{ws} being defined as $0.75[(1 + 2\lambda)/(1 + \lambda)][(b_w/L)\tan\Lambda]$;

K_{dwf} being equal to 0.774 for a delta wing, 1.0 otherwise;

K_{door} being equal to 1 for no cargo door, 1.06 for one side door, 1.12 for two doors or a single clam-shell door, and 1.25 for two doors and one clam-shell door;

K_{lg} being equal to 1.12 for fuselage mounted landing gear and 1.0 otherwise;

L being the length of the fuselage (f);

L_t being the fuselage tail length defined as the distance between the 1/4-m.a.c locations of the main wing and tail sections (f);

n being the design load factor;

q being the dynamic pressure at cruise conditions (lbs/f^2);

TABLE 11.4: Coefficients for Eq. [11.4].

	C_1	C_2	C_3	C_4	C_5	C_6	C_7	C_8	C_9
Fighter	0.499	K_{dwf}	1	0.35	0.25	0.50	0	0.849	0
Transport	0.328	K_{door}	K_{lg}	0.50	0.50	0.35	0	−0.100	0.302
Gen. Av.	0.052	1	1	0.177	0.177	−0.072	−0.051	0.072	1.086

	C_{10}	C_{11}	C_{12}	C_{13}
Fighter	0.685	0	0	0
Transport	0	0.04	0	0
Gen. Av.	0	0	0.241	W_p

S_f being the fuselage wetted area (f^2);

S_{vt} being the planform area of the vertical tail (f^2);

W being the structural width of the fuselage (f);

W_p being a weight penalty due to having a pressurized fuselage, $W_p = 11.9 + (V_{pr}\Delta P)^{0.271}$ (lbs), with V_{pr} being the volume of the pressurized section (f^3) and ΔP being the pressure difference ($\Delta P \simeq 8$ lb/f^2);

W_{dg} being the design gross weight (lbs);

Λ being the sweep angle of the maximum thickness line of the main wing ($°$);

λ being the taper ratio of the main wing.

11.1.5 Main Landing Gear Weight

The general formula for the weight of the main landing gear is

$$W_{\text{main lg}} = C_1 C_2 C_3 W_l^{C_4} n^{C_5} L_m^{C_6} N_{mw}^{C_7} N_{mss}^{C_8} V_s^{C_9}, \tag{11.5}$$

with coefficients given in Table 11.5.

The parameters used in Eq. [11.5] correspond to

K_{cb} being a coefficient equal to 2.25 for a cross-beam gear (similar to the F-111), or otherwise 1;

K_{mp} being a coefficient equal to 1.126 for a kneeling gear, or otherwise 1;

K_{tpg} being a coefficient equal to 0.826 for a tripod gear (similar to the A-7), or otherwise 1;

L_m being the length of the main landing gear (in.);

n being the design load factor;

N_{mw} being the number of main wheels;

N_{mss} being the number of main gear shock struts;

V_s being the stall velocity (f/s);

W_l being the landing design gross weight (lbs).

11.1.6 Nose Landing Gear Weight

The general formula for the weight of the nose landing gear is

$$W_{\text{nose lg}} = C_1 C_2 W_l^{C_3} n^{C_4} L_n^{C_5} N_{nw}^{C_6}, \tag{11.6}$$

with coefficients given in Table 11.6.

TABLE 11.5: Coefficients for Eq. [11.5].

	C_1	C_2	C_3	C_4	C_5	C_6	C_7	C_8	C_9
Fighter	1	K_{cb}	K_{tpg}	0.250	0.25	0.973	0	0	0
Transport	0.0106	1	K_{mp}	0.888	0.25	0.400	0.321	-0.5	0.1
Gen. Av.	0.0344	1	1	0.768	0.768	0.409	0	0	0

TABLE 11.6: Coefficients for Eq. [11.6].

	C_1	C_2	C_3	C_4	C_5	C_6
Fighter	1	1	0.290	0.290	0.500	0.525
Transport	0.032	K_{np}	0.646	0.200	0.500	0.450
Gen. Av.	1	0.0153	0.566	0.566	0.845	0

TABLE 11.7: Weight multipliers for different groups of aircraft.

Type	Engine W_{inst}/W_{uninst}	W_{remain}/W_{TO}
Fighter	1.3	0.17
Transport	1.3	0.17
Gen. Av.	1.4	0.14

The parameters used in Eq. [11.6] correspond to

K_{np} being a coefficient equal to 1.15 for a kneeling gear, or otherwise 1;

L_n being the length of the nose landing gear (in.);

n being the design load factor;

N_{nw} being the number of nose wheels;

W_l being the landing design gross weight (lbs).

The other significant weight is due to the engine and fuel. As in the previous chapter, the installed engine weight is estimated as the uninstalled weight times a factor. These factors are reproduced in Table 11.7.

The total weights of all the other components of the aircraft are simply estimated as a fixed fraction of the take-off weight. These fractions are also listed in Table 11.7 for the three different groups of aircraft. They range from 14 to 17 percent of the take-off weight. The components making up this weight are distributed all over the aircraft. In order to refine the estimate of the location of the center of gravity, the resultant load of the remaining weight can be assumed to act at the fuselage $x/L = 0.5$.

11.2 STATIC STABILITY

In the design, it is generally important that the aircraft be statically stable in flight. However, excessive stability can have adverse effects on maneuverability and performance. The static stability of the aircraft is presented by considering each of the three directions of motion separately. These are defined in Figure 11.1, along with the sign convention.

11.2.1 Longitudinal (Pitch) Stability

The longitudinal stability is the measure of the response of the aircraft due to a changing pitch angle condition. As discussed in the previous chapter, positive stability in this case

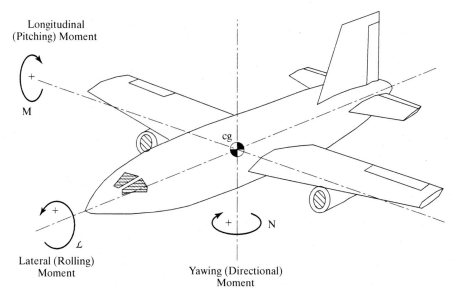

FIGURE 11.1: Illustration showing aircraft body axes and direction sign convention.

means that without control, the nose of the aircraft will pitch down. This is desirable because it prevents a precipitous increase in the angle of attack, which can lead to stall.

Previously, the static margin (SM) was used as an estimate of the longitudinal stability of the aircraft. In this section, the coefficient of the longitudinal stability, C_{M_α}, will be derived. The two are, however, related with

$$(C_{M_\alpha})_W = -(\text{SM})(C_{L_\alpha})_W, \qquad (11.7)$$

where W refers to the main wing and $(C_{L_\alpha})_W = dC_L/d\alpha$.

Figure 11.2 shows a schematic of the forces and moment arms with respect to the center of lift for a general aircraft configuration. The aircraft is presumed to be at an angle of attack, α. The Cartesian coordinate system is taken to be through the centerline of the fuselage (x-direction) and in the direction normal to the centerline (y-direction). Therefore, the lift and drag forces are projected onto that coordinate frame so that

$$F_y = L \cos\alpha + D \sin\alpha \qquad (11.8)$$

and

$$F_x = D \cos\alpha - L \sin\alpha, \qquad (11.9)$$

where L and D refer to the lift and drag forces, respectively.

In order to derive the longitudinal stability coefficient, we start by writing an equation that represents the sum of the moments about the aircraft center of gravity (cg). For Figure 11.2, this is indicated as the dimensionless moment coefficient, $C_{M_{cg}}$, where

$$C_{M_{cg}} = \left(C_L \frac{x_W}{\bar{c}} + C_D \frac{z}{\bar{c}} + C_{M_{ac}}\right)_W + \frac{T z_T}{q_\infty S_W \bar{c}} - C_{L_{HT}} \frac{l_{HT} S_{HT}}{S_W \bar{c}} \frac{q_{\infty_T}}{q_\infty} - \left(C_{M_{cg}}\right)_I. \quad (11.10)$$

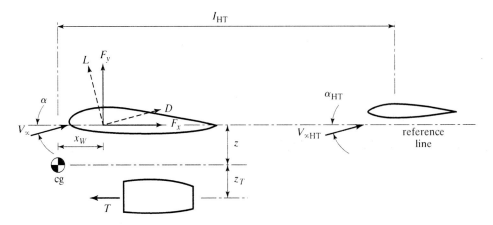

FIGURE 11.2: Illustration showing forces, moments, and lengths governing pitching motion in a general aft tail layout.

Equation [11.10] is generally referred to as the "trim equation." Here, the forces have been normalized by the product of the wing dynamic pressure and planform area, $q S_W$, and the moment arms have been normalized by the wing mean aerodynamic chord, $\bar{c}$.

The quantity $C_{L_{HT}}$ refers to the 3-D lift coefficient for the horizontal tail. The effective velocity or angle of attack of the air stream at the location of the horizontal tail can be affected by any upstream components such as the fuselage or the main wing. A measure of the upstream influence is the "tail efficiency," η_{HT}, where

$$\eta_{HT} = \frac{q_{\infty HT}}{q_\infty} \le 1. \tag{11.11}$$

For a canard, $\eta_{HT} = 1$. For an aft tail, η_{HT} is generally less than one. The only difference in the trim equation (Eq. [11.10]) for a canard configuration is that the sign on the moment produced by the horizontal tail is reversed (+ changed to a −).

The lift coefficient produced by the horizontal tail depends on if it uses a flap, or variable pitch angle, to vary the horizontal tail lift. In the case of the former (flapped) configuration,

$$C_{L_{HT}} = \frac{dC_L}{d\alpha}\left[\left(1 - \frac{de}{d\alpha}\right)\alpha - \alpha_{0_L}\right], \tag{11.12}$$

where $de/d\alpha$ is a factor that changes the effective angle of attack as a result of downwash from the main wing.

The quantity, $de/d\alpha$, can be estimated using Figure 11.3 (from Silverstein and Katzoff, NACA TR 648). It is a function of the main wing aspect ratio, A; the taper ratio, λ, of the main wing; and the placement of the horizontal stabilizer with respect to the main wing, as shown in the illustration at the top of the figure.

The effective angle of attack of the horizontal tail referred to in Figure 11.2 is then

$$\alpha_{HT} = \left(1 - \frac{de}{d\alpha}\right)\alpha. \tag{11.13}$$

For a canard, $\alpha_{HT} = \alpha$, namely, the main wing has no effect.

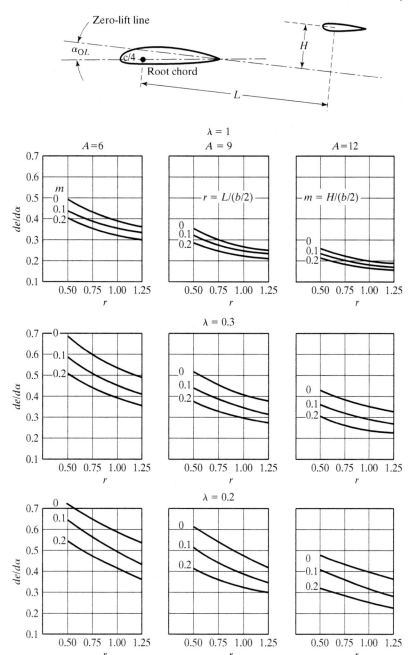

FIGURE 11.3: Downwash factor, $de/d\alpha$, for horizontal stabilizers. (From NACA TR 648.)

In Eq. [11.12], α_{0_L} is the zero-lift angle of attack. This depends on the flap deflection, which is equivalent to adding camber. The analysis is identical to what was done for the flapped main wing, which was presented in Chapter 9.

In a horizontal tail arrangement that uses a variable pitch angle to vary the lift (a so-called "flying tail"),

$$C_{L_{HT}} = \frac{dC_L}{d\alpha} \left[\alpha_{HT} + \alpha_{cs}\right], \tag{11.14}$$

where α_{cs} is the "control stick" angle of attack of the horizontal stabilizer.

The only other term that requires some explanation is the moment coefficient, $(C_{M_{cg}})_I$. This represents a force in the y-direction, which is produced by the turning of the air at the inlet of an engine. This is illustrated in Figure 11.4. In this case, streamlines with velocity, V_∞, are turned by an angle β in order to enter along the engine axial centerline. This produces a y-component of velocity, V_y, and a force F_{y_I} given as

$$(F_y)_I = \dot{m}_I V_\infty \tan \beta. \tag{11.15}$$

For small turning angles,

$$(F_y)_I \simeq \dot{m}_I V_\infty \beta. \tag{11.16}$$

The moment, M_I, produced by this force is

$$M_I = (F_y)_I l_I. \tag{11.17}$$

The normalized moment coefficient is then

$$(C_M)_I = \frac{(F_y)_I l_I}{q\,S_W \bar{c}} \simeq \frac{2\dot{m}l_I \beta}{\rho V_\infty S_W \bar{c}}. \tag{11.18}$$

The coefficient of longitudinal stability, C_{M_α}, is found by differentiating the terms in the trim equation with respect to angle of attack, namely, $C_{M_\alpha} = dC_{M_{cg}}/d\alpha$. Thus,

$$C_{M_\alpha} = \frac{dC_L}{d\alpha}\frac{x_W}{\bar{c}} + \frac{dC_D}{d\alpha}\frac{z}{\bar{c}} - \frac{dC_{L_{HT}}}{d\alpha}\left(1 - \frac{de}{d\alpha}\right)\eta_T \bar{V}_{HT} - \left(\frac{dC_{M_{cg}}}{d\alpha}\right)_I, \tag{11.19}$$

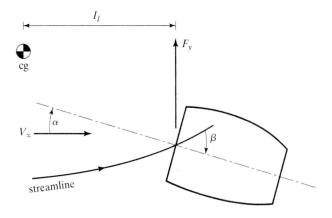

FIGURE 11.4: Illustration of engine inlet turning angle effect, which results in the force F_y.

where

$$\bar{V}_{HT} = \frac{l_{HT} S_{HT}}{S_W \bar{c}} \tag{11.20}$$

is called the horizontal tail "volume coefficient."

Within the drag bucket, $dC_D/d\alpha \simeq 0$. Therefore, the second term in Eq. [11.19] can be neglected with respect to the others. Given this, Eq. [11.19] becomes

$$C_{M_\alpha} = (C_{L_\alpha})_W \frac{x_W}{\bar{c}} - (C_{L_\alpha})_{HT} \left(1 - \frac{de}{d\alpha}\right) \eta_T \bar{V}_{HT} - \left(\frac{dC_{M_{cg}}}{d\alpha}\right)_I. \tag{11.21}$$

The first two terms in Eq. [11.21], depend on the 3-D lift characteristics of the main wing and horizontal tail. The latter has a correction for angle of attack based on any downwash from the wing.

Following Eq. [11.18], the effect of the engine inlet on the longitudinal stability coefficient is

$$\left(\frac{dC_{M_{cg}}}{d\alpha}\right)_I = \frac{2\dot{m} l_I}{\rho V_\infty S_W \bar{c}} \frac{d\beta}{d\alpha}. \tag{11.22}$$

For this, the value of $d\beta/d\alpha$ depends on the Mach number and the location of the engine inlet as follows:

1. For supersonic flow,

$$\frac{d\beta}{d\alpha} = 1. \tag{11.23}$$

2. For subsonic flow, the following are true:

   (a) For engine inlets located *downstream* of the trailing edge of the main wing,

   $$\frac{d\beta}{d\alpha} \simeq \left(1 - \frac{de}{d\alpha}\right) \frac{x_I}{l_{HT}}, \tag{11.24}$$

   where x_I is the distance from the wing trailing edge to the engine inlet and l_{HT} is the distance between the wing trailing edge to the horizontal tail.

   (b) For engine inlets located *upstream* of the wing leading edge, $d\beta/d\alpha$ is given by Figure 11.5.

Given this, all the terms in Eq. [11.21] can be computed in order to estimate the longitudinal stability coefficient.

Static stability is the tendency to return to equilibrium if perturbed. Based on this criterion,

$$\text{for pitch stability,} \quad C_{M_\alpha} < 0. \tag{11.25}$$

An appropriate range of values for C_{M_α} is likely to be

$$-1.5 \le C_{M_\alpha} \le -0.16. \tag{11.26}$$

Too large of a negative coefficient will result in an excessive amount of trim drag in aft tail designs. The impact C_{M_α} has on performance will be discussed later section of this chapter.

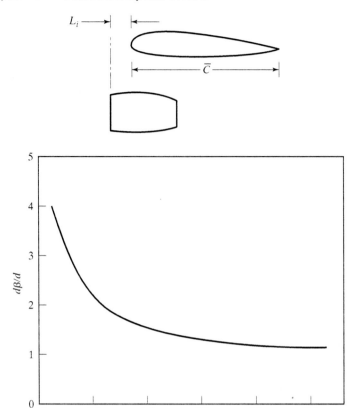

FIGURE 11.5: Values of $d\beta/d\alpha$ as a function engine placement for subsonic aircraft. (NACA TM 1036.)

11.2.2   Lateral Stability

The lateral motion is the rolling motion about the fuselage longitudinal centerline. The notation convention to describe the motion is shown in Figure 11.6. The top drawing shows a front view of an aircraft. R and L refer to the right and left wing tips. The rolling moment is defined as $\mathcal{L}$, with a positive value corresponding to the right wing tip moving down. The dimensionless rolling moment coefficient is defined as

$$C_{\mathcal{L}} = \frac{\mathcal{L}}{q S_w b}. \tag{11.27}$$

The process of rotating the right wing tip down will result in a side slip motion of the aircraft. As shown in the lower part of Figure 11.6, the side slip angle is denoted as β, which is positive for a positive $\mathcal{L}$ (right wing down).

Conversely, the rolling motion of the aircraft in response to a gust at an angle β with respect to the flight direction is a measure of the lateral stability. Therefore, the

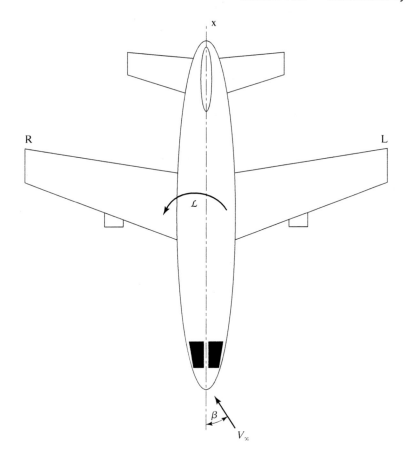

FIGURE 11.6: Illustration defining the convention for lateral (rolling) motion and the rolling moment, $\mathcal{L}$, for an aircraft.

lateral stability coefficient is defined as

$$C_{\mathcal{L}_\beta} = \frac{dC_{\mathcal{L}}}{d\beta}. \tag{11.28}$$

For lateral stability, the right wing tip should rotate up to counter a positive side slip angle. Therefore,

$$\text{for lateral stability,} \quad C_{\mathcal{L}_\beta} < 0. \tag{11.29}$$

The total lateral stability is primarily considered to be the sum of the contributions of the main wing, vertical stabilizer, and the wing-fuselage. Therefore,

$$C_{\mathcal{L}_\beta} = \left(C_{\mathcal{L}_\beta}\right)_W + \left(C_{\mathcal{L}_\beta}\right)_{VS} + \left(C_{\mathcal{L}_\beta}\right)_{W-F}. \tag{11.30}$$

The contribution of the main wing is further subdivided into three elements consisting of the basic wing, sweep angle, and dihedral angle. Therefore,

$$\left(C_{\mathcal{L}_\beta}\right)_W = \left(C_{\mathcal{L}_\beta}\right)_{BW} + \left(C_{\mathcal{L}_\beta}\right)_\Lambda + \left(C_{\mathcal{L}_\beta}\right)_\Gamma. \tag{11.31}$$

The basic wing only considers the effect of aspect ratio and taper ratio on the lateral stability. Generally, the basic wing is stabilizing to lateral motions $((C_{\mathcal{L}_\beta})_{BW} < 0)$. Values normalized by the 3-D lift coefficient can fall in the range of

$$-0.9 \leq (C_{\mathcal{L}_\beta})_{BW}/C_L \leq -0.05, \tag{11.32}$$

with greater stability occurring with smaller aspect ratios.

The effect of the taper ratio on $(C_{\mathcal{L}_\beta})_{BW}$ is relatively small, so that aspect ratio is the dominant effect on the lateral stability of the basic wing.

If large aspect ratios are used for the main wing, the lowered lateral stability can be increased by applying a positive sweep angle, Λ. In addition, adding a positive (upward) dihedral angle, Γ, greatly improves the lateral stability. This in fact is one of the most commonly used approaches for improving the lateral stability in aircraft designs. Note however that with dihedral, the vertical lift is reduced according to $\cos(\Gamma)$.

Besides the wing, the other components of the aircraft that influence the lateral stability are the wing-fuselage and vertical tail. The wing-fuselage refers to the vertical placement of the wing on the fuselage. A "high wing" configuration mounts the wing above the horizontal centerline of the fuselage. This arrangement is stabilizing to lateral motions. A "low-wing" places the wing below the fuselage centerline. This is destabilizing to lateral motions. Finally, the "mid-wing" locates the wing on the fuselage centerline. This has a neutral effect on the lateral stability.

The horizontal lift on the vertical stabilizer caused by a side slip angle β naturally acts to turn the aircraft in a direction that reduces the side slip. Therefore, by definition, it has a stabilizing effect to lateral motions. The magnitude of the coefficient $(C_{\mathcal{L}_\beta})_{VS}$ increases as $(C_{L_\alpha})_{VT}$, S_{VT}, and z_{VT} increase. The latter of these is the moment arm consisting of the distance between the aerodynamic center of the vertical stabilizer and the aircraft center of gravity.

Ultimately, it is a relatively difficult task to accurately estimate the lateral stability coefficient for a conceptual aircraft. Being conservative with too large a value is not necessarily desirable. Therefore, as a **first approximation**, the approach is to take

$$C_{\mathcal{L}_\beta} = -C_{n_\beta}, \tag{11.33}$$

where C_{n_β} is the coefficient of directional stability, which is presented in the next section.

11.2.3 Directional Stability

The directional motion of an aircraft is a rotation about its vertical axis. This is illustrated in Figure 11.7. Here again, the left (L) and right (R) wing tips are denoted. The directional moment is about the center of gravity location shown as N. It is defined as positive in the clockwise direction (right wing back).

The forces acting to rotate the aircraft are lateral forces produced by the fuselage and the wing due to a side slip vector with angle β. The vertical tail opposes this motion by a lateral lift force that produces a counter moment, which acts to reduce the side slip angle and maintain the original flight direction.

Given that the directional moment coefficient is defined as

$$C_n = \frac{N}{q\,S_W b} \tag{11.34}$$

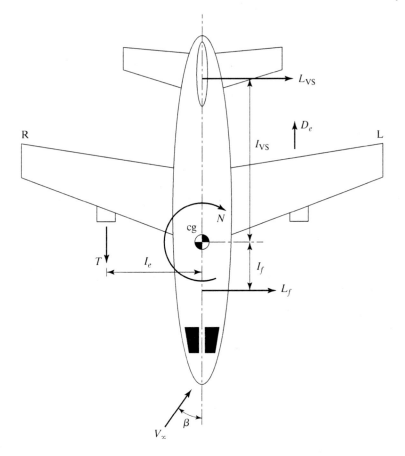

FIGURE 11.7: Illustration defining the convention for directional motion and the directional moment, N, of an aircraft.

and accounting for the fuselage, wing, and vertical stabilizer, then

$$C_n = -\frac{l_F L_F}{q S_W b} + \frac{N_W}{q S_W b} + \frac{l_{VS} L_{VS}}{b S_W},$$ (11.35)

where F, W, and VS refer to the fuselage, wing, and vertical stabilizer, respectively.

For the vertical stabilizer, L_{VS} represents its lateral lift force. This is given as

$$L_{VS} = (C_{L_\alpha})_{VS} \, \alpha_{VS} \frac{q_{VS}}{q} S_{VS},$$ (11.36)

where q_{VS} and α_{VS} account for any upstream influence of the fuselage on the flow approaching the vertical stabilizer.

The effective angle of attack of the vertical stabilizer, α_{VS}, is given as

$$\alpha_{VS} = \left(1 + \frac{d\sigma}{d\beta}\right) \beta.$$ (11.37)

The directional stability coefficient reflects the change in the directional moment coefficient, C_n, with respect to the side slip angle, β, or

$$C_{n_\beta} = \frac{dC_n}{d\beta} = (C_{n_\beta})_F + (C_{n_\beta})_W + (C_{n_\beta})_{VS}. \tag{11.38}$$

This again emphasizes the separate contributions of the fuselage, wing, and vertical stabilizer.

For directional stability, the aircraft should rotate such that the side slip angle is reduced. In terms of the directional coefficient,

$$\text{for directional stability, } \quad C_{n_\beta} > 0. \tag{11.39}$$

The fuselage, wing, and vertical stabilizer all contribute to the directional stability. For the vertical stabilizer,

$$(C_{n_\beta})_{VS} = \bar{V}_{VS} (C_{L_\alpha})_{VS} \left(1 + \frac{d\sigma}{d\beta}\right) \frac{q_{VS}}{q}, \tag{11.40}$$

where $\bar{V}_{VS}$ is the vertical tail volume coefficient, which is similar to the horizontal tail volume coefficient V_{HT} used in the pitch stability.

For the vertical tail, the volume coefficient is defined as

$$\bar{V}_{VS} = \frac{l_{VS} S_{VS}}{b S_W}. \tag{11.41}$$

The quantity $\left(1 + \frac{d\sigma}{d\beta}\right) \frac{q_{VS}}{q}$ in Eq. [11.40] represents the influence of the fuselage and wing on the effectiveness of the vertical stabilizer. An estimate can be obtained from the empirical relation given by Ellison (1968), which is

$$\left(1 + \frac{d\sigma}{d\beta}\right) \frac{q_{VS}}{q} = 0.724 + \frac{3.06 S_{VS}/S_W}{1 + \cos \Lambda_{VS}} + 0.4 \frac{z_W}{h} + 0.009 A_W. \tag{11.42}$$

where S_{VS} is the planform area of the vertical stabilizer measured down to the fuselage centerline, z_W is the height of the wing root chord with respect to the fuselage centerline, and h is the total height of the fuselage.

The effect of the main wing on the directional stability is due to an asymmetric spanwise lift and drag distribution that is caused by side slip. The effect is a function of the main wing aspect ratio and sweep angle. An empirical relation, which is also from Ellison (1968), can be used to provide an estimate. This is given as

$$(C_{n_\beta})_W = C_L^2 \left[\frac{1}{4\pi A} - \frac{\tan \Lambda}{\pi A(A + 4\cos \Lambda)} \left(\cos \Lambda - \frac{A}{2} - \frac{A^2}{8\cos \Lambda} + 6\frac{x}{\bar{c}} \frac{\sin \Lambda}{A}\right)\right]. \tag{11.43}$$

where all of the parameters refer to the main wing, with x being the distance from the aircraft c.g. to the wing aerodynamic center (i.e., x_W in Figure 11.2). Again the coefficient, $(C_{n_\beta})_W$, has units of $[\text{rad}]^{-1}$.

Although Eq. [11.43] is a relatively complex function, we note that the directional stability increases as the sweep angle, Λ, increases and the aspect ratio, A, decreases.

As a result of a side slip, the fuselage is at an angle of attack to the free stream flow. This results in a lift distribution along the length of the fuselage that produces a moment, which counters the stable restoring moment produced by the vertical tail.

An empirical estimate of the fuselage directional stability coefficient is

$$\left(C_{n_\beta}\right)_F = -1.3 \frac{(\text{VOL})_F}{S_W b} \frac{h}{w},$$ (11.44)

where $(\text{VOL})_F$ is the total volume of the fuselage and h and w are the average height and width of the fuselage, respectively.

Accounting for the destabilizing effects of the wing and fuselage on the directional stability, the vertical tail characteristics can be adjusted to achieve a desired positive stability coefficient. A reasonable range is

$$0.08 \le C_{n_\beta} \le 0.28,$$ (11.45)

where again, C_{n_β} has units of $[\text{rad}]^{-1}$. Figure 11.8 also provides a suggested trend in C_{n_β} as a function of the design cruise Mach number.

11.2.4   Aileron Sizing

As discussed in the previous sections, in order for an aircraft to be responsive to roll maneuvers, the roll stability coefficient cannot be too large, and as an initial estimate, it would be taken to be the negative of the directional stability coefficient or

$$C_{\mathcal{L}_\beta} = -C_{n_\beta}.$$ (11.46)

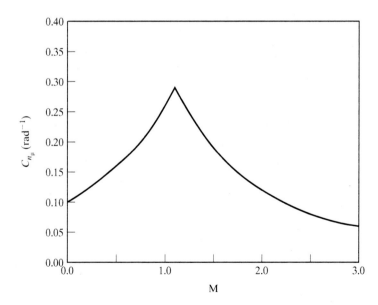

FIGURE 11.8: Recommended directional stability coefficient, C_{n_β}, as a function of the cruise Mach number. (NASA TN D-423.)

This approach will slightly over-design $C_{\mathcal{L}_\beta}$ for subsonic aircraft, but will give a fairly close estimate for transonic aircraft designs, such as for commercial air transports.

Steady Roll Requirements.   The rolling motion of an aircraft is controlled by ailerons on the main wing. The ailerons acts like flaps, so that as they deflect, they change the effective camber of the wing section. This then changes α_{0_L} for that section. The ailerons are usually located outboard on the wings, near the wing tips, so that they can produce the largest rolling moment. They deflect in opposite directions on opposite wing tips to produce a coordinated motion.

When the ailerons are deflected, they produce a moment about the fuselage longitudinal axis. If the ailerons remain deflected, the aircraft will roll. The rate of roll will increase until the moment produced by the aileron deflection is balanced by the damping in the roll moment. At this point, a steady roll condition is reached.

In the past, roll rate requirements were based on a wing helix angle, $Pb/2V$, which is equivalent to the pitch angle on a screw and represents the trajectory of a wing tip in a steady roll with forward flight. In this definition, P is the roll rate in radians per second, V is the forward flight speed, and b is the wing span. For a flight speed and wing span previously set in the design, the necessary roll helix angle then determines the steady roll rate.

Flight tests compiled by NACA (report NACA 715) determined that a helix angle of at least 0.07 radians is considered good by pilots. Table 11.8 also compiles values of P based on MIL-F-8785 requirements. These are given in terms of the time required to roll the aircraft by a specified number of degrees. Since the time to accelerate to a steady roll rate is relatively short, the rotation angles can be divided by the time for the roll, to approximate the steady roll rate. This is done in the third column in the table.

11.2.5   Rudder Area Sizing

The rudder area is sized in order to provide directional control capable of holding a zero side slip angle ($\beta = 0$) based on a worst-case condition consisting of one of the following scenarios:

1. an asymmetrical power condition caused by having one engine out, at a velocity of $1.2V_{TO}$ (equal to $1.44V_s$);

TABLE 11.8: Roll requirements based on MIL-F-8785.

Aircraft Type	Required Roll	Approx. Equiv. Roll Rate (rad/s)
Light Utility, Observation, Primary Trainer	60° in 1.3 s	0.81
Medium Bomber and Cargo Transport	45° in 1.4 s	0.60
Heavy Bomber and Cargo Transport	30° in 1.5 s	0.35
Attack Fighter and Intercepter	90° in 1.3 s	1.21
Air-to-Air Combat	90° in 1.0 s	1.57
Fighter with Air-to-Ground Stores	90° in 1.7 s	0.92

2. Landing and take-off in a cross-wind of $0.2V_{TO}$ (equal to $0.24V_s$) and a side slip angle of $\beta = 11.5°$.

Considering first the asymmetric power condition, the engine that is out not only does not supply thrust, but also adds drag, acting to increase the side slip angle.

If this is only offset by the rudder deflection, then at equilibrium

$$C_n = 0 = -\frac{(T + D_e)}{q S_w b} + C_{n_{\delta_R}} \delta_R, \tag{11.47}$$

where D_e is the drag on the engine, δ_R is the rudder deflection, and $C_{n_{\delta_R}}$ is the rudder control power defined as

$$C_{n_{\delta_R}} = \frac{dC_n}{d\delta_R}. \tag{11.48}$$

The drag on the engine can be estimated as

$$D_e = C_d q S_e, \tag{11.49}$$

where C_d is a drag coefficient, which is approximately 1.2, and S_e is the frontal area of the engine. The dynamic pressure, q, would be based on take-off conditions.

Considering the cross-wind condition, at equilibrium

$$C_n = 0 = C_{n_\beta}\beta + C_{n_{\delta_R}} \delta_R. \tag{11.50}$$

In this case, $\beta = 11.5°$. In both cases, a rudder deflection angle of $\delta_R = 20°$ is used.

The rudder control power is proportional to the transverse lift produced by the rudder deflection times the moment arm. An estimate of this is

$$C_{n_{\delta_R}} \simeq 0.9 \left(C_{L_\alpha}\right)_{VS} \bar{V}_{VS} \frac{d\alpha_{0_L}}{d\delta_R}. \tag{11.51}$$

Here again $\bar{V}$ is the vertical tail volume coefficient given by Eq. [11.41]. The quantity $d\alpha_{0_L}/d\delta_R$ is the change on the zero-lift angle of attack due to the rudder deflection. This can be easily found by the same analysis used in Chapter 9 to predict the effect of trailing-edge flaps.

The rudder can be analyzed as a plane flap. Therefore, the change in α_{0_L} with rudder deflection is

$$\frac{d\alpha_{0_L}}{d\delta_R} = -\frac{dC_l}{d\delta_R} \frac{1}{C_{l_\alpha}} K'. \tag{11.52}$$

For this, the 2-D section lift coefficient slope, C_{l_α} is approximately 2π (rad)$^{-1}$. The quantities K' and $dC_l/d\delta_R$ are found from Figures 9.3 and 9.4 (in Chapter 9), where δ_f now refers to δ_R. The nonlinear correction, K', is a function of δ_R and C_R/C_{VS}. For $\delta_R = 20°$, this correction ranges from $0.7 \leq K' \leq 0.9$, for $0.1 \leq C_R/C \leq 0.5$.

The quantity $dC_l/d\delta_R$ is primarily a function of C_R/C and a weak function of the thickness-to-chord ratio. Assuming that $t/c = 0.12$, the data from Figures 9.3 and 9.4 can be combined to generate a plot of $d\alpha_{0_L}/d\delta_R$ versus C_R/C_{VS}. This is shown in Figure 11.9.

If we assume that the ratio of the rudder chord to the total chord exactly scales with the ratio of the area of the rudder to the total area of the vertical stabilizer, then

$$\frac{C_R}{C_{VS}} = \frac{S_R}{S_{VS}}.$$
(11.53)

The equations for the directional control requirements and Figure 11.9 then lead to a procedure for sizing the rudder in the vertical stabilizer. In either of two cases consisting of the asymmetric power condition given by Eq. [11.47], or the cross-wind condition given by Eq. [11.50], the procedure consists of the following three steps:

1. Solve for $C_{n_{\delta_R}}$ in both Eq. [11.47] and Eq. [11.50]. The units on $C_{n_{\delta_R}}$ are $[\text{rad}]^{-1}$.
2. Take the case with the largest value of $C_{n_{\delta_R}}$ from the previous step. Use Eq. [11.51] to put it in terms of $d\alpha_{0_L}/d\delta_R$, and solve for that quantity. Care needs to be taken with units here, where $d\alpha_{0_L}/d\delta_R$ is dimensionless.
3. Use Figure 11.9, and based on the value of $d\alpha_{0_L}/d\delta_R$, find C_R/C_{VS}. Assuming that the ratio is equivalent to S_R/S_{VS}, solve for the necessary rudder area, S_R, where S_{VS} is known from a previous design step in Chapter 6.

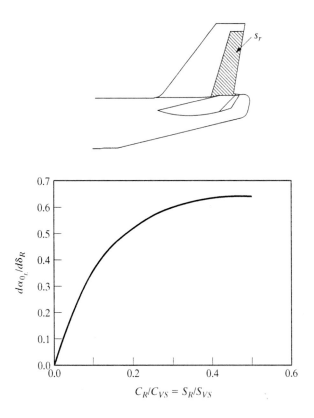

FIGURE 11.9: Rudder effectiveness based on $t/c = 0.12$, $\delta_R = 20°$.

11.2.6 Longitudinal Stability Effect on Performance

As discussed in Section 11.2.1, longitudinal (pitch) stability requires a negative pitch stability coefficient, $C_{M_\alpha} < 0$. In this section, we examine the implications of this on the performance of the aircraft, especially the drag.

In this discussion, it is possibly more convenient to consider the static margin as a measure of the longitudinal stability. Recall that the longitudinal stability coefficient is related to the static margin by Eq. [11.7], which is reproduced in

$$(C_{M_\alpha})_W = -(SM)(C_{L_\alpha})_W. \tag{11.54}$$

The static margin is the normalized difference between the locations of the neutral point (x_{np}) and the center of gravity (x_{cg}). This is defined as

$$SM = \frac{x_{np} - x_{cg}}{\bar{c}} > 0 \text{ for stability.} \tag{11.55}$$

Neglecting any lift force induced by the fuselage or horizontal control surfaces (so-called stick-free condition), the neutral point corresponds to the neutral lift point of the main wing. A schematic that illustrates this is shown in Figure 11.10.

Aft Tail Design. If we consider for the moment an aft tail design, the stipulation of having a positive static margin requires that the horizontal tail provide some downward lift in order to maintain a level attitude of the fuselage. This condition is illustrated in Figure 11.11. This force was determined in the previous chapter to calculate the loads on the fuselage.

Assuming that the fuselage does not generate lift, the net lift force is the sum of that produced by the main wing and horizontal tail, namely,

$$\text{Net Lift} = L_W + L_{HT}. \tag{11.56}$$

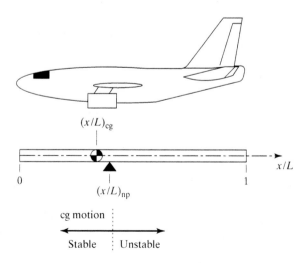

FIGURE 11.10: Illustration showing the relation between the neutral point and center of gravity on the static margin.

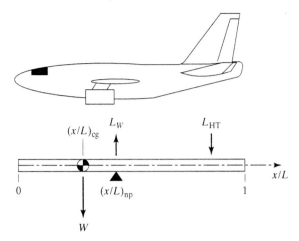

FIGURE 11.11: Illustration of the compensating lift force from an aft horizontal tail to maintain level flight with a positive static margin.

With a positive static margin, the lift on the aft tail is directed downward, opposite to the main wing. Since the net lift must equal the weight at cruise, the lift produced by the main wing has to increase to compensate for the downward lift of the tail. This leads to higher lift-induced drag. If the static margin is too large, this can lead to an excessive drag increase.

In discussing the design of the horizontal tail in Chapter 6, the recommendation was to use an uncambered profile shape. As a result, unless an elevator deflection is applied, no lift is generated. The tail lift that is required to balance the static margin, therefore, requires a fixed elevator deflection. This is referred to as elevator "trim." The lift in this case produces a lift-induced drag, which in this instance is referred to as "trim drag."

The trim drag is easily estimated using the same methods that were used for the main wing. Therefore,

$$D_{\text{trim}} = q_\infty S_{\text{HT}} k_{\text{HT}} (C_L)_{\text{HT}}^2,$$ (11.57)

where

$$k = \frac{1}{\pi A_{\text{HT}} e}$$ (11.58)

and A_{HT} is the aspect ratio of the horizontal tail.

To put this in terms of the lift,

$$(C_L)_{\text{HT}} = \frac{L_{\text{HT}}}{q_\infty S_{\text{HT}}}.$$ (11.59)

Therefore, substituting into Eq. [11.57],

$$D_{\text{trim}} = \frac{k L_{\text{HT}}^2}{q_\infty S_{\text{HT}}}.$$ (11.60)

The lift on the horizontal tail, L_{HT}, is determined based on balancing the aircraft about the center of gravity in level cruise. That value substituted into Eq. [11.60], along with the conditions at cruise, gives a value for the trim drag.

It is important to compare the trim drag to the overall drag at cruise, which was found as part of the engine selection in Chapter 7. As a rule of thumb,

$$D_{trim} \leq 0.1 D_{tot}. \tag{11.61}$$

If this is not the case, there are three possible solutions:

1. The static margin needs to be reduced. This requires either moving the center of gravity aft or moving the center of lift forward.
2. The moment arm of the horizontal tail, l_{HT}, needs to be increased. This requires increasing the distance between the horizontal tail and main wing m.a.c.
3. The aspect ratio of the horizontal tail needs be increased. This will reduce the lift-induced drag by reducing k. The use of an h-tail design, which increases the effective aspect ratio of the horizontal tail through an end-plate (winglet) effect, offers another possible alternative.

Considering the total drag from the main wing and horizontal tail,

$$D = q S_W (C_{D_0} + k C_L^2) + D_{trim}. \tag{11.62}$$

Substituting for C_L for the main wing and substituting for D_{trim} from Eq. [11.60],

$$D = q S_W C_{D_0} + \frac{k_W L_W^2}{q_\infty S_W} + \frac{k L_{HT}^2}{q_\infty S_{HT}}. \tag{11.63}$$

Rearranging terms,

$$D - q S_W C_{D_0} = \frac{k_W}{S_W q_\infty} \left[L_W^2 + \frac{k_{HT} S_W}{k_W S_{HT}} L_{HT}^2 \right]. \tag{11.64}$$

At cruise, the total lift generated equals the weight; therefore,

$$L_W + L_{HT} = W. \tag{11.65}$$

Substituting for L_W and k in Eq. [11.64] then gives

$$D - q S_W C_{D_0} = \frac{k_W}{S_W q_\infty} \left[(W - L_{HT})^2 + \frac{A_W S_W}{A_{HT} S_{HT}} L_{HT}^2 \right]. \tag{11.66}$$

Eq. [11.66] allows us to examine the effect of the value of the static margin on the drag. As the static margin changes, the lift that has to be produced by the horizontal tail will have to change in order to balance the aircraft in level flight. This will affect the overall drag, based on Eq. [11.66].

TABLE 11.9: Fuselage load summary for schematic in (Chapter 10) Figure 10.10.

Load Type	Magnitude (lbf)	x/L_{start}	x/L_{end}	Resultant	M @ x_{CL} (f-lbs)
Fuel	10,000	0.3	0.5	0.4	−1000
Payload	5,000	0.1	0.6	0.35	−750
Fuse. Struct.	9,000	0	1	0.5	0
Engine	3,000	0.7	0.7	0.7	600
Wing Struct.	2,000	0.3	0.6	0.45	−100
Tail Struct.	3,000	0.8	1	0.9	1200
Tail Lift	−125	0.9	0.9	0.9	−50

TABLE 11.10: Variation of L_{HT} with changing static margin for the conditions in Table 11.9.

SM	L_{HT} (lbs)
−17.2	120,000
−13.6	95,000
−10.0	70,000
−6.4	45,000
−2.9	20,000
−1.1	7,500
0.7	−5,000
4.3	−30,000
6.1	−42,000
11.4	−80,000

To demonstrate this, we use the same conditions for a fictitious aircraft that were presented in Table 10.6. These conditions are again listed in Table 11.9.

In this case, the neutral point was located at $x/L = 0.5$, where L is the length of the fuselage. The static margin was varied by concentrating the payload to a particular x/L position and varying that position. The static margin and L_{HT} were then tabulated. The result is given in Table 11.10.

Note in Table 11.10, L_{HT} is positive in the upward direction, so that it augments the lift of the main wing with negative (unstable) static margins. The values of L_{HT} were substituted into Eq. [11.66] in order to calculate the drag. For this, $W = 32,000$ lbs, and $(A_W S_W)/(A_{HT} S_{HT})$ was taken to be 1. The result is plotted in Figure 11.12. In this case, the drag is normalized by the minimum value.

It is clear from Figure 11.12 that the minimum drag occurs for a *negative* static margin. In addition, having too large of a positive static margin can significantly increase the drag. In the range of SM values most typically used (3 to 12); in this example, the drag could be from approximately 5 to 30 times larger than the minimum value.

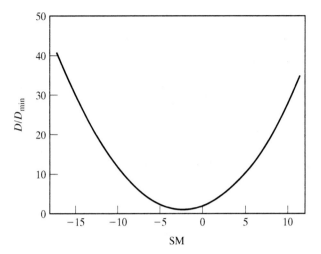

FIGURE 11.12: Effect of the static margin on drag for an aft tail design based on conditions in Table 11.9.

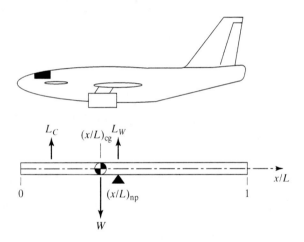

FIGURE 11.13: Illustration of the compensating lift force from a canard to maintain level flight with a positive static margin.

Forward (Canard) Tail Design. If we now consider a forward (canard) tail design, the stipulation of having a positive static margin requires that the horizontal tail provide some upward lift in order to maintain a level attitude of the fuselage. This condition is illustrated in Figure 11.13. This lift now augments the lift produced by the main wing. This is favorable with respect to the overall drag because it then reduces the lift-induced drag on the main wing.

To demonstrate this, we use the same conditions for a fictitious aircraft that were presented in Table 11.9. The position of the canard is taken to be the same distance

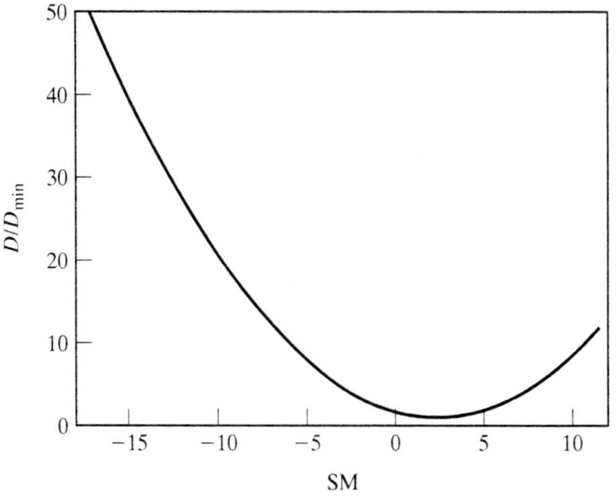

FIGURE 11.14: Effect of the static margin on drag for a forward (canard) tail design based on conditions in Table 11.9.

ahead of the neutral point as it was behind the neutral point for the aft tail. For the sake of comparison, the tail structure weight and its location are assumed to be the same as before (considered to be the vertical tail alone). Under these conditions, the lift force on the canard is the same magnitude, but opposite sign, as with the previous aft tail case. These values were then substituted into Eq. [11.66], and the result is plotted in Figure 11.14.

It is clear from Figure 11.14 that with a canard, the minimum drag now occurs for a *positive* (stable) static margin. In addition, the drag penalty for having too large of a positive static margin is not as severe as with the aft tail. For example, in the range of SM values from 3 to 12, the drag ranges from the minimum to only approximately 10 times larger. Therefore, the canard has a distinct advantage over the aft tail from a performance aspect.

11.3   SPREADSHEET FOR REFINED WEIGHT AND STABILITY ANALYSIS

The topics and equations in this chapter have been put into two spreadsheets that perform a refined weight estimate and a static stability analysis. These are named **refwt.xls** and **stab.xls**, respectively. These require input from many of the previous spreadsheet calculations and design parameters, and as such, depend heavily on the previous design steps. A discussion of these is given in the following two sections.

11.3.1   Refined Weight Analysis

The formulas for estimating the weights of the major components consisting of

1. the main wing,
2. the horizontal tail,
3. the vertical tail,

4. the fuselage,

5. the main landing gear, and

6. the nose landing gear.

have been incorporated into the spreadsheet named **refwt.xls**. A sample is shown in Figures 11.15 through 11.17, with parameters that are relevant to the conceptual SSBJ that was proposed in Chapter 1. The formulas are based on Eqs. [11.1–11.6] and their accompanying tables of coefficients. The coefficients include the three general types of aircraft consisting of combat/fighter, transport, and general aviation.

Each aircraft component is analyzed in a separate part of the spreadsheet. The general format is to list the equation coefficients in the three left-most columns, corresponding to the three types of aircraft. The parameters that are required as input to the formulas are listed in the columns to the right of the coefficients. These consist of the parameter designation (e.g., the wing aspect ratio, A) and the parameter value. Because many of the same parameters are needed in the formulas for the different aircraft components, many of the values are given global names, for example, K_{dw} is assigned the name KDW. The global names are then referred to in the formulas. The parameters appear in a **bold font** if they require values. Others that are not in a bold font receive their values by a link to a global value, or by an internal calculation.

The weights of each component are listed at the bottom of each section, below the coefficient columns. These are given for each of the three general types of aircraft. The very bottom of the spreadsheet contains a summary table of the component weights for an aircraft selected from one of the three general types. This includes the component weight, as well as its fraction of the total structure weight, W_{str}, which is input to the spreadsheet based on the initial take-off weight estimate taken from the spreadsheet **itertow.xls**. The total percentage is summed at the bottom of the table. The remaining percentage and corresponding weight needed to bring the total to 100 percent are listed at the bottom of the table. This is labeled as "other" and compared to the "target" value, which corresponds to a percentage of the take-off weight, based on Table 11.7. Finally, the component weight percentages for the selected aircraft type are plotted next to the table.

The estimate for the weight of the wing is based on Eq. [11.1]. Most of the parameters used in the formula come from the wing design in the **wing.xls** spreadsheet. The two parameters K_{dw} and K_{vs} have different values based on whether the main wing has a delta planform shape ($\lambda = 0$). These are linked to coefficients C_2 and C_3 in the table for the fighter aircraft. The ratio of the flapped area to the total area of the wing, S_f/S_W, is taken from the enhanced lift design in the spreadsheet **flaps.xls**. The design gross weight, W_{dg}, and the weight of fuel stored in the wings, W_{fw}, can be obtained from the take-off weight estimate in the spreadsheet **itertow.xls**, where the former is the weight at take-off and the latter is the weight of fuel at take-off, assuming that all of it is stored inside the wing.

The estimate of the weight of the horizontal tail is based on Eq. [11.2]. Here most of the parameters come from the tail design spreadsheet **tail.xls**. The parameter L_{ht} is based on the position of the horizontal tail on the fuselage. Although this parameter is used in the tail design spreadsheet, the static stability analysis may require some adjustment of the position. Therefore, it is possible that it may need to be updated.

Wing Weight					
Fighter	*Transport*	*Gen. Av.*	**A**	2	
0.0103	0.0051	0.0360	**K_dw**	0.77	
0.7680	1.0000	1.0000	**K_vs**	1	
1.0000	1.0000	1.0000	**n**	4.5	
0.5000	0.5570	0.4900	**q**	531	lb/f^2
0.5000	0.5770	0.4900	**S_w**	519	f^2
0.6220	0.6490	0.7580	**(Sf/Sw)**	0.6	
0.7850	0.5000	0.6000	**S_f**	311.4	f^2
−0.4000	−0.4000	−0.3000	**t/c**	0.04	
1.0000	1.0000	0.0000	**W_dg**	90523	lbs
0.0500	0.1000	0.0040	**W_fw**	41261	lbs
−1.0000	−1.0000	0.3000	**Λ**	62	deg
0.0400	0.1000	0.0000	**λ**	0	
0.0000	0.0000	0.0060			
0.0000	0.0000	0.0035			

Fighter	4,122.67	lbs
Transport	7,843.28	lbs
Gen. Av.	0.00	lbs

Horizontal Tail Weight					
Fighter	*Transport*	*Gen. Av.*			
0.5503	0.0379	0.0092	**A_ht**	2	
−2.0000	−0.2500	0.0000	**b_ht**	12.2	f
0.2600	0.6390	0.4140	**F_w**	6	f
0.2600	0.1000	0.4140	**K_y**	10.08	f
0.8060	0.7500	0.8960	**L_ht**	33.6	f
0.0000	−1.0000	0.0000	**n**	4.5	
0.0000	0.7040	0.0000	**q**	531	lbs/f^2
0.0000	−1.0000	0.0340	**S_ht**	74	f^2
0.0000	0.1160	0.0430	**(t/c)ht**	0.04	
0.0000	0.0000	−0.1200	**W_dg**	90523	lbs
0.0000	0.0000	−0.0200	**(Λ)ht**	62	deg
0.0000	0.0000	0.1680	**(λ)ht**	0.35	

Fighter	228.23	lbs
Transport	516.16	lbs
Gen. Av.	395.94	lbs

Vertical Tail Weight					
Fighter	*Transport*	*Gen. Av.*			
0.4520	0.0026	0.0076	**A_vt**	1.1	
1.0000	1.0000	0.2000	**H_ht**	0	f
0.5000	0.2250	1.0000	**H_vt**	8.1	f
0.4880	0.5560	0.3760	**K_z**	40	f
0.4880	0.5360	0.3760	**K_rht**	1	
0.7180	0.5000	0.8730	**L_vt**	40	f
0.3410	0.0000	0.0000	**M**	2.1	
−1.0000	−0.5000	0.0000	**n**	4.5	

FIGURE 11.15: Spreadsheet for estimating the weights of the major aircraft components based on Eqs. [11.1–11.6] (Part 1).

0.3480	0.0000	0.0000		q	531	lbs/f^2
0.2230	0.3500	0.3570		S_r/S_vt	0.3	
1.0000	0.0000	0.0000		S_r	30	f^2
0.2500	0.0000	0.0390		S_vt	100	f^2
−0.3230	−1.0000	−0.2240		(t/c)_vt	0.04	
0.0000	−0.5000	−0.4900		W_dg	90523	lbs
0.0000	0.8750	0.0000		(Λ)vt	63	deg
0.0000	0.0000	0.1220		(λ)vt	0.3	

Fighter	334.67	lbs
Transport	1,505.78	lbs
Gen. Av	667.91	lbs

Fuselage Weight

Fighter	Transport	Gen. Av.				
0.4990	0.3280	0.0520		b_w	40.8	f
0.7740	1.0600	1.0000		D	9	f
1.0000	1.1200	1.0000		K_ws	−0.27	
0.3500	0.5000	0.1770		K_dwf	0.77	
0.2500	0.5000	0.1770		K_door	1.06	
0.5000	0.3500	−0.0720		K_lg	1.12	
0.0000	0.0000	−0.0510		L_f	126	f
0.8490	−0.1000	0.0720		L_t	33.6	f
0.0000	0.3020	1.0860		n	4.5	
0.6850	0.0000	0.0000		q	531	
0.0000	0.0400	0.0000		S_f	2284	f^2
0.0000	0.0000	0.2410		S_vt	100	f^2
0.0000	0.0000	28.5189		W_f	9	f
				V_pr	3990	f^3
				W_p	28.52	
				W_dg	90523	lbs
				Λ	62	deg
				λ	0	

Fighter	9,977.08	lbs
Transport	11,067.01	lbs
Gen. Av	7,155.33	lbs

Main Landing Gear Weight

Fighter	Transport	Gen. Av.				
1.0000	0.0106	0.0344		K_cb	1	
1.0000	1.0000	1.0000		K_mp	1	
1.0000	1.0000	1.0000		K_tpg	1	
0.2500	0.8880	0.7680		L_m	40	in
0.2500	0.2500	0.7680		n	4.5	
0.9730	0.4000	0.4090		N_mw	4	wheels
0.0000	0.3210	0.0000		N_mss	2	struts
0.0000	−0.5000	0.0000		V_s	194	f/s
0.0000	0.1000	0.0000		W_l	51,597	lbs

FIGURE 11.16: Spreadsheet for estimating the weights of the major aircraft components based on Eqs. [11.1–11.6] (Part 2).

Fighter	794.81								
Transport	1930.95								
Gen. Av.	2054.77								
	Nose Landing Gear Weight								
Fighter	*Transport*	*Gen. Av.*							
1.0000	0.0320	1.0000		**K_np**		1			
1.0000	1.0000	0.0153		**L_n**		50	in		
0.2900	0.6460	0.5660		n		4.5			
0.2900	0.2000	0.5660		**N_nw**		2	wheels		
0.5000	0.5000	0.8450		**W_l**		51597	lbs		
0.5250	0.4500	0.0000							
Fighter	366.12	lbs							
Transport	462.48	lbs							
Gen. Av.	454.34	lbs							
	Summary								
	Fighter	Transport	Gen. Av.						
Wing	4,122.67	7,843.28	0.00						
Horiz. Tail	228.23	516.16	395.94						
Vert. Tail	334.67	1,505.78	667.91						
Fuselage	9,977.08	11,067.01	7,155.33						
Main Gear	794.81	1930.95	2054.77						
Nose Gear	366.12	462.48	454.34						
Total	15,823.57	23,325.66	10,728.30						

	Case Study			
Wstr	45,000	lbs		
Parts	**Weights**	**W/Wstr**		
Wing	4,122.67	9.16	%	
Horiz. Tail	228.23	0.51	%	
Vert. Tail	334.67	0.74	%	
Fuselage	9,977.08	22.17	%	
Main Gear	794.81	1.77	%	
Nose Gear	366.12	0.81	%	
Ins. Eng.	12,974.00	28.83	%	
Total		**63.99**	%	
Other	16,202.43	36.01	%	
Target	15,388.91	34.2	%	

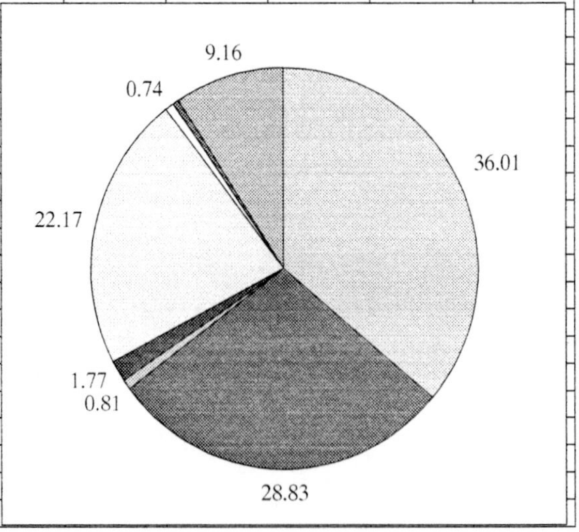

FIGURE 11.17: Spreadsheet for estimating the weights of the major aircraft components based on Eqs. [11.1–11.6] (Part 3).

The estimate of the weight of the vertical tail is based on Eq. [11.3]. It also uses parameters that primarily come from the tail design spreadsheet **tail.xls**. The parameter K_{rht} has one of two values (1.047 or 1) depending on whether it is an all-moving design. This has to be input. Again, the distance L_{vt} may have to be updated from the earlier tail design based on the stability analysis.

The estimate of the fuselage weight is based on Eq. [11.4]. This involves parameters that primarily come from the spreadsheet named **fuse.xls**. The coefficient K_{ws} is based on parameters of the wing and is calculated. K_{dwf} is 0.774 for a delta wing, or otherwise 1. K_{door} ranges from 1 to 1.25 depending on the number and type of cargo doors in the fuselage. K_{lg} depends on whether the landing gear is mounted to the fuselage. The weight penalty for having a pressurized cabin, W_p, involves the internal volume of the fuselage. This can be estimated by using the fuselage spreadsheet, which has at the bottom a table of the local diameter at successive x/L stations. Assuming a stepwise change in the diameter, the local volume is then the cross-section area times the distance between stations.

The weight of the main landing gear is estimated based on Eq. [11.5]. This has a number of parameters that can have one of two values depending on the design of the landing gear. These include K_{cb}, K_{mp}, and K_{tpg}. The complete description of these is given in Section 11.1.5.

Finally the weight of the nose landing gear is based on Eq. [11.6]. The parameters for this are similar to those for the main landing gear. In addition, it has a coefficient, K_{np}, which depends on if a kneeling gear is used.

All of the component weights, for each of the aircraft types, are automatically copied to the summary table at the bottom of the spreadsheet. These are then summed to give the total weight. To use this for a specific design, the weights for the most relevant aircraft type have to be copied to the lower table where the weight fraction percentages are determined. With this, the weight of the installed engines also needs to be supplied to the table. The installed weight is a factor times the uninstalled engine weight. The factor depends on the category of aircraft, according to Table 11.7. The uninstalled engine weight comes from the engine design spreadsheet, **engine.xls**. For multiple- engine aircraft, this weight should include all the engines.

The difference between the sum of the weight fraction percentages and 100 percent is listed under the category of "other." Also based on Table 11.7, this should fall between 14 to 17 percent of the take-off weight, based on the aircraft type. In terms of the structure factor, $K_{str} = W_{str}/W_{TO}$, the "target" structure weight remaining is

$$W_{remain} = \frac{W_{remain}}{W_{TO}} \frac{W_{str}}{K_{str}}. \qquad (11.67)$$

Ideally, we want the weight in the "other" category to match that in the "target" category. An allowable difference is probably ± 5 percent. If the difference is greater, we would hope that the "other" exceed the "target," since this indicates that the weights of the major components have been **underestimated**. If the weight in the "other" category is less than that in the "target" category, it indicates that a weight reduction is necessary or that the structure factor has to be increased. Since the structure factor was set in the first estimate of the take-off weight, such a change has implications on every step of the design.

The pie-chart plot of the weight fraction percentages provides a good graphical presentation of how the component weights are distributed for the design. Generally speaking, the largest weights correspond to the fuselage and engine. Next follows the vertical tail and wing. Of course, the exact order of component weights is specific to each design.

11.3.2 Static Stability Analysis

The formulas for estimating the static stability coefficients for the aircraft design are implemented in the spreadsheet named **stab.xls**. This includes analysis of

1. the static margin, SM;
2. the longitudinal (pitch) stability coefficient, C_{M_α}; and
3. the yawing (directional) stability coefficient, C_{n_β}.

The lateral (rolling) stability coefficient is estimated by taking $C_{L_\beta} = -C_{n_\beta}$. In addition, the last part of the spreadsheet calculates the area of the vertical stabilizer rudder based on asymmetric power or cross-wind conditions.

A sample of the spreadsheet is shown in Figures 11.18 through 11.20. This includes parameters that are relevant to the conceptual SSBJ proposed in Chapter 1. The formulas are based on equations from Section 11.2 and accompanying graphs and tables.

Longitudinal Stability. The spreadsheet is organized so that the top half deals with the longitudinal stability and the bottom half deals with the directional stability. The longitudinal stability is further divided into two parts consisting of the static margin in the top part and the longitudinal stability coefficient in the following part.

The analysis of the static margin is very important because it is used to move integral components along the fuselage in order to "balance" the aircraft and to make the SM > 0. Some of the general input that is needed are the length of the fuselage, L; the x/L-location of the center of lift, L_{ctr}; and the mean aerodynamic chord (m.a.c.) of the main wing. These are entered at the top of the spreadsheet and used in various formulas. The next step in the spreadsheet is to input the component weights made up of

1. fuel,
2. payload,
3. fuselage structure,
4. engines,
5. wing structure,
6. horizontal tail,
7. vertical tail, and
8. the weight in the "other" category from the refined weight analysis.

All these weights should come from the **refwt.xls** spreadsheet. These are entered in the second column of the "Load Summary (fuselage)" table. In addition to the component weights, the distance over which they act needs to be entered. This is designated as x/L_{start} and x/L_{end}. If x/L_{start} is less than x/L_{end}, the component weight is assumed to be uniformly distributed over that region. If x/L_{start} equals x/L_{end}, the component weight is treated as a point load. In either case, the location where the equivalent resultant

	Longitudinal Stability					
Fuselage Length						
L (f)	126					
Wing Center of Lift						
L_ctr (x/L)	0.6					
m.a.c. (ft)	27.2					
Load Summary (fuselage)						
Load Type	Magnitude	x/L_start	x/L_end	resultant	M @ C_lift	dw
	(lbs)			x/L	f-lb (+ cw)	
Fuel	41261	0.4	0.6	0.5	−519888.6	8252.2
Payload	4000	0.1	0.8	0.45	−75600	266.67
Fus.Struct.	9977	0	1	0.5	−125710.2	475.1
Engine(s)	13000	0.7	0.7	0.7	163800	13000
Wing Struct.	4122	0.4	0.8	0.6	0	458
Horiz. Tail	228	0.15	0.2	0.18	−12209.4	114
Vert. Tail	334	0.8	1	0.9	12625.2	66.8
Other	16202	0	1	0.5	−204145.2	771.52
ΣL	89124			ΣM	−761.1E+3	
Tail Lift (req)	14213.41	0.15	0.2	0.18	−761.1E+3	7106.71
Center of Gravity						
X_cg / L	0.53					
X_cg (ft)	67.06	f				
Static Margin						
S.M.	0.31	stable				
Longitudinal Stability Coefficient:						
Wing Parameters:						
S_w	831	f^2				
(C_L_α)_w	0.04	(deg)^−1				
x_w	−8.54	f				
cbar	27.2	f				
Horiz. Tail Paramters:						
(C_L_α)_ht	0.03	(deg)^−1				
de/dα	0	Fig. 11.3				
η_ht	1					
l_ht	−45.01	f				
S_ht	74	f^2				
Engine Parameters						
mdot	135.4	lbm/s				
l_i	1.6	f				

FIGURE 11.18: Spreadsheet for estimating the static margin and static stability coefficients for the three directions of motion of an aircraft (Part 1).

rho	0.01	lbm/f^3					
V	1925.7	f/s					
dβ/dα	1						
Calculations							
V_bar_hs	−0.15						
inlet effect	0	unstable					
wing effect	−0.79	stable	check: C_M_α = −S.M.*C_L_α			−0.79	
h. tail effect	−0.25	stable					
C_M_α	**−0.54**	stable					
Directional Stability Coefficient:							
Wing Parameters:							
A_w	2						
Λ	62	deg					
λ	0						
S_w	831	f^2					
b	40.8	f					
z_w	−4	f					
C_L (cruise)	0.2						
Fuselage Parameters:							
h	9	f					
w	9	f					
Vol_f	3990	f^3					
Vertical Tail Parameters:							
(C_L_α)_vs	0.02	(deg)^−1					
l_vs	46.34	f					
S_vs	100	f^2					
Λ_vs	63	deg					
Calculations							
V_bar_vs	0.14						
(1+dσ/dβ)q/q	0.82	Eq[11.42]					
v. tail effect	0.1472	Eq[11.40]	stable				
fuse. effect	−0.1530	Eq[11.44]	unstable				
wing effect	0.0092	Eq[11.43]	stable				
C_n_β	**0.0034**	stable					
C_L_β	**−0.0034**	stable					
	Rudder Sizing						
Input Parameters							
δ_r	20	deg					
β	11.5	deg					
Asym. T	10200	lbs					

FIGURE 11.19: Spreadsheet for estimating the static margin and static stability coefficients for the three directions of motion of an aircraft (Part 2).

S_w	831	f^2				
b	40.8	f				
C_n_β	0					
diam_e	2.79	f				
V_T-O	303	f/s				
rho_T-O	0.07	lbm/f^3				
Calculations						
1.2V_T-O	363.6	f/s				
0.2V_T-O	60.6	f/s				
q	139.6	lbs/f^2				
D_e	1024.12	lbs				
C_n_δR:						
Asy. Power	0.01	[rad]^−1	Eq[11.47]			
Cross Wind	0	[rad]^−1	Eq[11.50]			
dα_0L/dδ_r	0.04		Eq[11.51]			
C_R/C_VS	4	%	Fig. 11.9			

FIGURE 11.20: Spreadsheet for estimating the static margin and static stability coefficients for the three directions of motion of an aircraft (Part 3).

point load acts is calculated in the next column. The resultant is then used to calculate the moment about the location of the center of lift produced by each component load. This moment has two functions. The first is to find the longitudinal center of gravity (cg) for the aircraft. The second is to find the lift that must be produced by the horizontal tail in order to balance the aircraft for level flight. This is referred to as the "Tail Lift (req)" in the spreadsheet. As pointed out in Section 11.2.6, for an aft-tail design, requiring excessive tail lift dramatically increases the drag on the aircraft.

Having determined the location of the cg and specified the location of the center of lift, the static margin is calculated. A label appears in the column next to the SM value. Based on the condition of SM > 0 or SM < 0, this label displays the word "stable" or "unstable," respectively.

As mentioned, the static margin is a good indication of the longitudinal stability of the aircraft, and the longitudinal stability coefficient of the wing is exactly the negative of the product of the static margin and $(C_{L_\alpha})_W$ (Eq. [11.7]). However, a better estimate comes by including the effect of the horizontal tail and engine.

The longitudinal stability coefficient is found by solving Eq. [11.21]. This includes the effects of the main wing (W), horizontal tail (HT), and the turning of the flow at the inlet of the engines (I). For each of these, input parameters are required. Most of the wing parameters come from the wing design spreadsheet, **wing.xls**. The distance x_W is calculated from knowing the location of the center of gravity, calculated earlier, and the center of lift, previously input.

The horizontal tail parameters are generally found from the tail design spreadsheet, **tail.xls**. The distance l_{HT} is used in the horizontal tail volume coefficient, $\bar{V}_{HT}$. In the spreadsheet, l_{HT} is calculated as the distance from where the resultant lift load acts to the cg location. The quantity $de/d\alpha$ is found from Figure 11.3. This represents the downstream effect of the main wing on the horizontal tail. For the case study aircraft that uses a canard, $de/d\alpha = 0$.

The engine parameters are generally found from the engine design spreadsheet, **engine.xls**. The effect of the engine on the longitudinal stability coefficient is based on Eq. [11.22]. This involves the term $d\beta/d\alpha$. The value depends on whether the cruise speed is supersonic or subsonic. For supersonic cruise, $d\beta/d\alpha = 1$. Otherwise, it depends on the location of the engine with respect to the main wing. If the engine is downstream of the main wing, $d\beta/d\alpha$ is based on Eq. [11.24]. Otherwise, it is based on Figure 11.5. In either of these cases, the value needs to be input to the spread-sheet.

The "calculations" portion of the spreadsheet dealing with the longitudinal stability lists the values of the contributions of each of the three components (wing, horizontal tail, and engine inlet) to the stability coefficient. Along with the magnitude, the sign is analyzed and the label next to the value indicates whether the components have "stable" or "unstable" contributions. The sum of the three contributions, corresponding to the total C_{M_α} is then listed at the bottom of the calculations table. Again a label based on the sign of the coefficient indicates if the condition is stable or unstable.

Directional Stability. The directional stability is next analyzed in the bottom half of the spreadsheet. The calculation of the directional stability coefficient, C_{n_β}, is based on Eq. [11.38]. This includes the contributions of the main wing (W), the fuselage (F), and the vertical stabilizer (VS).

The contribution of the wing to the coefficient is based on Eq. [11.43]. This again involves many of the parameters that can be found in the wing design spreadsheet, **wing.xls**.

The contribution of the fuselage to the coefficient is based on Eq. [11.44]. This involves parameters that can be obtained from the fuselage design spreadsheet, **fuse.xls**. The volume of the fuselage, VOL_F, can be estimated as was done before with the fuselage weight. This assumes a stepwise change in the diameter and computes the local volume as the local cross-section area times the distance between x-stations.

The contribution of the vertical stabilizer to the coefficient is based on Eq. [11.40]. Again many of the parameters can be obtained from the tail design spreadsheet, **tail.xls**. As with the horizontal tail, the distance l_{VT} used in the vertical tail volume coefficient, $\bar{V}_{VT}$, is calculated as the distance between the location of the vertical tail resultant load to the cg location. The term $(1 + \frac{d\sigma}{d\beta})\frac{q_{VS}}{q}$ is calculated based on Eq. [11.42].

The "calculations" portion of the spreadsheet dealing with the directional stability lists the values of the contributions of each of the three components (wing, fuselage, and vertical tail) to the stability coefficient. Along with the magnitude, the sign is again analyzed, with the label next to the value indicating whether the components make "stable" or "unstable" contributions. The sum of the three contributions, corresponding to the total C_{n_β} is then listed at the bottom of the calculations table. Again the label based on the sign of the coefficient indicates whether the condition is stable or unstable.

Finally, the lateral stability coefficient is estimated as $C_{\mathcal{L}_\beta} = -C_{n_\beta}$.

Rudder Area Sizing. The last part of the spreadsheet deals with the sizing of the rudder on the vertical stabilizer. This is based on either of the two conditions consisting of asymmetric power or a cross-wind at landing. These conditions were described in Section 11.2.5. The first step is to determine which of the conditions results in the largest control power, $C_{n_{\delta_R}}$. For the asymmetric power condition, it is based on Eq. [11.47]. For the cross-wind condition, it is based on Eq. [11.50]. Both of these involve a number of parameters that are found in the various other spreadsheets or have been previously calculated in this spreadsheet. The asymmetric thrust should correspond to the **unbalanced** thrust that would occur for a one-engine-out condition. For example, with four engines, it would correspond to the thrust of only one engine, since the other two would balance each other.

The drag on the engine that is out is based on Eq. [11.48]. It uses the velocity at landing corresponding to $1.2V_{TO}$ and a drag coefficient value of 1.2.

The largest of the $C_{n_{\delta_R}}$ values (regardless of sign) is used according to Eq. [11.51] to solve for the quantity $d\alpha_{0_L}/d\delta_R$. This value is then used with Figure 11.9 to find the chord ratio of the vertical stabilizer rudder, C_R/C_{VS}. This has to be read manually from the figure. If we assume that the area ratio scales directly with the chord ratio, then the result gives S_R/S_{VS}.

11.4   CASE STUDY: REFINED WEIGHT AND STATIC STABILITY ANALYSIS

We continue the case study corresponding to a conceptual supersonic business jet. The values in the spreadsheets discussed in the previous sections represent characteristics of this design that have evolved in the previous chapters.

11.4.1   Refined Weight Analysis

The spreadsheets corresponding to the refined weight analysis were presented in Figures 11.15 through 11.17. These address the weight estimates for the primary components on the aircraft.

For the main wing, the planform has a delta shape ($\lambda = 0$) so that $K_{dw} = 0.768$. In addition, it is a fixed geometry so that $K_{VS} = 1$. The design gross weight corresponds to the take-off weight, so that $W_{dg} = W_{TO} = 90,523$ lbs. All of the weight of fuel at take-off is assumed to be stored inside the wings, so that $W_{fw} = W_{fuel_{TO}} = 41,261$ lbs.

With these parameters, as well as the general characteristics of the main wing, the weights of the main wing are estimated to be 4122, 7843, and 0 pounds for the fighter, transport, and general aviation aircraft, respectively. The value of 0 lbs for the general aviation aircraft results from the taper ratio being zero. For the other aircraft, the coefficient exponent on λ is zero, making them independent of the taper ratio. Therefore, for this formula to be useful for estimating the weight of the wing on a general aviation class aircraft, the taper ratio cannot be zero. This is probably not a serious restriction since such aircraft would not likely use a delta wing planform.

The horizontal tail for this design is a canard. Most of its characteristics were developed in the **tail.xls** spreadsheet. The weight that is estimated by this spreadsheet has values of 228, 516, and 395 pounds for the fighter, transport, and general aviation class aircraft, respectively.

The vertical tail is a standard design. Again most of its characteristics were developed in the **tail.xls** spreadsheet. The vertical tail has a fixed geometry, with a rudder for control, so that $K_{rht} = 1$. We note that the estimated weights are higher than those of the horizontal tail. These have values of 334, 1505, and 667 pounds for the fighter, transport, and general aviation class aircraft, respectively.

Because of the supersonic cruise speed, the fuselage has a large l/d ratio to minimize wave drag. This ultimately gives it a relatively large weight. The design of the fuselage was done in the **fuse.xls** spreadsheet. It had been based on a Von-Karman Ogive shape, so that it has a circular cross section. Therefore, the structural depth (D) and width (W) are equal and set to be the maximum fuselage diameter. The coefficient K_{dwf} equals 0.774 because the main wing is a delta planform. Also the fuselage is designed to have one side cargo door, so that $K_{door} = 1.06$. Because of the small aspect ratio and small t/c of the main wing, the main landing gear is designed to mount to the fuselage. Therefore, the coefficient K_{lg} equals 1.12.

The wetted area of the fuselage, S_f, was calculated in the **fuse.xls** spreadsheet. The internal volume was estimated by summing the volume in 10 (12.6-foot long) constant diameter segments. Given these parameters, the estimated weight of the fuselage is 9977, 11,067, and 7155 pounds for the fighter, transport, and general aviation class aircraft, respectively.

The landing gear is envisioned to be very conventional. It will not be a cross-beam design so that $K_{cb} = 1$. It also will not be a kneeling gear so that $K_{mp} = 1$. Finally, it will not use a tripod configuration (unlike the A-7) so that $K_{tpg} = 1$. The design will have two struts and four main wheels (two on each strut), which is appropriate for its design gross weight. The stall velocity can be based on take-off, where the weight of the aircraft is a maximum, or based on landing, where it is a minimum. In these extremes, V_s ranges from 113 to 252 f/s.

In an extreme case in the landing analysis, the aircraft was assumed to land with 50 percent of its fuel remaining. This results in an approximate 30 percent reduction in the landing stall velocity to $V_s \simeq 194$ f/s. This is the value used for the landing gear weight estimate.

Based on these parameters, the weight of the landing gear is estimated to be 794, 1930, and 2054 pounds for the fighter, transport, and general aviation class aircraft, respectively.

The nose landing gear is also envisioned to be very conventional. It will not be a kneeling type so that $K_{np} = 1$. It will have a single strut with two nose wheels, which is again appropriate for its design gross weight. Based on this design, the weight of the nose landing gear is estimated to be 366, 462, and 454 pounds for the fighter, transport, and general aviation class aircraft, respectively.

By virtue of its cruise Mach number and size, the conceptual aircraft is possibly closer to a fighter class of aircraft. Therefore, the component weights based on the fighter aircraft equations are summarized in the table at the bottom of the spreadsheet. Added to this list is the weight of the *installed* engines. For transport aircraft, based on Table 11.7, this is 1.3 times the total uninstalled engine weight.

The component weights are presented as a percentage of the total structure weight in the third column of the table and in the pie chart. Here we observe that the largest component weight is the engines at 28 percent, followed by the fuselage at 22 percent, and the main wing at 9 percent. The other components individually correspond to less than 2 percent of the structure weight. The largest of these is the main landing gear.

The difference between the combined weight of these components and the total structure weight corresponds to approximately 36 percent of the structure weight. The "target" amount for the remaining structure based on 17 percent of W_{TO} (from Table 11.7) corresponds to 34 percent of the structure weight. That is, the total of the component weights should not exceed 66 percent of the overall structure weight. Based on these estimates, the design is within this limit.

11.4.2 Static Stability Analysis

The component weights were employed in the analysis of the static stability in the **stab.xls** spreadsheet. These are given in the Load Summary table at the top of the spreadsheet (Figure 11.18). The payload represents passengers and is, therefore, distributed over most of the length of the fuselage. The weight of the wing structure is distributed in the portion of the fuselage, which is slightly aft of the mid-point. The wing will carry all of the fuel. This will be primarily in the forward portion so that its resultant will be forward of the center of lift. The engines will be mounted under the wings, aft of the leading edge. Therefore, their position is centered at $x/L = 0.7$. The vertical tail is nearly to the aft end of the fuselage. The horizontal tail is placed near the most forward part of the fuselage. These positions maximize their moment arms. The remaining weight of the "other" components are assumed to be uniformly distributed over the rest of the fuselage.

With this distribution of weights, the center of gravity was found to be at $x_{cg}/L = 0.53$. This gave a static margin of 0.31, which is greater than zero and, therefore, statically **stable**. With this static margin, the lift force that the canard needs to generate for level flight is approximately 14,213 pounds. However, as pointed out in Section 11.2.6, such a canard configuration reduces the overall drag. Based on Figure 11.14, this static margin is close to producing the minimum drag.

Another important check is to verify that the aircraft remains statically stable as the fuel is used. When the fuel weight was zeroed, the static margin was reduced from 0.31 to 0.19; therefore, the aircraft still remains statically stable.

The analysis of the longitudinal stability coefficient showed that the contributions of the wing and canard were stabilizing. The engine inlet effect was destabilizing, but its magnitude was extremely small owing to the supersonic Mach number. Therefore, the total longitudinal stability coefficient was essentially the sum of those for the wing and canard, which gave a stable value of $C_{M_\alpha} = -0.54$.

The analysis of the directional stability showed that the contributions of the wing and vertical tail were stabilizing and that of the fuselage was destabilizing. In this case, because of its large l/d ratio, the directional coefficient of the fuselage was the largest of the three. The contribution of the wing was relatively small, so that the main component to offset the fuselage was the vertical tail. As a result, the area of the vertical tail was increased over the initial empirical estimate made in the tail (**tail.xls**) spreadsheet. The end result was a slightly stable directional stability coefficient of $C_{n_\beta} = 0.0034$.

Following our approach, the lateral stability coefficient was simply estimated to be the negative of the directional stability coefficient or $C_{\mathcal{L}_\beta} = -0.0034$.

For the rudder area sizing, the asymmetric power condition was found to require the largest rudder power coefficient. Based on the larger value, $d\alpha_{0_L}/d\delta_r = 0.04$. Using

Figure 11.9, this gave a required rudder area of only approximately 4 percent of the vertical stabilizer area. The reasons for such a small value are the relatively small directional stability coefficient and small asymmetric thrust. In the case of the latter, with four engines, the one-engine-out condition gave an asymmetric thrust of one engine.

Overall, all the conditions of static stability were met in the design. At this point, the conceptual design is essentially complete, except for an estimate of the cost. This topic is covered in the next chapter.

11.5 PROBLEMS

11.1. Consider the case study aircraft with an aft horizontal tail instead of a canard. Keeping the weight of the horizontal tail the same, place it at the same location as the vertical tail ($0.8 \leq x/L \leq 1$). Determine the fuel placement that gives the same static margin and longitudinal stability coefficient as for the canard configuration. Check to see that the static margin remains positive as the fuel weight goes to zero. Can you see any disadvantages to using an aft horizontal tail in this case?

11.2. For the case study aircraft, plot the weight of the main wing as a function of the following independent parameters:

1. t/c;
2. taper ratio, λ;
3. sweep angle, Λ.

Based on how the wing weight changes with these parameters, what is the shape of the optimum wing to minimize weight?

11.3. For the case study aircraft, plot the weight of the fuselage as a function of the following independent parameters:

1. ratio of the length to structural depth, L/D;
2. load factor, n;
3. design gross weight, W_{dg}.

Which has the greatest effect on the fuselage weight?

11.4. The weight of the fuselage also depends on the following properties of the main wing:

1. wing span, b_w;
2. wing sweep angle, Λ;
3. wing taper ratio, λ.

Vary these independently for the case study aircraft, and plot how the fuselage weight changes. Explain why you think the wing design is linked to the fuselage weight?

11.5. The effect of the main wing on the longitudinal (pitch) stability is through $dC_L/d\alpha$. Discuss how the parameters of the main wing (taper ratio, sweep angle, and aspect ratio) then affect the longitudinal stability. You may want to use the **wing.xls** spreadsheet in this discussion. Indicate how they change to *increase* the stability.

11.6. The effect of the main wing on the directional stability is given by Eq. [11.43]. Plot how the directional stability coefficient for the wing varies with aspect ratio and sweep angle. Based on this, what is the wing configuration that will maintain the flight direction the best?

11.7. The directional stability increases for aircraft with winglets on the main wing. Based on the results of Problem 11.6, explain why this happens.

11.8. The effect of the fuselage on the directional stability is given by Eq. [11.44], where it is a function of the fuselage volume, height, and width. Based on this, plot the directional stability coefficient for the fuselage as a function of the fineness ratio, d/l. What is the optimum shape of the fuselage from the point of view of the directional stability. Are there any disadvantages to that shape? Explain.

11.9. For the case study aircraft, compare the component weights given by the spreadsheet for the three classes of aircraft, to those estimated in the previous chapter using the factors in Table 10.3. What might be the source of differences in the two estimates?

11.10. To obtain Figure 11.12, the term in the brackets in Eq. [11.66] was solved for the conditions given in Table 11.9. The static margin was changed by moving the location of the resultant of the payload. This then changed the value of the lift on the horizontal stabilizer that was needed to balance the aircraft for level flight. Generate a plot similar to Figure 11.12 for the conditions of the case study aircraft. How far is the design from the minimum drag optimum?

11.11. Using Eq. [11.60], calculate the trim drag on the case study aircraft. Compare it to the total drag used for the engine selection in **engine.xls**. Does the trim drag satisfy the criterion given by Eq. [11.61]? What can be done if it does not?

11.12. In Eq. [11.66], what is the optimum static margin to minimize drag if the quantity $(A_W S_W)/(A_{HT} S_{HT})$ is large?

CHAPTER 12

Cost Estimate

12.1 COST ESTIMATING RELATIONSHIPS
12.2 UNIT PRICE
12.3 SPREADSHEET FOR COST ESTIMATION
12.4 CASE STUDY: COST ESTIMATE
12.5 PROBLEMS

Photograph of the Northrop Grumman Corporation B-2 bomber. The aircraft has a take-off weight of approximately 336,000 pounds, a high subsonic maximum speed, and a maximum altitude of approximately 50,000 feet. The unit cost of this aircraft is approximately $1.3 billion making it one of the most expensive ever built. There are 21 in service and one test aircraft. (U.S. Air Force Photograph)

This chapter presents an approach for obtaining an estimate of the cost of building an aircraft based on a conceptual design. These costs are grouped into two parts consisting of

1. research, development, test and evaluation (RDT&E);
2. acquisition (A).

The sum of these defines the purchase price of a new aircraft.

The costs of RDT&E and acquisition are two of the three parts that make up the total life-cycle cost (LCC) of an aircraft. The final part is the cost of operations and maintenance (O&M). Although this chapter will not attempt to estimate O&M costs, it is important to the purchaser, and because of the long lifetime of most aircraft, it generally represents the largest part of the life-cycle cost.

In estimating the cost of a conceptual design, RDT&E includes all of the technology research, development, engineering, fabrication, and flight testing of prototype aircrafts prior to committing to full production. The number of development aircraft fabricated in this phase is denoted as N_D. It usually varies from 1 to 6, with the lower being the more common.

RDT&E is broken into multiple elements for which costs are derived. These include

1. airframe engineering;
2. development support;
3. flight test aircraft, which is further broken down into

   (a) engine and avionics,
   (b) manufacturing labor,
   (c) manufacturing materials,
   (d) tooling, and
   (e) quality control;

4. flight test operations;
5. profit.

The acquisition (A) phase refers to the fabrication of N_P production aircraft. It includes many of the elements that go into the production of the flight test aircraft in the RDT&E phase. These are

1. engine and avionics,
2. manufacturing labor,
3. manufacturing materials,
4. airframe engineering,
5. tooling,
6. quality control, and
7. profit.

In this phase, airframe engineering supports modifications and upgrades that occur during production.

12.1 COST ESTIMATING RELATIONSHIPS

The basis for estimating the cost of manufacturing development or production aircraft is in cost estimating relationships (CERs). These attempt to correlate a few important characteristics of a large group of aircraft to produce simple model equations for estimating the cost. The equations utilize coefficients that are found by a best fit of these characteristics to the cost of existing aircraft in the data set. One of the earlier CERs is from a 1971 Rand Corporation Report (R-761-PR). This included 29 aircraft built between 1945 and 1970. It has been updated in a 1987 version (R-3255-AF) to include more aircraft of more modern design. The conclusion from these reports is that the primary characteristics that drive the cost of an aircraft are

1. the structure weight, W_s (units of pounds will be used);
2. the maximum speed at best altitude, V_{max} (units of knots will be used);
3. the quantity of aircraft produced, N_D or N_P;
4. in one element, the production rate, R.

It is reasonable to assume that the structure weight and maximum speed are related to the complexity of the aircraft design and thereby to the ultimate cost. The correlation with the quantity produced represents the "learning curve" effect on the time and cost of the different elements. The production rate only enters into the cost of tooling, since tools designed for low production rates do not have to be as well engineered as those for high production rates and, thereby, cost less.

The following will present the CERs for the different elements of RDT&E and acquisition listed earlier. For this, coefficients from both the 1971 and 1987 reports will be listed. In addition, the costs of labor in 1970 and 1986 from these reports will be listed in order to provide a means of extrapolation to the present year. Each of the models estimate the cost in their respective years. These need to be converted to present dollars using an appropriate economic escalation factor. A logical choice for this is the consumer price index (CPI), which is readily available from the U.S. Department of Labor Bureau of Labor Statistics Web site. For example, based on this, the conversion from 1970 and 1986 dollars to 1998 dollars is

$$\frac{\$_{\cdot 70}}{\$_{\cdot 98}} = 3.95 \tag{12.1}$$

and

$$\frac{\$_{\cdot 86}}{\$_{\cdot 98}} = 1.49. \tag{12.2}$$

12.1.1 Airframe Engineering

Airframe engineering relates to both RDT&E and acquisition. It involves the airframe design and analysis, wind tunnel testing, mock-ups, test engineering, and evaluation. During the acquisition phase, it also includes analysis and incorporation of modifications, material and process specifications, and reliability analysis.

The airframe engineering cost is first expressed as the total hours that are associated with this element. The hours are then converted to a cost based on an hourly rate for engineering. The CER for airframe engineering hours is

$$H_E = C_1 W_s^{C_2} V_{max}^{C_3} N^{C_4}, \tag{12.3}$$

TABLE 12.1: Coefficients for Eq. [12.3].

Year	C_1	C_2	C_3	C_4
1970	0.027	0.791	1.526	0.183
1986	4.860	0.777	0.894	0.163

where the coefficient values from the 1970 and 1986 reports are given in Table 12.1. In the RDT&E phase, $N = N_D$. For the sustained engineering in the acquisition phase, $N = N_P$.

It should be pointed out that in the 1971 CERs, the aircraft structure weight used in the formula was an AMPR (or DCPR) weight, which is the empty weight less the weight of components such as wheels, engines, instruments, electrical equipment, etc. The AMPR weight is approximately $0.62 W_s$. For consistency between the 1971 and 1987 relations, the factor relating the AMPR and structure weight has been absorbed into the coefficient C_1.

12.1.2   Development Support Cost

Development support is defined as the nonrecurring manufacturing effort to support engineering during the RDT&E phase. This involves the labor and material required to produce mock-ups, test parts, and other items needed for the airframe design and development. The CER for the cost of this element is

$$C_D = C_1 W_s^{C_2} V_{max}^{C_3} N^{C_4}. \tag{12.4}$$

Since this cost is only relevant to the RDT&E phase, $N = N_D$ only. The coefficients for Eq. [12.4] are given in Table 12.2.

12.1.3   Engine and Avionics Cost

The engine and avionics are assumed to be "off-the-shelf" items so that the cost is presumably known from the manufacturer. For the engines, an estimate based on 1971 data for turbo-jet engines relates the cost to the maximum sea-level thrust, namely,

$$C_{EN} = C_1 T_{max-SL}^{C_2}, \tag{12.5}$$

where the coefficient values are given in Table 12.3.

Alternatively, the 1987 report includes a different relation for the engine cost which is based on the maximum thrust at sea level, T_{max-SL}, as well as the maximum Mach

TABLE 12.2: Coefficients for Eq. [12.4].

Year	C_1	C_2	C_3	C_4
1970	0.00549	0.873	1.890	0.346
1986	45.42	0.630	1.300	0

TABLE 12.3: Coefficients for Eq. [12.5].

Year	C_1	C_2
1970	109 (turbo-jet) 130 (turbofan)	0.836
1986	—	—

TABLE 12.4: Coefficients for Eq. [12.6].

Year	C_1	C_2	C_3	C_4	C_5
1970	—	—	—	—	—
1986	1548.0	0.043	243.25	0.969	2228.0

number, M_{max}, and the turbine inlet temperature, Θ_I, with units of absolute temperature (Rankine). The equation for the 1986 engine cost is

$$C_{EN} = C_1 \left[C_2 T_{max-SL} + C_3 M_{max} + C_4 \Theta_I - C_5 \right].$$ (12.6)

where the coefficient values are given in Table 12.4.

12.1.4 Manufacturing Labor Cost

The manufacturing labor costs relate to both the RDT&E and acquisition phases. It is based on the number of hours that are needed to fabricate and assemble the major structural elements of the aircraft. It includes the labor associated with the installation of off-site or purchased manufactured components and the labor costs of manufacturing performed by subcontractors. The CER for the hours associated with the manufacturing labor cost is

$$H_{ML} = C_1 W_s^{C_2} V_{max}^{C_3} N^{C_4},$$ (12.7)

where the coefficients are given in Table 12.5. Again for the RDT&E phase, $N = N_D$, and for the acquisition phase, $N = N_P$. The hours are converted to a cost based on an hourly rate for manufacturing labor.

12.1.5 Manufacturing Materials Cost

The manufacturing materials costs also relate to both the RDT&E and acquisition phases. It includes the raw materials, hardware, and equipment required for the fabrication and

TABLE 12.5: Coefficients for Eq. [12.7].

Year	C_1	C_2	C_3	C_4
1970	20.348	0.740	0.543	0.524
1986	7.370	0.820	0.484	0.641

TABLE 12.6: Coefficients for Eq. [12.8].

Year	C_1	C_2	C_3	C_4
1970	18.47	0.689	0.624	0.792
1986	11.00	0.921	0.621	0.799

assembly of the airframe. The CER for this cost is

$$C_{MM} = C_1 W_s^{C_2} V_{max}^{C_3} N^{C_4}, \tag{12.8}$$

where the coefficients are given in Table 12.6. For the RDT&E, $N = N_D$, and for the Acquisition, $N = N_P$.

12.1.6 Tooling Cost

Tooling refers to jigs, fixtures, dies, and other special equipment that are used in the fabrication of the aircraft. The cost of tooling is first expressed in hours required for tool design, fabrication, and maintenance. It also includes production planning since the tool design depends on the production rate, where tool wear and breakage are factors. The CER for the hours related to tooling is

$$H_T = C_1 W_s^{C_2} V_{max}^{C_3} N^{C_4} R^{C_5}, \tag{12.9}$$

where R is the production rate defined as aircraft per month. The coefficients for Eq. [12.9] are given in Table 12.7.

Tooling is relevant to both RDT&E and acquisition. Therefore, $N = N_D$ or N_P for the respective phases. In either, the hours for tooling are converted to a cost based on an hourly rate for tooling labor.

12.1.7 Quality Control

Quality control is the task of inspecting (1) fabricated and purchased parts and (2) assemblies for defects and adherence to specifications. The time associated with quality control is related to the total number of labor hours. The CER for quality-control hours is

$$H_{QC} = C_1 H_{ML}, \tag{12.10}$$

where the coefficient is given in Table 12.8. We note that the 1987 CER associates approximately half as many hours to quality control for cargo aircraft. Again the total hours are converted to a cost using an appropriate hourly wage rate. This relates to both RDT&E and acquisition phases.

TABLE 12.7: Coefficients for Eq. [12.9].

Year	C_1	C_2	C_3	C_4	C_5
1970	2.79	0.764	0.899	0.178	0.066
1986	5.99	0.777	0.696	0.263	0

TABLE 12.8: Coefficients for Eq. [12.10].

Year	C_1
1970	0.13
1986	0.13 (0.076 for cargo)

12.1.8 Flight Test Cost

The flight test costs include all the elements involved in conducting aircraft flight tests. It includes flight test engineering planning, data analysis, instrumentation, fuel, test pilot salary, facilities, and insurance. The flight tests are essential for establishing the aircrafts operation envelope and flying characteristics. The CER cost estimate for this is

$$C_{FT} = C_1 W_s^{C_2} V_{max}^{C_3} N^{C_4},$$ (12.11)

where the coefficients are given in Table 12.9. Flight testing only relates to the RDT&E phase, so that $N = N_D$.

12.1.9 Profit

Profit is based on a fixed percentage of the total cost of all of the elements in RDT&E and acquisition phases. A typical profit value, PV, is 10 percent. The total profit is then

$$C_P = PV\Sigma(C_{RDT\&E} + C_A).$$ (12.12)

12.1.10 Hourly Rates

As mentioned, the hours estimated in some of the CERs are converted to a cost based on an appropriate hourly rate for labor. These hourly rates include the total cost made up of salaries, overhead, benefits, administrative expenses, and miscellaneous direct charges. Table 12.10 lists the hourly rates for airframe engineering, tooling, manufacturing, and

TABLE 12.9: Coefficients for Eq. [12.11].

Year	C_1	C_2	C_3	C_4
1970	0.000714	1.160	1.371	1.281
1986	1243.03	0.325	0.822	1.210

TABLE 12.10: Hourly rates for engineering, tooling, manufacturing, and quality control for the years 1970 and 1986.

Year	Engineering	Tooling	Manufacturing	Quality Control
1970	16.00	11.50	10.00	10.00
1986	59.10	61.70	55.40	50.10

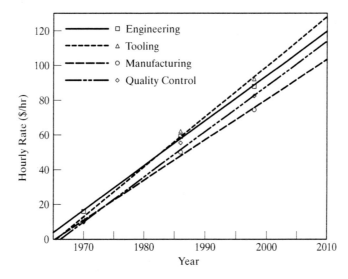

FIGURE 12.1: Trends in hourly rates in aircraft construction for engineering, tooling, manufacturing, and quality control.

quality control for the years 1970 and 1986. These have also been plotted in Figure 12.1. The figure also includes values for 1998 based on the Consumer Price Index conversion (Eq. [12.2]) of the 1986 hourly rates. A linear curve fit has been applied to the data in order to aid in making projections to future years. For reference, the linear relations are

$$\mathrm{CR}_E = -5058 + 2.576Y; \tag{12.13a}$$

$$\mathrm{CR}_T = -5666 + 2.883Y; \tag{12.13b}$$

$$\mathrm{CR}_{\mathrm{ML}} = -4552 + 2.316Y; \tag{12.13c}$$

$$\mathrm{CR}_{\mathrm{QC}} = -5112 + 2.600Y; \tag{12.13d}$$

where Y is the year (e.g., 1989) and the cost rate, CR, is in dollars in year Y.

In general, the hourly rates follow a linear trend reasonably well. At the present time, and based on their projections, tooling has the highest hourly rate, followed by engineering, quality control, and manufacturing. These hourly rates are then used to convert the hour values associated with their respective elements to the cost in dollars.

12.2   UNIT PRICE

The unit price is made up of the costs of RDT&E and acquisition elements in the construction. In general, the cost of RDT&E is amortized over a period of time, t_{am}. This time can be put in terms of a number of aircraft built, based on the production rate, R. Therefore, the number of aircraft built during the amortization period is $N_{\mathrm{am}} = Rt_{\mathrm{am}}$. For example, if the production rate is 10 aircraft per year, and the amortized period is 4 years, then the cost of RDT&E will be recovered after 40 aircraft are built and sold.

In general, including an amortization period over which RDT&E costs are distributed, the unit price of an aircraft is then as follows:

For $t \le t_{am}$, Unit Price ($)

$$= \frac{1}{N_P} \sum \left(C_E + C_{ML} + C_{MM} + C_T + C_{QC} + C_{EN} + C_P \right)_A$$

$$+ \frac{1}{N_{am}} \sum \left(C_E + C_D + C_{ML} + C_{MM} + C_T + C_{QC} + C_{FT} + C_{EN} + C_P \right)_{RDT\&E}. \qquad (12.14)$$

For $t > t_{am}$, Unit Price ($)

$$= \frac{1}{N_P} \sum \left(C_E + C_{ML} + C_{MM} + C_T + C_{QC} + C_{EN} + C_P \right)_A. \qquad (12.15)$$

Remember that the difference between the costs of elements that are common to both RDT&E and acquisition is in the number of aircraft produced, N_D versus N_P.

12.3 SPREADSHEET FOR COST ESTIMATION

The CERs for the cost estimation of RDT&E and acquisition for a conceptual design were put into the spreadsheet named **cost.xls**. A sample of the spreadsheet with input parameters relevant to the case study aircraft is shown in Figures 12.2 to 12.5.

The top of the spreadsheet contains various parameters that are needed for the CER relations. These include

1. the structure weight, W_s (in pounds);
2. the maximum velocity, V_s (in feet per second which is converted to knots);
3. the number of RDT&E aircraft to be built, N_D;
4. the number of acquisition aircraft to be produced, N_P;
5. the production rate of $RDT\&E$ aircraft (aircraft/month), R_D;
6. the production rate of acquisition aircraft (aircraft/month), R_P;
7. the year, Y, to which dollar amounts are to be referenced.

Because these parameters are used in all the CER formulas, they were given global names throughout the spreadsheet.

The top right portion of the spreadsheet lists CPI rates that are used to convert dollars from 1970 to 1998, 1986 to 1998, and 1998 to 2000. The former two are historic and need not be changed. The latter is an estimate to the present time and would need to be updated in future years. The latter CPI must be consistent with the value given as year (Y) in the list of inputs. That is, the last CPI is used to convert from 1998 in either of the two models (1971 or 1987) to the year Y. The three CPI rates are also given global names (CPI1, CPI2, and CPI3, respectively) in the spreadsheet.

The cost rates in dollars per hour are computed for year Y based on Eqs. [12.13a–12.13d], (Figure 12.1). These also appear in the top portion of the spreadsheet. This includes the cost rates for engineering, CR_E; tooling, CR_T; manufacturing labor, CR_{ML}; and Quality Control, CR_{QC}.

		Cost Analysis			
Input Parameters:				**Constants:**	
W_s	45,261	lbs		CPI (70–98)	3.950
V_max	1926	f/s		CPI (86–98)	1.490
N_D	2	aircraft		CPI (98–00)	1.056
N_P	250	aircraft			
R_D	0.08	aircraft/mo.			
R_P	1	aircraft/mo.			
Year	2000				
V_max	1140.37	knots			
CR_E	94	$/hr			
CR_T	100	$/hr			
CR_ML	80	$/hr			
CR_QC	88	$/hr			
Research, Development, Test and Evaluation					
		Based on 1970	**Based on 1986**		
Airframe Engineering:					
	H_E (hrs)	15,188,522.88	19,484,287.81		
	C_E ($)	$1,427,721,150	$1,831,523,054		
Development Support:					
	C_D ($)	$545,051,187	$1,140,923,947		
Manufacturing Labor:					
	H_ML (hrs)	4,955,251.20	2,936,940.55		
	C_ML ($)	$396,420,096	$234,955,244		
Manufacturing Materials:					
	C_MM	$24,130,322	$64,041,180		
Tooling:					
	H_T (hrs)	8,664,149.58	5,757,264.00		
	C_T ($)	$866,414,958	$575,726,400		
Quality Control:					
	(C or O)			O	Cargo or Other Aircraft
	H_QC	644,182.66	381,802.27		
	C_QC	$56,688,074	$33,598,600		
Flight Test:					
	C_FT	$58,007,269	$73,905,354		

FIGURE 12.2: Spreadsheet for estimating the unit price of a conceptual aircraft (Part 1).

Engine:					
	T_max (lbs)	10,200			
	(TJ or TF)	TJ		turbofan or turbojet	
	N_eng	4			
	C_EN ($)	$8,165,208	$8,165,208		
Subtotal (RDT&E):		$3,382,598,265	$3,962,838,987		
Profit:					
	Percent	10	10		
	C_P	$338,259,827	$396,283,899		
Total (RDT&E):		$3,720,858,092	$4,359,122,886		
Acquisition:					
		Based on 1970	Based on 1986		
Airframe Engineering:					
	H_E (hrs)	36,749,384.70	42,803,656.30		
	C_E ($)	$3,454,442,162	$4,023,543,693		
Manufacturing Labor:					
	H_ML (hrs)	62,208,040.61	64,865,605.02		
	C_ML ($)	$4,976,643,249	$5,189,248,402		
Manufacturing Materials:					
	C_MM	$1,104,881,883	$3,033,125,833		
Tooling:					
	H_T (hrs)	20,463,311.91	20,497,621.92		
	C_T ($)	$2,046,331,191	$2,049,762,192		
Quality Control:					
	(C or O)		O	Cargo or Other Aircraft	
	H_QC	8,087,045.28	8,432,528.65		
	C_QC	$711,659,985	$742,062,521		
Engine:					
	T_max (lbs)	10,200			
	(TJ or TF)	TJ		turbofan or turbojet	
	N_eng	4			
	C_EN ($)	$1,020,650,976	$1,020,650,976		
Subtotal (A):		$13,314,609,445	$16,058,393,617		
Profit:					
	Percent	10	10		
	C_P	$1,331,460,945	$1,605,839,362		
Total (A):		$14,646,070,390	$17,664,232,978		

FIGURE 12.3: Spreadsheet for estimating the unit price of a conceptual aircraft (Part 2).

Total Profit:		$1,669,720,771	$2,002,123,260		
Aircraft Unit Cost:					
RDT&E amortized					
unit no.:		200	200		
Cost / Prod. Aircraft		$18,604,290	$21,795,614		
Unit cost:					
1	200	**$77,188,572**	**$92,452,546**		
201	250	**$58,584,282**	**$70,656,932**		
Summary Totals:		**Based on 1970**	**Based on 1986**		
	C_E ($)	$4,882,163,312	$5,855,066,747		
	C_D ($)	$545,051,187	$1,140,923,947		
	C_ML ($)	$5,373,063,345	$5,424,203,645		
	C_MM ($)	$1,129,012,206	$3,097,167,013		
	C_T ($)	$2,912,746,149	$2,625,488,592		
	C_QC ($)	$768,348,058	$775,661,121		
	C_EN ($)	$1,028,816,184	$1,028,816,184		
	C_P ($)	$1,669,720,771	$2,002,123,260		

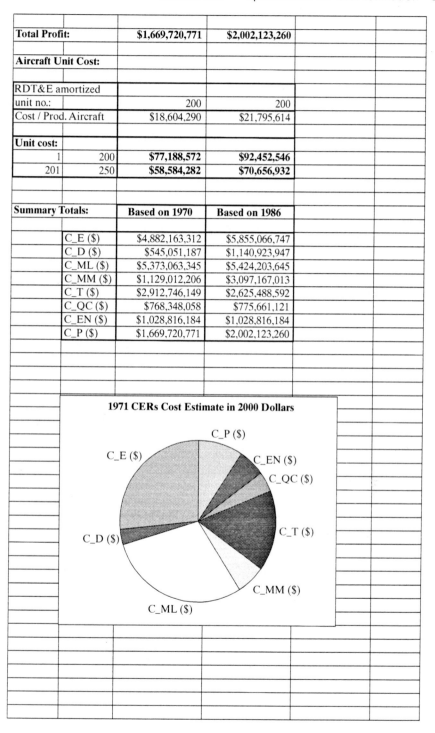

FIGURE 12.4: Spreadsheet for estimating the unit price of a conceptual aircraft (Part 3).

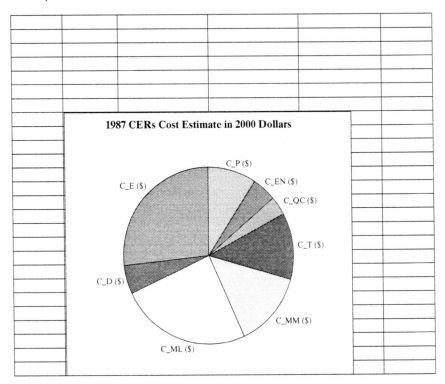

FIGURE 12.5: Spreadsheet for estimating the unit price of a conceptual aircraft (Part 4).

Finally, the velocity that is used in the CERs, V_{max}, is converted to knots, which is the desired unit for the correlation relations. The value is shown just below the input parameters on the spreadsheet.

The cost estimates are divided into two parts: RDT&E and acquisition. Each of these are further broken down into the relevant elements. The costs in each of these are associated with the respective number and production rate of development (D) or production (P) aircraft. In some cases, the CER predicts the number of hours associated with an element. This is converted to a year Y dollar amount using the appropriate cost rate. If the CER makes a direct estimate in the cost of an element, it is converted to year Y dollars by multiplying by the appropriate CPI values.

Since the CERs are empirical, the spreadsheet includes the estimates based on **both** the 1971 and 1987 formulas and correlation constants. The costs predicted in each of these are originally in 1970 and 1986 dollars. However, as previously mentioned, these are automatically converted to year Y dollars. It is reasonable to assume that the 1987 CERs would provide better estimates, since they should include more aircraft of more modern construction. However, comparing the estimates from the two CERs can help to provide an additional perspective on the trends for the cost of aircraft construction and possibly give a quantitative uncertainty to the estimates.

The costs of each of the elements of RDT&E and acquisition are computed based on the parameters input at the top of the spreadsheet. Besides these, the estimate of the engine cost and profit require additional input values. The cost estimate for the engine is based on Eq. [12.5], which is from the 1971 formulas. This requires values for the maximum thrust, $T_{\mathrm{max-SL}}$; the number of engines, N_{eng}; and a designation of TJ or TF, which signify "turbo-jet" or "turbofan."

The estimate for the cost of quality control using the 1987 CER requires designating whether the design is a cargo aircraft. This is specified by a C for "cargo" or an O for "other" in the sections of the spreadsheet labeled "Quality Control."

The costs of each of the elements, before profit, are added together to form a subtotal. The profit cost requires a value for the profit rate in percent. The profit is computed based on the subtotal and then added to the subtotal to give the respective cost associated with RDT&E and acquisition.

The unit cost of the aircraft is based on Eqs. [12.14] & [12.15]. This requires specifying an amortizing period, given as the number of production aircraft. Based on this, the total cost of RDT&E is divided by the number of aircraft produced in the amortizing period, to give a RDT&E cost per aircraft.

The unit cost in the amortizing period, $1 \leq N \leq N_{\mathrm{am}}$, is the total acquisition cost divided by the number of production aircraft, N_P, plus the RDT&E cost divided over N_{am} aircraft. For $N > N_{\mathrm{am}}$, the unit cost is only the total acquisition cost divided by the number of production aircraft.

The bottom of the spreadsheet gives a summary of the cost in each of the smaller elements of RDT&E and acquisition. For elements that were common to both of these, the sum of both is listed. These are then plotted as a percentage of the total cost in the form of a pie chart. Both of the estimates based on the 1971 and 1987 CERs are plotted for comparison.

The pie charts provide a nice way of viewing the cost breakdown and comparing the two cost estimates. In both models, the engineering and manufacturing labor represented the largest costs, making up nearly 60 percent of the total cost of the aircraft. The next highest costs were associated with tooling and manufacturing materials. In the 1971 CERs, the cost of tooling was more than twice that of the other. In the 1987 CERs, the two costs were nearly the same, with manufacturing materials increasing in percentage by more than a factor of two.

Overall, the two models were relatively close in their cost breakdown. The largest difference occurred with the cost of development in RDT&E. In the 1971 model, C_D depended on the number of development aircraft. In the 1987 model, the cost of development was not a function of N_D. As a result, for a small number of development aircraft, the difference could be quite large. For example, for the other input parameters remaining the same, it would require building approximately 17 development aircraft to have an equal amount of development cost between the two CER models. Otherwise, with only the intended two development aircraft being built, the difference between the two development cost estimates corresponds to an additional $2.5M on the cost of each of the first 200 aircraft.

12.4   CASE STUDY: COST ESTIMATE

We continue the case study corresponding to a conceptual supersonic business jet (SSBJ) by obtaining an estimate of the cost per aircraft. As mentioned, the spread-sheet shown in Figures 12.2 to 12.5 includes all of the parameters for this design.

The structure weight, W_s, came from the first spreadsheet, **itertow.xls**. The maximum velocity corresponded to the highest cruise velocity, which occurs at the start of cruise, at the design cruise Mach number and altitude. This was taken from the wing loading spreadsheet, **wing.xls**.

Since the aircraft was expected to be expensive, the number of development aircraft was set to be a relatively low value of two. The production rate for the development aircraft was estimated to be one per year.

In order to keep the unit price down, the number of production aircraft needed to be relatively high. In the case study, this was set to be 250 aircraft. The production rate was set to be 12 aircraft per year, or one per month. Finally, Y was set to give the cost in year 2000 dollars.

In the estimate of the cost of quality control, the aircraft type was signified as "O" to indicate that it is *other* than a cargo aircraft.

In the estimate for the engine cost, the maximum thrust, T_{max-SL}, was taken from the engine spreadsheet, **engine.xls**. The selected engine is a GE-F404-100D, which would have a scaled thrust of 10,200 lbs. This is a low bypass engine so that "TJ" was input to signify a *turbo-jet*.

The profit value was set to be a nominal 10 percent. Depending on the CERs used, this amounted to approximately $1.7 to 2.0 billion.

In order to keep the cost of the aircraft at a reasonable level, the RDT&E cost was amortized over the first 200 aircraft. Therefore, based on the two CERs for the first 200 aircraft, the cost per aircraft would range from $77 to $92 million. This would drop to approximately $58 to $70 million after the 250th aircraft was produced.

How does this price compare to the competition? Of the six comparison aircraft listed in the design proposal in Chapter 1, the two that are closest in characteristics (Sukhoi S-21 and Dassault SSBJ) have not been built. The others, including the Mig-31, TU-22M, and Dassault Falcon 900B, do not have published prices. A published estimate for the Dassault SSBJ, which was pictured in Figure 1.10, is from $70 to 80 million which is close to our estimate.

To provide some additional reference, the Learjet 31A is considered for comparison. Its published price in year 2000 dollars is approximately $6 million. This is approximately 13 times less that the proposed supersonic business jet! However, the Learjet has a maximum Mach number of 0.81 versus 2.1 for the SSBJ. In addition, it carries half the payload and has half the range. Therefore, the case study SSBJ has particular advantages that could justify the cost difference.

12.5 PROBLEMS

12.1. Compare the cost of the engine for the case study aircraft using Eq. [12.6] to that in the spreadsheet that was based on Eq. [12.5].

12.2. Keeping everything else the same, plot the unit price of the case study aircraft as an independent function of
  1. the maximum velocity,
  2. the structure weight, and
  3. the number of production aircraft.
  Describe the trends.

12.3. Explain what you think accounts for the changes in the hourly rate costs seen in Figure 12.1, from the 1970s to the present. Based on these, how best can the cost of an aircraft be reduced?

12.4. In 1974, the cost of a Cessna Citation was approximately $840,000. For this aircraft, the structure weight was approximately 6200 lbs and the design cruise Mach number was 0.7 at an altitude of 30,000 feet. It was powered by two PW JT15D turbofan engines, each having a maximum thrust of 2500 lbs.

Using the spreadsheet, compare the estimated cost from the 1971 CERs based on having three (3) development aircraft and 250 production aircraft. Assume that the cost of RDT&E will be amortized over 500 aircraft.

12.5. The Boeing 777-200 had an estimated development cost of $4 billion in 1991. This involved the construction of nine development aircraft. The maximum take-off weight is 515,000 lbs, and the maximum payload is 121,000 lbs. Its maximum Mach number is 0.83 at an altitude of 37,000 feet. It uses two engines that each produce 73,000 lbs of thrust (static-sea level). The maximum fuel amount is 44,000 U.S. gallons. Its empty weight is listed as 298,900 lbs.

Based on this information, compare the RDT&E cost you obtain from the CERs in the spreadsheet to the listed development cost. Indicate any assumptions that you need to make.

12.6. The McDonnell Douglas F-15E had a cost of $35 million in 1992. This aircraft has an empty weight of 32,000 lbs and a maximum Mach number of 2.5. It is powered by two F110 engines, each producing 29,100 lbs of thrust (static-sea level). A total of 209 aircraft were built from 1986 through 1992.

Based on this information, compare the total cost you obtain from the CERs relations in the spreadsheet to the listed price. Indicate any assumptions that you need to make.

12.7. The Northrop B-2 bomber is one of the most expensive aircraft ever built, with a 1992 unit cost of approximately $2.2 billion. Initially, in 1987, the total cost for 75 aircraft was to have been $64.7 billion, giving a unit cost of approximately $860 million. The 1992 cost was based on only producing 20 aircraft.

1. Assuming that the RDT&E costs were the same, what accounts for the increase in the cost of the B-2 in 1992 versus 1987?

2. The B-2 has an empty weight of 110,000 lbs and is powered by four F118-GE engines, each producing 19,000 lbs of thrust. The maximum speed is unlisted. Given this information, estimate the RDT&E costs based on the 1987 CERs in the spreadsheet.

12.8. The Gulfstream V is a twin turbofan long-range business transport. Its empty weight is 45,000 lbs. It has a cruise speed of 459 knots. Its two engines have a combined maximum thrust of 29,360 lbs. The unit cost for the first 24 aircraft was $29.5 million in 1993.

Use the CERs in the spreadsheet, and compare the estimated cost to the price in 1993.

12.9. The Swearingen SJ30 is a small business jet that carries four passengers and two crew. Its empty weight is 6210 lbs. Its maximum velocity is 470 knots, and it is powered by twin turbofan jet engines that produce a combined maximum thrust of 3800 lbs. The unit price for this aircraft was $2.595 million in 1989.

Use the CERs in the spreadsheet, and compare the cost per aircraft to the cited price.

CHAPTER 13

Design Summary and Trade Study

13.1 TRADE STUDY
13.2 PROBLEMS

Photograph that gives the appearance of a "what if" that exemplifies the end-of-design trade study. Actually it is a turboprop experimental version of the XF-88 Voodoo, which was built to conduct propeller research for supersonic planes. (Courtesy of the Boeing Company.)

Following the steps in the previous chapters, a complete conceptual design is reached. The process started with a proposal for a type of aircraft, which was specified by a few characteristics such as its range, payload, cruise Mach number and altitude, wing aspect ratio, engine thrust-specific fuel consumption, and structure factor. These were chosen by considering the characteristics of existing aircraft, which were of a similar class as the proposed design. Based on these, along with some theoretical relations, with gaps filled by historical trends, we were able to estimate the weight of the conceptual aircraft at every step in a flight plan from take-off to landing.

A crucial aspect of the design was the selection of "principle design drivers." This step acknowledged that although our objective was to optimize every part of the performance of the aircraft, optimizing one often had an adverse effect on another. There were many examples where this occurred, such as having a high wing loading to improve cruise efficiency also increased take-off and landing distances; or having a large wing sweep angle to lower the drag in high-speed flight also greatly reduced the effectiveness of lift-enhancing devices (flaps and slats). Therefore, the final design has many features that are the outcome of the initial selection of a *few* principle design drivers.

The first step in the conceptual design was a proposal. This defined the type of aircraft and its performance objectives. Accompanying this was a list of comparison aircraft, whose characteristics could be used as both a check on aspects of the conceptual design, as well as a *motivation* to exceed the capabilities of close existing aircraft.

In the end, it is possible that some of the original mission requirements were not met. On the other hand, some may have been exceeded. The latter is often the case as more of the inter-relationships between parameters are better appreciated. Therefore, it is ultimately important to examine the final characteristics for a new design. As an example, a detailed summary for the conceptual SSBJ that was proposed in Chapter 1, along with those of its comparison aircraft, is given in Table 13.1.

The characteristics of the conceptual SSBJ most closely resemble those of the Mig-31. In terms of the weights, the take-off, empty, fuel, and payload are all quite comparable. An important parameter in estimating the take-off weight is the structure factor, W_e/W_{TO}. These were nearly the same for the two aircraft.

With regards to the wing design, the wing area of the SSBJ was lower, giving a higher wing loading compared to the Mig. This is appropriate given the different design drivers for the two aircraft. For the Mig, a lower wing loading is appropriate for better maneuverability, whereas, for the SSBJ, a higher wing loading is more appropriate for long range cruise.

The aspect ratio of the SSBJ is also smaller than on the Mig or the proposed S-21. This primarily affects the 3-D lift distribution and the lift-induced drag. However, the difference between aspect ratios of 2 to 3 on these is relatively insignificant. The advantage of having a smaller aspect ratio is towards lowering the wing weight. This was important to this design because it then lowered the take-off weight, which ultimately affected the unit cost.

The wing span of the case study aircraft is about 25 percent shorter than that of the Mig, which is consistent with the smaller wing area. The proposed S-21 has twice the wing span; however, its wing planform is considerably different, with a taper ratio that is closer to one.

The fuselage lengths of the conceptual SSBJ and the S-21 are relatively close. Both have fineness ratios, H/L, which are small to minimize the wave drag in supersonic flight. The SSBJ fineness ratio is the optimum $H/L = 1/14$. The proposed S-21 would have

TABLE 13.1: Detailed summary of the characteristics of the conceptual SSBJ and other comparison aircraft.

	SSBJ Final	Sukhoi* S-21	Mig-31	Tu-22M	Dassault Falcon 900B
Weights:					
W_{TO} (lbs)	90,523	106,000	90,390	273,370	45,500
W_e (lbs)	45,261	—	48,115	—	22,575
W_f (lbs)	41,261	—	36,045	110,230	19,158
W_p (lbs)	4,000	—	—	52,910	4,817
W_e/W_{TO}	0.50	—	0.53	0.40	0.50
Wing:					
S (f^2)	519	—	663	1892	527
b (f)	32.2	65.3	44	76.5	63.5
A	2.0	3.1	2.94	3.09	7.62
$(W/S)_c$ (lbs/f^2)	157	—	132	144	83
Fuselage:					
L (f)	126	132.9	—	139	66
H (f)	9	6	—	—	8
Engines					
No.	4	2	2	2	3
T_{max-SL} (lbs)	11,000	27,560	34,170	48,500	4,750
$(T/W)_{max-SL}$	0.54	0.52	0.76	0.36	0.31
Take-off/Landing					
s_{TO} (f)	7,566	6,500	3,940	6,700	4,970
s_L (f)	3,906	6,500	2,625	4,000	2,300

a value of 1/22. There are certainly other considerations for the selection of a higher fineness ratio for the S-21, but the trade-off is a larger amount of viscous drag.

The SSBJ was fairly optimally designed to minimize drag. As a result, its total thrust was the least of the group of comparison aircraft. The next closest was the proposed S-21, which would have a total maximum thrust of 55,120 lbs. The S-21 would use two engines to produce this thrust. The SSBJ proposed four engines. The rationale for this design was to minimize the engine diameter, since the engines would mount below the wing, and the smaller engine would reduce the ground clearance height. However, like the S-21, it is possible to modify the SSBJ design and use two scaled turbo-jet engines, which can supply the necessary thrust.

The take-off and landing distances of the SSBJ are somewhat mixed with regard to the comparison aircraft. Those of the Mig-31 are shorter, although the landing distance is comparable. The take-off distance of the Mig-31 is significantly shorter, but this is due to its higher thrust-to-weight ratio. The SSBJ take-off distance is comparable to that projected for the S-21, but the landing distance is considerably less. These distances are somewhat longer than that of the Dassault, which is representative of subsonic business jets; however, they are short enough to give it access to the same air fields.

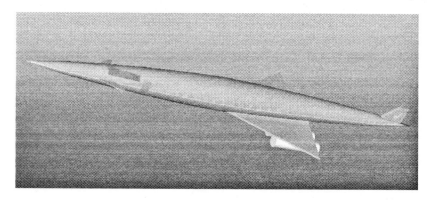

FIGURE 13.1: Final 3-D Drawing of supersonic business jet case study proposed in Chapter 1.

A final 3-D drawing of the SSBJ case study is shown in Figure 13.1. This illustrates all of the features of the design. The similarity to the Dassault design in Figure 1.10 reflects the fact that the two aircraft had similar design drivers.

13.1 TRADE STUDY

It is possible that a designer may want to examine variations in the original design. Such an examination of the impact of individual parameters on the total design is referred to as a "trade study." An example of this would be to evaluate how wing loading, range, or payload might affect the unit cost of a conceptual aircraft. The following lists some commonly made design trades and their respective outcomes to be evaluated by the study.

Trade Study		Outcome to Be Evaluated
T/W and W/S	$\longrightarrow$	Range and Payload
A and Λ	$\longrightarrow$	Range & Loiter Time
t/c, λ, and Airfoil Type	$\longrightarrow$	Cruise speed
$C_{L_{max}}$	$\longrightarrow$	$\dot{\psi}$ & s_{TO} and s_L
TSFC and T/W	$\longrightarrow$	s_{TO} and s_L
Static Margin	$\longrightarrow$	Range & Payload
Configuration	$\longrightarrow$	Cost

Examining each of these, the thrust-to-weight ratio (T/W) affects the range and payload because it affects the thrust-specific fuel consumption (TSFC). A higher T/W increases the amount of fuel used for a mission. For the same take-off weight, the added fuel weight has to be traded for a reduced range or a reduced payload. In either case, the outcome will be very dramatic, owing to an exponential dependence of both of these on the TSFC.

For a specified take-off weight, the value of the wing loading sets the wing area, S. As the wing area increases, the viscous drag increases. For a high aspect-ratio wing, this can be a significant part of the total drag on the wing. The higher drag then requires an increase in the required thrust and, thereby, an increase in the fuel weight, which then must be traded off with payload or range.

Both the aspect ratio, A, and the sweep angle, Λ, affect the drag on the wing. In the initial estimate of the take-off weight (**itertow.xls**), the aspect ratio affects the lift-to-drag ratio (L/D), which appears in the range equation for the fuel fraction. For supersonic Mach numbers however, L/D is independent of aspect ratio.

The sweep angle affects the effective Mach number over the wing. This most directly affects the viscous drag on the wing (**wing.xls**). This effect will be most greatly felt with high aspect-ratio wings, in which the lift-induced drag is minimal.

The choice of the wing characteristics affects the drag and, therefore, the excess power needed for high-speed intercept. For subsonic aircraft, the optimum t/c was between 12 to 14 percent. In supersonic aircraft, smaller values are needed to reduce wave drag. For efficient cruise, a cambered airfoil is usually selected so that the design lift coefficient falls within the "drag bucket." In an optimum design, the full range of the lift coefficient, from the start to the end of cruise, remains in the drag bucket. A trade study on the choice of the airfoil characteristics would examine this aspect on the performance.

Lift-enhancing devices augment the maximum lift coefficient for the wing. This is primarily important for take-off and landing, and high maneuverability. However, flaps and slats add complexity, weight, and cost to the main wing. Therefore, a trade study that examines the sensitivity of the performance and cost to $\Delta C_{L_{\max}}$ is beneficial.

In addition to $C_{L_{\max}}$, the other important parameter affecting take-off and landing distances is T/W. The choice of a too high T/W is detrimental to efficient cruise, since it requires throttling down the engine, which is less efficient, with a higher TSFC. However, a trade study could be made that examines the effect of the number of engines, engine bypass and pressure ratios, and possibly the propulsion concept, namely, turbo-jet versus turbofan versus turboprop.

It is desirable to have a positive static margin for pitch stability. However, for an aft tail aircraft, having too large of a static margin results in excessive trim drag, which reduces the cruise range and payload. A reasonable trade study would then be to examine the sensitivity of the static margin on these characteristics for a particular design.

Finally, in a design, many configuration choices that are made could later be evaluated in trade studies. These include the vertical tail design, the horizontal tail design and placement (forward or aft), wing placement on the fuselage, and fuselage height and width. These impact the performance, weight, and cost of the aircraft.

Such studies can be quite involved, linking many parameters. For many, the use of automated calculations, such as the spreadsheets in this text, are essential. This process is demonstrated in the next section, by examining some simple trade studies for the conceptual SSBJ.

13.1.1   SSBJ Trade Study

With the conceptual SSBJ, two of the most important design drivers were the supersonic cruise Mach number and the long (4000-nm) range. These led to an aircraft whose unit price was significantly higher than *subsonic* business jets with which it might compete in the market. Therefore, the following trade study examines how the range and supersonic cruise Mach number affected the unit cost of this aircraft.

TABLE 13.2: Range and Mach number versus cost trade study for the conceptual SSBJ. (Bold are selected conditions.)

Mach No.	Range (nm)	W_{TO}	Unit Cost (year 2000 M$)
1.9	3000	38,218	45.6
1.9	3500	55,715	60.0
1.9	4000	111,635	101.3
1.9	4200	174,330	143.0
1.9	4500	967,002	558.6
2.1	3000	36,232	46.9
2.1	3500	52,334	61.3
2.1	**4000**	**90,523**	**92.5**
2.1	4200	125,926	119.1
2.1	4500	294,617	231.5
2.3	3000	34,861	48.5
2.3	3500	48,757	61.9
2.3	4000	78,553	88.4
2.3	4200	102,827	108.5
2.3	4500	187,693	173.0

For this, the **itertow.xls** and **cost.xls** spreadsheets were used. The range and Mach number affect the overall take-off weight. The structure weight, which is a fixed percentage of the take-off weight, and the design cruise velocity affect the cost. The results are compiled in Table 13.2.

The values in Table 13.2 are plotted in Figure 13.2. This shows the cost, in millions of year 2000 dollars, versus the range, as a function of the cruise Mach number. Note that the cost is shown on a log axis. The large circle indicates the condition that was selected for the design. We observe that at any Mach number, the cost goes up rapidly with increasing range. We also observe that over this range of cruise Mach numbers, the cost decreases with increasing Mach number. This is a result of a predicted decrease in the take-off weight, which comes from an increase in the L/D ratio with increasing Mach number, greater than one.

An important aspect to note from Figure 13.2 is that the cost sensitivity to Mach number becomes more significant at the higher range values. However, below a range of 4000 nm, the effect of the cruise Mach number on cost is not very significant.

Trade studies often examine the effect of various performance characteristics on thrust-to-weight and wing loading. Two of the characteristics that depend on both of these are the take-off distance, s_{TO}, and the sustained turning rate, $\dot{\psi}$.

In order to show the interdependence of T/W and W/S on the performance of the conceptual SSBJ, the wing area was assumed to be fixed at the selected value of $S = 519$ f^2. The change in the take-off weight given in Table 13.3 was converted to a change in the wing loading for the fixed wing area. Given these wing loadings, the T/W was calculated based on

1. the T/W needed to have a take-off distance, which was equal to the case study value of 3667 feet, or

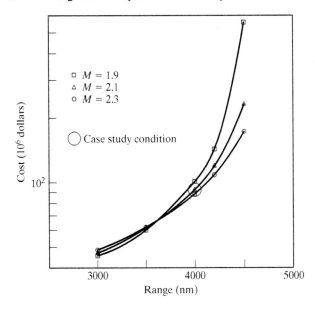

FIGURE 13.2: Cost trade study on range and cruise Mach number for conceptual SSBJ.

TABLE 13.3: Wing loading/thrust-to-weight trade study for take-off and turn rate. (Bold are case study conditions.)

W/S (lbs/f^2)	T/W for $s_{TO} = 3667$ f	T/W for $\dot{\psi} = 0.78°$
73.6	0.323	0.298
111.2	0.511	0.236
215.1	1.096	0.220
336.0	1.890	0.265
1863.0	38.4	1.170
69.8	0.304	0.309
100.8	0.457	0.247
174.0	**0.855**	**0.215**
242.0	1.266	0.228
567.0	3.81	0.386
67.1	0.292	0.318
93.9	0.423	0.256
151.4	0.726	0.217
198.1	0.994	0.217
361.6	2.082	0.277

2. the T/W needed to have a sustained turn rate equal to the case study value of 0.78°/s.

The former was based on solving Eq. [3.5], which used the take-off parameter, and the latter was based on Eq. [3.28] in which the load factor was $n = 1$. The results are tabulated in Table 13.3.

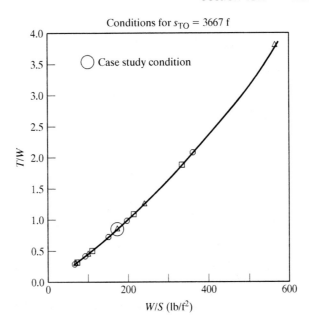

FIGURE 13.3: Take-off distance trade study on wing loading and thrust-to-weight ratio for conceptual SSBJ.

The values in Table 13.3 are plotted in Figures 13.3 and 13.4. The symbols correspond to the three Mach numbers in Figure 13.2, from which the wing loadings were derived. The larger open circle again corresponds to the conditions selected for the conceptual SSBJ.

In Figure 13.3, the curve through the points represents the required T/W that will give a take-off distance of 3667 feet for the range of wing loadings. This illustrates that if the wing loading were to increase in the case study, the T/W would also have to increase in order to maintain the same take-off distance. By choosing different take-off distances, a family of curves could be generated. The local slopes of these curves would give an indication of the sensitivity of these parameters to the take-off distance.

In Figure 13.4, the curve through the points represents the T/W that is required to give a sustained turning rate of 0.78°/s for the range of wing loadings. Note that the T/W for the selected conditions is a minimum. For higher or lower wing loadings, the required T/W increases. Here again, a family of curves could be generated for different turning rates that would indicate its sensitivity to these parameters.

The purpose of this discussion is to illustrate a methodology for evaluating and improving upon the overall conceptual design. There are many degrees of freedom, and so possibly no single optimum exists. However, it is possible to optimize on a few principle design drivers. The challenge is to achieve this with as few changes as possible. This requires a full understanding of how parameters interact and the effect they have on the **total design**.

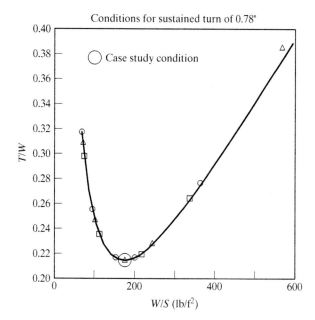

FIGURE 13.4: Sustained turning rate trade study on wing loading and thrust-to-weight ratio for conceptual SSBJ.

13.2 PROBLEMS

13.1. For the same wing loading values in Table 13.3, generate curves that are similar to Figure 13.3 that show the relation between T/W and W/S for $s_{TO} = 3000$ feet and 4500 feet. From these curves and Figure 13.3, plot the take-off distance as a function of wing loading, for $T/W = 0.8$.

13.2. For the same wing loading values in Table 13.3, generate curves that are similar to Figure 13.4 that show the relation between T/W and W/S for $\dot{\psi} = 0.5°/s$ and $1.0°/s$. From these curves and Figure 13.4, plot the sustained turn rate as a function of wing loading, for $T/W = 0.25$.

13.3. Equation 3.14 gives a relation between wing loading and thrust-to-weight ratio for different climb gradients. Generate a table, and plot for the case study aircraft T/W versus W/S for climb gradients of 8, 10, and 12 degrees. Describe any sensitivities.

13.4. The Dassault Falcon 900B was used as a generic subsonic business jet for comparison to the case study aircraft. The characteristics of the Falcon are given in Table 13.1. Using the two spreadsheets **itertow.xls** and **cost.xls** do the following:

   1. Input the appropriate values into the spreadsheets for the Falcon, and first verify how closely the estimated take-off weight agrees with the published value.

   2. Using the estimated take-off weight from the spreadsheet, estimate the unit cost, assuming the same production schedule as for the case study.

   3. Vary the range for the Falcon, and determine how that changes the unit price. Plot the result in a fashion similar to Figure 13.2.

   4. Describe any differences between the cost sensitivity to range in this case, to that of the conceptual SSBJ.

13.5. A paper by Fellers and Patierno (1970) concluded that it is more efficient in terms of W_{TO} to improve $\dot{\psi}$ by decreasing the wing loading, rather than by increasing the T/W. Using the results in Tables 13.2 and 13.3, and Figure 13.4, for the case study aircraft, discuss if you agree or disagree with their conclusion.

13.6. In this chapter, an argument was given for using four engines, rather than two in the case study design. Describe a trade study that would examine the impact of that choice on such things as cost, take-off weight, and fuel efficiency. Be specific, and use formulas, tables, and graphs.

13.7. The excess engine power, P_s, determines the time to accelerate to high speed. This can be given as

$$P_s = V \frac{T}{W} - V q S C_{D_0} - q \mathrm{Sk} \left(\frac{W}{S} \frac{n}{q} \right)^2 . \tag{13.1}$$

  1. For the case study aircraft, determine the excess power ((f-lbf)/s) at the end of take-off.
  2. Generate a plot of T/W versus W/S such that P_s remains the same as in the previous step.
  3. Based on this result, which is better to maximize acceleration, larger T/W or lower W/S? Are there any other factors in the *total design* that should also be considered?

13.8. Repeat Problem [13.7], except generate curves of constant P_s for different the wing aspect ratios, A. Based on this, what would you conclude is better for a high-speed intercept aircraft, high or low A? Is the choice very sensitive?

13.9. In the initial estimate of the take-off weight in the spreadsheet **itertow.xls**, the only dependence on thrust comes from combat. Since the characteristics of the case study business jet were similar to the Mig-31, we can use it as a combat model, with the addition of 10 minutes of combat time. With this one addition, do the following:
  1. Determine the new take-off weight.
  2. Assuming that the wing area remains the same ($S = 519$ f^2), plot T/W versus W/S, which keeps W_{TO} at the same value.
  3. Repeat [2] for two other wing areas. Discuss how the take-off weight is or is not sensitive to the choice of wing loading in this instance.

APPENDIX A

1976 Standard Atmosphere Data

Altitude (feet)	Temperature (°R)	Pressure lbs/f²	σ^*	a (f/s)
0	518.6700	2116.2281	1	1116.4503
1000	515.1038	2040.8640	0.9711	1112.6056
2000	511.5377	1967.6881	0.9428	1108.7475
3000	507.9715	1896.6514	0.9151	1104.8760
4000	504.4054	1827.7057	0.8881	1100.9908
5000	500.8392	1760.8036	0.8617	1097.0919
6000	497.2730	1695.8984	0.8359	1093.1791
7000	493.7069	1632.9442	0.8106	1089.2522
8000	490.1407	1571.8959	0.7860	1085.3111
9000	486.5746	1512.7089	0.7620	1081.3556
10000	483.0084	1455.3396	0.7385	1077.3857
11000	479.4422	1399.7449	0.7156	1073.4010
12000	475.8761	1345.8825	0.6932	1069.4015
13000	472.3099	1293.7108	0.6713	1065.3870
14000	468.7438	1243.1889	0.6500	1061.3573
15000	465.1776	1194.2766	0.6292	1057.3122
16000	461.6114	1146.9344	0.6090	1053.2516
17000	458.0453	1101.1234	0.5892	1049.1753
18000	454.4791	1056.8055	0.5699	1045.0831
19000	450.9130	1013.9430	0.5511	1040.9748
20000	447.3468	972.4993	0.5328	1036.8502
21000	443.7806	932.4380	0.5150	1032.7091
22000	440.2145	893.7236	0.4976	1028.5514
23000	436.6483	856.3211	0.4807	1024.3768
24000	433.0822	820.1964	0.4642	1020.1851
25000	429.5160	785.3157	0.4481	1015.9762
26000	425.9498	751.6460	0.4325	1011.7497
27000	422.3837	719.1549	0.4173	1007.5055
28000	418.8175	687.8104	0.4025	1003.2433
29000	415.2514	657.5815	0.3881	998.9629
30000	411.6852	628.4375	0.3741	994.6642

Altitude (f)	Temperature (°R)	Pressure lbs/f^2	σ^*	a (f/s)
31000	408.1190	600.3483	0.3605	990.3467
32000	404.5529	573.2845	0.3473	986.0104
33000	400.9867	547.2172	0.3345	981.6549
34000	397.4206	522.1181	0.3220	977.2800
35000	393.8544	497.9594	0.3099	972.8854
36000	390.2882	474.7140	0.2981	968.4709
37000	389.9700	452.4380	0.2844	968.0760
38000	389.9700	431.2066	0.2710	968.0760
39000	389.9700	410.9715	0.2583	968.0760
40000	389.9700	391.6860	0.2462	968.0760
45000	389.9700	308.0134	0.1936	968.0760
50000	389.9700	242.2151	0.1522	968.0760
55000	389.9700	190.4727	0.1197	968.0760
60000	389.9700	149.7836	0.0941	968.0760
65000	389.9700	117.7866	0.0740	968.0760
70000	392.3748	92.6848	0.0579	971.0563
75000	395.1180	73.0534	0.0453	974.4448
80000	397.8612	57.6751	0.0355	977.8216
85000	400.6044	45.6080	0.0279	981.1868
90000	403.3476	36.1234	0.0220	984.5405
95000	406.0908	28.6565	0.0173	987.8828
100000	408.8340	22.7686	0.0136	991.2138
110000	419.2711	14.4591	0.0085	1003.7864
120000	434.6330	9.3210	0.0053	1022.0102
130000	449.9950	6.1011	0.0033	1039.9146
140000	465.3569	4.0508	0.0021	1057.5159
150000	480.7188	2.7255	0.0014	1074.8291
160000	487.1700	1.8530	0.0009	1082.0171
170000	483.0574	1.2607	0.0006	1077.4403
180000	467.6954	0.8499	0.0004	1060.1698
190000	452.3335	0.5654	0.0003	1042.6132
200000	436.9716	0.3709	0.0002	1024.7560

$^*\rho_{SL} = 0.0024$ slugs/f^3.

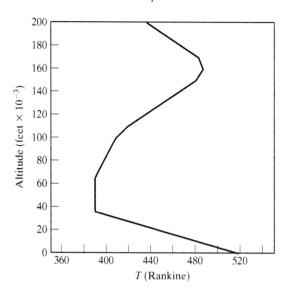

FIGURE A.1: Temperature variation with altitude in 1976 Standard Atmosphere.

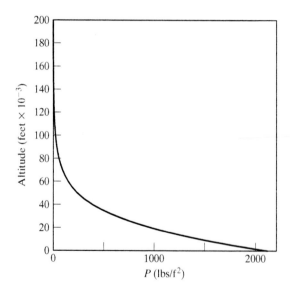

FIGURE A.2: Pressure variation with altitude in 1976 Standard Atmosphere.

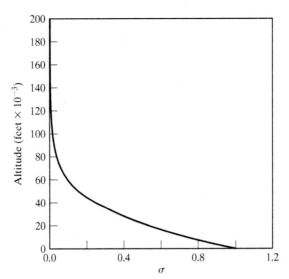

FIGURE A.3: Variation of $\sigma = \rho/\rho_{SL}$ with altitude in 1976 Standard Atmosphere.

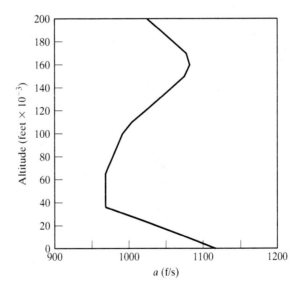

FIGURE A.4: Variation of the speed of sound with altitude in 1976 Standard Atmosphere.

APPENDIX B

Case Study: High-Performance Kit Aircraft (HPKA)

B.1 TAKE-OFF WEIGHT ESTIMATE
B.2 WING LOADING SELECTION
B.3 WING DESIGN
B.4 FUSELAGE DESIGN
B.5 TAIL DESIGN
B.6 PROPULSION SYSTEM DESIGN
B.7 TAKE-OFF AND LANDING ANALYSIS
B.8 ENHANCED LIFT DESIGN
B.9 REFINED WEIGHT ANALYSIS
B.10 STATIC STABILITY ANALYSIS
B.11 COST ANALYSIS
B.12 SUMMARY

DESIGN PROPOSAL: HIGH-PERFORMANCE KIT AIRCRAFT (HPKA)

We propose to design a propeller-driven, low-wing mono-plane of composite construction that will combine the features of a large engine, extended range, high cruise and maximum speeds, large take-off weight, and high flight ceiling. It will carry a pilot and one passenger, and have relatively low initial and operating costs to provide a weekend flier with an inexpensive higher performance aircraft. It is intended to be developed into a kit.

The proposed characteristics, along with those of some comparison aircraft are given in the following table. The principle design drivers will be longer range, higher cruise speed, and moderate cost.

HPKA and Aircraft with Similar Characteristics

	HPKA	Diamond Katana	Pulsar Sport 150	Dyn'Aero MCR01 VLA
Max. Speed (kts)	150	80	118	100
Cruise Speed (kts)	250	157	135	197
Max. T-O Weight (lbs)	1800	1610	1200	992
Empty Weight (lbs)	1000	1085	680	485
Fuel Capacity (gal)	30	20.1	20	23.2
Range (nm)	1000	500	521	894
T-O Distance (ft)	1500	1560	—	1395
Landing Distance (ft)	1500	1490	—	—
Aspect Ratio	7	10	6	8.45

B.1 TAKE-OFF WEIGHT ESTIMATE

The take-off weight is estimated using the **itertow.xls** spreadsheet. The design drivers that heavily affect the take-off weight are operating range, cruise speed and altitude, thrust-specific fuel consumption, and aspect ratio. The proposed range of 1000 nm (500-nm operating radius) would allow the aircraft to make cross-country flights without frequent refueling. Based on historic data (Figure 2.5), the average structure factor should be approximately 0.6. Because this design is intended to make extensive use of composite materials, the estimated structure factor will be a relatively low 0.5. This would be comparable to the Dyn'Aero comparison aircraft.

The payload weight is 450 lbs. This consists of the pilot, one passenger, and baggage for both.

The cruise altitude is specified as 17,000 feet. At the cruise velocity of 250 knots, this gives a cruise Mach number of 0.35. An acceleration to a higher Mach number of 0.4 is included in the flight plan. There is no combat time, so that the maximum TSFC and engine thrust are not used in the take-off weight estimate. The cruise TSFC is estimated to be 0.4. A loiter time of 30 minutes is listed. This would allow less-experienced pilots generous time to position the aircraft prior to the landing approach.

For the estimate of the fuel-weight fraction used in cruise and loiter, the governing equations were modified from the original spreadsheet to include the propeller efficiency (Eqs. [2.6b & 2.19b]). The efficiency is listed at the bottom of the input portion of the spreadsheet. It was given a typical value of 0.8.

Given these input values, the resulting spreadsheet is shown in Figure B.1. The final take-off weight given by the spreadsheet was 1142 lbs. Of this, 450 lbs is payload, 121 lbs is fuel, and 571 lbs is the empty or structure weight.

B.2 WING LOADING SELECTION

The effect of the wing loading on different performance characteristics was evaluated using the **wingld.xls** spreadsheet. In designs where range is an important design driver, the wing loading is generally based on having the optimum cruise. In the design of

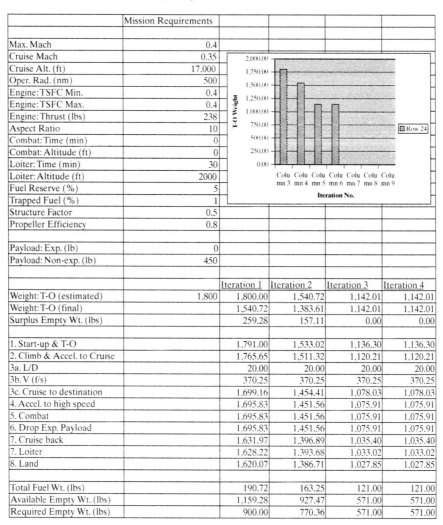

	Mission Requirements			
Max. Mach	0.4			
Cruise Mach	0.35			
Cruise Alt. (ft)	17,000			
Oper. Rad. (nm)	500			
Engine: TSFC Min.	0.4			
Engine: TSFC Max.	0.4			
Engine: Thrust (lbs)	238			
Aspect Ratio	10			
Combat: Time (min)	0			
Combat: Altitude (ft)	0			
Loiter: Time (min)	30			
Loiter: Altitude (ft)	2000			
Fuel Reserve (%)	5			
Trapped Fuel (%)	1			
Structure Factor	0.5			
Propeller Efficiency	0.8			
Payload: Exp. (lb)	0			
Payload: Non-exp. (lb)	450			
		Iteration 1	Iteration 2	Iteration 3	Iteration 4
Weight: T-O (estimated)	1,800	1,800.00	1,540.72	1,142.01	1,142.01
Weight: T-O (final)		1,540.72	1,383.61	1,142.01	1,142.01
Surplus Empty Wt. (lbs)		259.28	157.11	0.00	0.00
1. Start-up & T-O		1,791.00	1,533.02	1,136.30	1,136.30
2. Climb & Accel. to Cruise		1,765.65	1,511.32	1,120.21	1,120.21
3a. L/D		20.00	20.00	20.00	20.00
3b. V (f/s)		370.25	370.25	370.25	370.25
3c. Cruise to destination		1,699.16	1,454.41	1,078.03	1,078.03
4. Accel. to high speed		1,695.83	1,451.56	1,075.91	1,075.91
5. Combat		1,695.83	1,451.56	1,075.91	1,075.91
6. Drop Exp. Payload		1,695.83	1,451.56	1,075.91	1,075.91
7. Cruise back		1,631.97	1,396.89	1,035.40	1,035.40
7. Loiter		1,628.22	1,393.68	1,033.02	1,033.02
8. Land		1,620.07	1,386.71	1,027.85	1,027.85
Total Fuel Wt. (lbs)		190.72	163.25	121.00	121.00
Available Empty Wt. (lbs)		1,159.28	927.47	571.00	571.00
Required Empty Wt. (lbs)		900.00	770.36	571.00	571.00

FIGURE B.1: Take-off weight spreadsheet for HPKA case study.

the HPKA, a greater emphasis was placed on short take-off and landing distances and having a higher cruise speed. The spreadsheets for the wing loading analysis are shown in Figures B.2 and B.3.

In the normal operation of this spreadsheet, the wing loading is determined for most efficient cruise. In the section marked "Cruise Start," the optimum wing loading for cruise is determined to be 67 lbs/f^2. As indicated in Table B.1, except for one of the comparison aircraft, this would be from three to four times higher than those listed in the design proposal. Since cruise range is not a principle design driver, and the other performance objectives benefit from having a lower wing loading, the value based on cruise was not used.

Take-off			Landing	
H (f)	500		H (f)	500
CL(max)	2.08		CL(max)	2.08
T(max) (lb)	237		T(max) (lb)	237
W_TO (lb)	1142		W_L (lb)	1033
S (f^2)	57.1		W/S (lb/f^2)	18.09
W/S (lb/f^2)	20.00		S (f^2)	16.71
SIGMA	0.99		SIGMA	0.99
T/W	0.21		T/W	0.23
TOP	47.04		LP	8.83
S_TO (f)	1254.88		S_TO (f)	1441.92
Cruise Start			**Cruise End**	
CD_0	0.0350		CD_0	0.0350
A	10		A	10
H (f)	17,000		H (f)	20,000
Cruise Mach	0.35		Cruise Mach	0.35
W(lb)	1120		W (lb)	1,035
k	0.04		k	0.04
V (f/s)	370.25		V (f/s)	366.64
rho (lbm/f^3)	0.0475		rho (lbm/f^3)	0.04
q (lbf/f^2)	101.06		q (lbf/f^2)	91.05
W/S (lb/f^2)	67.02		W/S_optimum	60.38
S (f^2)	16.71		W/S_actual	61.94
Climb			**Acceleration**	
H (f)	1,000		H (f)	17,000
Climb Mach	0.125		Cruise Mach	0.26
dH/dt (f/min)	500		n	2.5
V (f/s)	137.58			
G (rad)	0.06		V (f/s)	272.01
Gamma (deg)	3.47		rho (lbm/f^3)	0.05
rho (lbm/f^3)	0.07		q (lbf/f^2)	54.55
q (lbf/f^2)	21.89		W/S (lb/f^2)	20.46
T/W min	0.135			
T/W min − G	0.07			
W/S_ + (lb/f^2)	20.53			
W/S_ − (lb/f^2)	20.53			

FIGURE B.2: Wing loading spreadsheet for HPKA case study (Part 1).

As a user note for this spreadsheet, the wing loading data used in other parts of the spreadsheet refer to the wing loading value based on cruise, which is given a global variable name, "WS." Therefore, if that value is not used, the number in the cells next to the W/S designation have to be inserted manually. In addition, the wing area, S, is

Turn-Inst.			Turn-Sustained	
H (f)	17.000		H (f)	17.000
Cruise Mach	0.35		Cruise Mach	0.35
CL_max	0.45		n	2.52
W/S (lb/f^2)	20		T/W_max	0.19
			W/S (lb/f^2)	20
V (f/s)	370.25			
rho (lbm/f^3)	0.05		V (f/s)	370.25
q (lbf/f^2)	101.06		rho (lbm/f^3)	0.05
psi_dot (rad/s)	0.18		q (lbf/f^2)	101.06
psi_dot (deg/s)	10.18		psi_dot (rad/s)	0.06
n	2.27		psi_dot (deg/s)	3.25
Ceiling			**Glide**	
W (lb)	1120		W (lb)	1120
S (f^2)	57		S (f^2)	57
H (f)	20.000		H (f)	17.000
Cruise Mach	0.35		Cruise Mach	0.35
W/S (lb/f^2)	19.65		W/S (lb/f^2)	19.65
V (f/s)	366.64		V (f/s)	370.25
rho (lbm/f^3)	0.04		rho (lbm/f^3)	0.05
q (lbf/f^2)	91.05		q (lbf/f^2)	101.06
CL_required	0.22		V_min_D (f/s)	168.58
			M_min_D	0.16

FIGURE B.3: Wing loading spreadsheet for HPKA case study (Part 2).

TABLE B.1: Comparison of wing loading related performance between HPKA and comparison aircraft.

	HPKA	Diamond Katana	Pulsar Sport 150	Dyn'Aero MCR01 VLA
Wing Loading (lbs/f^2)	20	12.9	80.4	17.6
Wing Area (f^2)	57	125	88	56
Wing Span (f)	23.9	35.7	21.9	23
Take-Off Dist. (f)	1255	1120	1000	935
Landing Dist. (f)	1441	1490	—	—
Ceiling (f)	20,000	14,000	17,000	—
Load Factor	+4.0/−2	+4.4/−2.2	—	+4/−2

automatically calculated from the wing loading and specified weight of the aircraft. This also is given a global variable name of "S." Therefore, if the value from cruise is not used, it needs to be specified.

In this case study, the wing loading was specified to achieve good take-off and landing distances and maneuvering characteristics. Based on comparison aircraft, a wing loading of 20 lbs/f^2 was chosen. As indicated in Table B.1, this gave performance that was very comparable to the other aircraft.

The HPKA is intended to be a "sport flyer" so that it is designed to be very maneuverable. This wing loading gives it a good climb rate of 500 feet per minute and sustained and instantaneous turn rates of 3.25 and 10.18 degrees per second. Its flight ceiling is estimated to be 20,000 feet.

B.3   WING DESIGN

The spreadsheet for the wing design is shown in Figures B.4 and B.5. This uses the wing area from the wing loading spreadsheet.

Because the cruise and maximum Mach numbers are relatively low, the planform shape of the wing did not require any leading-edge sweep. Since this aircraft is designed to be a homebuilt, the taper ratio was selected to be 1. This gives a simple rectangular wing that should be easy to construct. An aspect ratio of 10 was selected. This gives a wing span of 23.9 feet, which is comparable to the comparison aircraft listed in Table B.1.

The wing section selected for the design is the NACA $63_2 - 015$. This choice was based on having low drag in the range of the design C_l for cruise conditions, $0.180 \leq C_l \leq 0.194$. The lift coefficient versus angle of attack and drag polar for this airfoil are presented in Figure B.6. This airfoil has a maximum thickness-to-chord ratio of 15 percent, which is near the optimum to give the lowest drag coefficient.

Figure B.5 compares the 2-D and 3-D lift coefficient versus angle of attack. The relatively large aspect ratio and no leading-edge sweep makes this wing behave nearly two-dimensional. The plan view of the half-span of the wing illustrates its simple geometry.

The L/D ratio for the wing is 15.94. This is slightly less than the value of 20 that was estimated in the first spreadsheet for the take-off weight estimate. The difference is because the first spreadsheet assumed a wing design for most efficient cruise.

B.4   FUSELAGE DESIGN

The fuselage shape is constructed as a series of ellipses that are blended and tapered to allow enough room for the crew of two and to provide room for baggage behind the seats and space for the engine. A fineness ratio of 0.20 (l/d = 5) was selected to minimize the total drag on the fuselage. The overall length is 24 feet, and the diameter is 4.9 feet.

The spreadsheet for the fuselage design is shown in Figure B.7. Of particular interest are the lack of any steps in the perimeter along the fuselage length that could lead to flow separations and higher drag. The diagram shows it to be smooth. The total viscous drag is estimated to be 82 lbs. Because this is a subsonic aircraft, the wave drag is zero. The equivalent drag coefficient, normalized by the wing area, is 0.0143.

B.5   TAIL DESIGN

The vertical and horizontal tail use the same family of airfoil as the main wing, with a 6 percent thickness-to-chord ratio, namely, a NACA $63_2 - 006$. For the vertical tail, $C_{VT} = 0.09$, based on historic data for aircraft of this type. The value of L_{VT} was taken as 60 percent of the fuselage length, which is also consistent with other similar aircraft. This corresponds to $L_{VT} = 14.5$ feet. Mainly for aesthetic purposes, the leading edge of the vertical tail was swept at a 30-degree angle. Other input values for the vertical tail are shown in the spreadsheet in Figure B.8. The total drag on the vertical tail is estimated to be 8.17 lbs. The planform shape is illustrated in the figure in the spreadsheet.

Design Parameters		
M	0.35	
S	57	ft²
A	10.0	
Λ_{LE}	0	deg
t/c	0.15	
λ	1.00	
W c-start	1.120	lbf/f^2
W c-end	1.035	lbf/f^2
q c-start	101.09	lbf/f^2
q c-end	91.08	lbf/f^2
Cl c-start	0.194	
Cl c-end	0.180	

Airfoil Data		
Name	NACA 63(2)–015	
Cl_{max}	1.5	
$C_{L\alpha}$	0.113	1/deg
a.c.	0.27	c
α_{OL}	0	deg
Cd0	0.005	
r_{le}	0.050	c
Cl_{minD}	−0.3 – 0.3	
(t/c)max	0.30	c

Calculations		
b	23.9	ft
M_{eff}	0.35	
c_r	2.4	ft
c_t	2.4	ft
m.a.c.	2.4	ft
β	0.94	
$C_{L\alpha}$	0.095	1/deg
C_{Lo}	−0.00	
α_{trim}	2.1	deg
C_{Ltrim}	0.194	
k	0.04	
C_D	0.012	
L/D	15.94	
Total Drag	68.7	lbf

Sweep Angles		
	x/c	$\Lambda_{x/c}$ (deg)
LE	0.00	0.0
1/4C	0.25	0.0
a.c	0.27	0.0
(t/c)max	0.30	0.0
TE	1.00	0.0

Viscous Drag	
V_eff	366.17 f/s
q_eff	98.85 lbf/f^2
Re_mac	3.88E+06
sqrt(Re)	1969.52
Cf	3.47E−03
S_wet	117.14 ft²
F	1.5
Q	1
C_{D0}	0.01069

Plotting:

Spanwise View

x	y
0	0
2.4	0
2.39	11.94
0	11.94
0	0

Lift Curves

α	Cl	α	CL
0	0	0	0
13.27	1.5	13.27	1.26

FIGURE B.4: Wing design spreadsheet for HPKA case study (Part 1).

The horizontal tail is located on top of the vertical tail in a "T-tail" arrangement. The tip chord of the vertical tail is the same as the root chord of the horizontal tail.

Based again on historical aircraft, C_{HT} was chosen to be 1.0. The distance, L_{HT}, was chosen to be 16.7 feet. The leading edge was swept by 15 degrees to match the

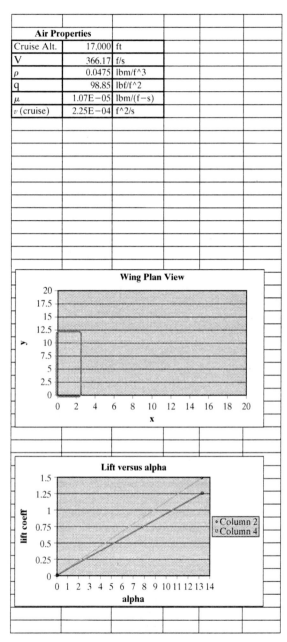

Air Properties

Cruise Alt.	17,000	ft
V	366.17	f/s
ρ	0.0475	lbm/f^3
q	98.85	lbf/f^2
μ	1.07E−05	lbm/(f−s)
v (cruise)	2.25E−04	f^2/s

FIGURE B.5: Wing design spreadsheet for HPKA case study (Part 2).

appearance of the vertical stabilizer. The aspect ratio was set to 2.0. Other input values for the horizontal tail are shown in the spreadsheet in Figure B.9. From this, the total drag was estimated to be 8.23 lbs. The planform shape is illustrated in the figure in the spreadsheet.

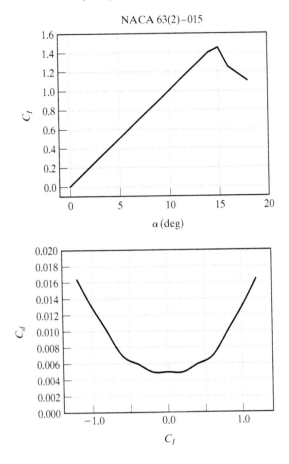

FIGURE B.6: Lift–drag characteristics of NACA 63_2–015 airfoil. (From NACA Report No. 824.)

With the wing, fuselage, and tail designed, the basic external shape of the aircraft is known. The exterior view of the HPKA is shown in a scale drawing in Figure B.10.

B.6   PROPULSION SYSTEM DESIGN

The Lycoming IO-320-B engine was selected as the propulsion system for the HPKA design. This is a four-cylinder, direct-drive reciprocating engine that produces 160 hp at 2700 rpm. A 6-foot diameter, two-bladed propeller was selected based on the airframe size and comparison aircraft. At this diameter, the propeller tip Mach number was below 0.85; therefore, strong compressibility effects were neglected.

The propeller spreadsheet **prop.xls** was used to determine the thrust produced by the engine and propeller. The spreadsheet is shown in Figure B.11. The conditions are based on the cruise Mach number and altitude. The total drag consisting of the sum of the drag forces on the wing, fuselage, and tail was 167.1 lbs. Based on the spreadsheet, the engine and propeller combination can produce 208.1 pounds of thrust. This gives a

Fuselage Design

Flight Regime Data:

Cruise Mach	0.35
Cruise Alt. (ft)	17,000
V (f/s)	370.25
ρ (lbm/f^3)	0.0475
q (lbf/f^2)	101.06
μ(lbm/(f−s))	1.1E−05
ν (cruise) (f^2/s)	2.3E−04

Dimension Data: | **Form Factors:**

D−max (ft)	4.8	F	1.49
L/D	5	Q	1
L (ft)	24	F*Q	1.49
S (f^2)	57		

Viscous Drag Calculations: *Von−Karman Ogive Fuselage Shape*

x/L	x (ft)	x−L/4 (ft)	D (ft)	P (ft)	Sw(ft^2)	Re_x	C_F	Drag (lbf)
0.00	0.00	0.00	0	0.0				
0.10	2.40	2.75	4	12.6	30.2	3.9E+06	3.46E−03	16
0.20	4.80	4.22	4.9	15.4	36.9	7.9E+06	3.08E−03	17
0.30	7.20	4.51	4.9	15.4	36.9	1.2E+07	2.89E−03	16
0.40	9.60	3.76	3.99	12.5	30.1	1.6E+07	2.76E−03	13
0.50	12.00	2.41	2.8	8.8	21.1	2.0E+07	2.67E−03	8
0.60	14.40	1.61	1.85	5.8	13.9	2.4E+07	2.60E−03	5
0.70	16.80	1.09	1.14	3.6	8.6	2.8E+07	2.54E−03	3
0.80	19.20	0.75	0.64	2.0	4.8	3.2E+07	2.49E−03	2
0.90	21.60	0.56	0.35	1.1	2.6	3.5E+07	2.44E−03	1
1	24.00	0.50	0.25	0.8	1.9	3.9E+07	2.40E−03	1
Totals:					**187.1**			**82**

Wave Drag Calculations:

A_max	18.1
CDW	0
Drag (lbf)	**0**

Total Drag:	**82**
(lbf)	

Equiv. CD	0.0143

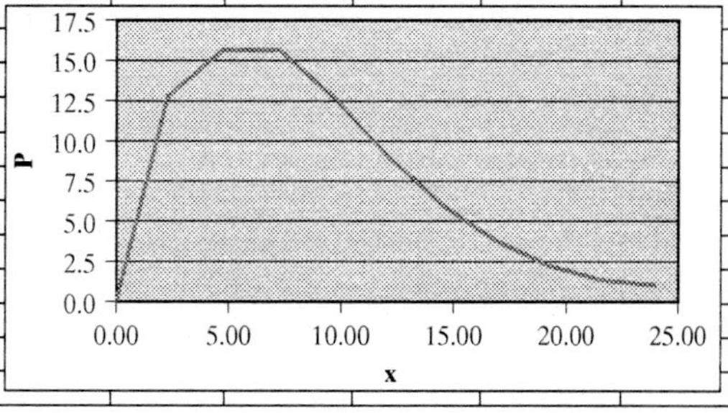

FIGURE B.7: Fuselage design spreadsheet for HPKA case study.

comfortable margin for this plane. The corrected static thrust is estimated to be 600.8 lbs. This will be used in the take-off analysis.

B.7   TAKE-OFF AND LANDING ANALYSIS

The take-off and landing analysis is a refined update of the earlier estimates listed in Table B.1 that were made using the take-off and landing parameters. The analysis was done using the spreadsheet **to-l.xls**. The spread sheet output is shown in Figures B.12

Main Wing Reference				Air Properties				
b	23.9	ft	Cruise Alt.	17,000	ft			
m.a.c.	2.4	ft	V	366.17	f/s			
S	57	ft²	ρ	0.05	lbm/f^3			
M	0.35		q	98.85	lbf/f^2			
Λ_{LE}	0	deg	μ	107.0E−7	lbm/(f−s)			
t/c	0.15		ν (cruise)	225.4E−6	f^2/s			
λ	1.00							
Vertical Tail								
Design Parameters			**Airfoil Data**					
Cvt	0.09		Name	NACA 63-006				
Lvt	14.4	ft	Cl_{max}	0.8				
Λ_{LE}	30	deg	Cl_{α}	0.11	1/deg			
t/c	0.06		a.c.	0.26	c			
λ	0.78		α_{0L}	0	deg			
Avt	1.20		Cd	0.004				
Calculations			**Sweep Angles**			**Viscous Drag**		
Svt	9	ft²		x/c	$\Lambda_{x/c}$ (deg)	V_eff	317.11	f/s
b	3.2	ft	LE	0.00	30.0	q_eff	74.14	lbf/f^2
c_r	3.0	ft	1/4 chord	0.25	25.4	M_eff	0.3	
c_t	2.3	ft	(t/c)max	0.35	23.4	Re_mac	376.7E+4	
m.a.c.	2.7	ft	TE	1.00	9.4	sqrt(Re)	1940.9	
β	0.94					Cf	3.50E−03	
$C_{L\alpha}$	0.030	1/deg				S_wet	17.1	ft²
						F	1.32	
						Q	1.05	
Total Drag	8.167	lbf				C_{D0}	**0.0097**	

Spanwise View	
x	y
0	0
3.0	0
3.26	3.2
0.92	3.2
0	0

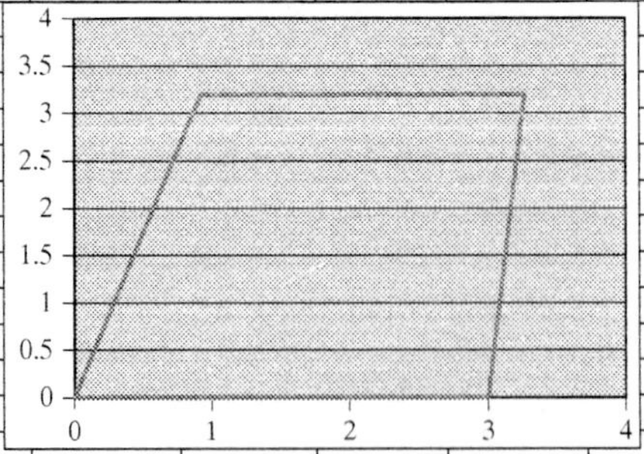

FIGURE B.8: Tail design spreadsheet for HPKA case study (Part 1).

Horizontal Tail									
Design Parameters			**Airfoil Data**						
Cht	1.00		Name	NACA 63-006					
Lht	16.7	ft	Cl_{max}	0.8					
Λ_{LE}	15	deg	Cl_{α}	0.11	1/deg				
t/c	0.06		a.c.	0.26	c				
λ	0.77		α_{0L}	0	deg				
Aht	2.00		Cd	0.004					
Calculations			**Sweep Angles**			**Viscous Drag**			
Sht	8.19	ft^2		x/c	$\Lambda_{x/c}$ (deg)	V_eff	353.69	f/s	
b	4.0	ft	LE	0.00	15.0	q_eff	92.23	lbf/f^2	
c_r	2.3	ft	1/4 chord	0.25	11.5	M_eff	0.34		
c_t	1.8	ft	(t/c)max	0.35	10.4	Re_mac	319.4E+4		
m.a.c.	2.0	ft	TE	1.00	0.5	sqrt(Re)	1787.18		
β	0.94					Cf	3.59E-03		
$C_{L\alpha}$	0.046	1/deg				S_wet	16.45	ft^2	
						F	1.34		
						Q	1.05		
Total Drag	8.227	lbf				C_{D0}	0.010		

Spanwise View	
x	y
0	0
2.3	0
2.3	2.0
0.54	2.0
0	0

FIGURE B.9: Tail design spreadsheet for HPKA case study (Part 2).

and B.13. The input parameters include characteristics of the design that were resolved in earlier steps. These include the wing aspect ratio, wing area, take-off and landing weights, and static engine thrust. The overall drag coefficient was taken as the sum of the coefficients for the wing, fuselage, and tail, based on previous spreadsheet calculations. The maximum lift coefficient, C_{LG}, corresponds to the flapped main wing. Analysis of lift enhancement, which is discussed in the next section, gave a value of approximately 2. The increase in drag due to the flaps was estimated to be $\Delta C_{D_{flaps}} = 0.1$. The actual value based on the enhanced lift analysis was 0.072. The rolling friction coefficient used in the analysis corresponds to a grass field, which is consistent with a recreational aircraft. Finally, the projected area of the landing gear was estimated to be 2 f^2 based on the comparison aircraft.

For take-off, a climb angle of 10 degrees and an obstacle height of 50 feet were used. For these, the total take-off distance was estimated to be 1340.8 feet. This is slightly more than the 1255 ft distance originally estimated in Table B.1.

For landing, a descent angle of 8 degrees was used. The other parameters remained the same as with take-off. For these, the total landing distance is 1580.5 feet. The

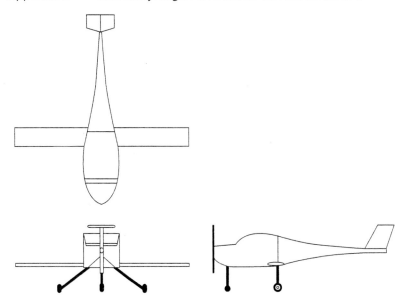

FIGURE B.10: Scale three-view drawing of the HPKA case study. Wing span is 23.9 f, fuselage length is 24 f, and maximum fuselage width is 4.9 f.

Cruise Conditions:				
Cruise Alt.	17.000	ft		
M	0.35			
V	370.30	f/s		
ρ	0.05	lbm/f^3		
Tot. Drag	**167.1**	lbf		
Propeller Design:				
V_tip	848.00	f/s		
M_tip	0.81			
D	6	f		
SHP	160	hp		
blade no.	2			
Calculations:				
n	44.99	rps	2699.27	rpm
J	1.37	1/rev		
C_P	0.0843			
η_P	0.85	(Fig. 7.5)		
η_P(COR)	0.88			
C_T	0.0538			
T_cruise	**208.06**	lbf		
Static Thrust:				
C_T/C_P	1.9	(Fig 7.6)		
T_static	619.43	lbf		
T_(COR)	**600.84**	lbf		

FIGURE B.11: Propulsion system design spreadsheet for HPKA case study.

Take-Off							
CD_0	0.05		mu_TO	0.05			
A	10		T_max (lb)	600.8			
H (f)	500		f_LG	3.48			
CL_G	2		A_LG (f^2)	2			
W_TO (lb)	1,142		deltCD_0_flap	0.1			
S (f^2)	57.06		gamma_CL (deg)	10			
			H_obstacle (f)	50			
k	0.04		T/W	0.53			
rho (lbm/f^3)	0.08		f1 (f/s^2)	15.33			
W/S (lb/f^2)	20.01		deltCD_0_LG	0.12			
S (f^2)	57.06		f2 (f^−1)	0			
V_T–O (f/s)	110.89		R_TR (f)	2545.7			
q_T_O (lb/f^2)	14.41		H_TR (f)	38.67			
			S_G (f)	501.84			
			S_R (f)	332.66			
			S_TR (f)	442.06			
			S_CL (f)	64.23			
			S_T–O (f)	**1340.78**			
Landing							
W_L (lb)	1027.8		D_50 (lb)	370.09			
W/S (lb/f^2)	18.01		gamma_A (deg)	−21.11			
V_50 (f/s)	113.96		gamma_A _act	−8			
V_TD (f/s)	100.81		R_TR (f)	1900.36			
q_50 (lb/f^2)	15.22		H_TR (f)	18.49			
q_TD (lb/f^2)	11.91		f1 (f/s^2)	−19.32			
mu _L	0.6		f2 (f^−1)	0.0017880			
T_L (lb)	0						
			S_A (f)	224.18			
			S_TR (f)	264.48			
			S_FR (f)	302.44			
			S_B (f)	789.41			
			S_L (f)	**1580.51**			
			1.6(S_L) (f)	2528.82			

FIGURE B.12: Take-off and landing spreadsheet for HPKA case study (Part 1).

60-percent higher value is not relevant to this design since it applies only to FAR regulations on commercial aircraft. The landing distance is also slightly longer than the initial 1441 ft estimate in Table B.1.

B.8   ENHANCED LIFT DESIGN

Slotted trailing-edge flaps were chosen for passive lift enhancement during take-off and landing. These flaps would have a length corresponding to 20 percent of the wing chord, have a maximum deflection angle of 60 degrees, and cover 48 percent of the wing span.

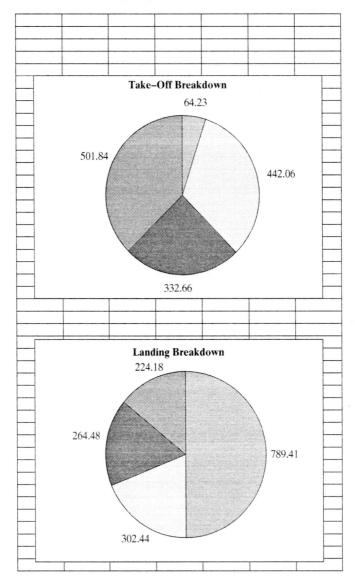

FIGURE B.13: Take-off and landing spreadsheet for HPKA case study (Part 2).

The analysis of the flapped wing was performed using the **flaps.xls** spreadsheet. This is shown in Figures B.14 to B.16. Most of the input parameters come from the wing design that was completed in an earlier design step using the spreadsheet **wing.xls**. This included C_{L_α} for the 2-D and 3-D wing.

In the spreadsheet, the flap type was designated as slotted by the key word "slot." The aspect ratio criterion designated the wing to be in the "high" category. Therefore, the low aspect-ratio basic wing results are to be ignored.

Wing Data:				
Airfoil	NACA	63(2)–15		
Λ_{LE}	0	deg		
λ	1.00			
t/c	0.15			
T-O Mach No.	0.1			
β	0.99			
A	10			
$\Lambda_{t/c}$	0.0	deg		
$C_{l\alpha}$ (no flap)	0.113	1/deg		
$C_{L\alpha}$ (no flap)	0.095	1/deg		
α_{0L}	0	deg		
C_{lmax}	1.5			
α_s	11	deg		
Trailing-edge Flap Design:				
Flap type	slot	slot, plane or split		
S_f/S_w	0.48			
δ_f	60	deg		
c_f/c	0.2			
Delta α 0L:				
	Plane Flap			
K'	0.51	Fig. 9.3		
$dC_l/d_f\delta$	0.38	Fig. 9.4		
$\Delta\alpha_{0l}$	−17.96	deg		
	Single Slotted & Fowler Flap			
$d\alpha/d\delta_f$	−0.2	Fig. 9.5		
$\Delta\alpha_{0l}$	−22.5	deg		
	Split Flaps			
k	1	Fig. 9.6		
ΔCl	1.35	Fig. 9.7		
$\Delta\alpha_{0l}$	−11.95	deg		
Aspect Ratio Criterion:				
C_1	0	Fig. 9.8		
High A criteria	4.00	High		
Basic Wing—High Aspect Ratio:	.			
Δy	3.28	%	Fig. 9.10	
C_{Lmax}/C_{lmax}	0.9	Fig. 9.9		
C_{Lmax}	1.35			
$\Delta\alpha_{CLmax}$	1.75	deg	Fig. 9.11	
α_s	15.99	deg		

FIGURE B.14: Enhanced lift design spreadsheet for HPKA case study.

The overall results are summarized in the table at the bottom of the spreadsheet in Figure B.15 and in the plots in the spreadsheet in Figure B.16. For the 2-D wing, the trailing-edge flap increased the maximum lift coefficient by an approximate factor of two, from 1.5 to 3.14. After accounting for the 3-D wing effect, the maximum lift coefficient is 2.07. The angle of attack at maximum lift is approximately 10.24 degrees.

Basic Wing—Low Aspect Ratio:

$(C1 + 1)$?	10.05		
$(C_{Lmax})_{base}$	1.2	Fig. 9.12	
C_2	0	Fig. 9.14	
$(C2 + 1)$?	0		
ΔC_{Lmax}	-0.02	Fig. 9.13	
C_{Lmax}	1.19		
$(\alpha_{CLmax})_{base}$	34	deg	Fig. 9.15
A cos(?	30		
$\Delta\alpha_{CLmax}$	1	deg	Fig. 9.16
α_s	35	deg	

Effect of Trailing-edge Flap:

Flap type	slot	slot, plane or split	
α_{0l}	-22.5		
Basic 3-D α_s	15.99	deg	
Basic 3-D C_L	1.35		
2-D $\Delta\alpha_s$	-5.75	deg	Fig. 9.18
2-D $\alpha_{s\ flapped}$	5.25	deg	
$(C_{lmax})_{flapped}$	3.14		
ΔC_{lmax}	1.64		
$K\Delta$	0.92		
ΔC_{Lmax}	0.72		
C_{Lmax}	2.07		
3-D $\alpha_{s\ flapped}$	10.24	deg	

Leading-edge flap CL Max:

ΔC_{lmax}	0	Table 9.1	
ΔC_{Lmax}	0		
C_{Lmax}	2.07		

Trailing-edge flap Added drag:

k_1	1	Fig. 9.20	
k_2	0.15	Fig. 9.21	
ΔC_{D0}	0.0720		

Lift Curve Plotting:

	2-D (no flaps)		2-D (flaps)		3-D (no flaps)		3-D (flaps)	
	α	C_1	α	C_1	α	C_L	α	C_L
	0	0	-22.5	0	0	0	-22.5	0
	10.62	1.2	-0.3	2.51	11.39	1.08	-5.01	1.66
	11	1.5	5.25	3.14	15.99	1.35	10.24	2.07
	11.38	1.2	10.8	2.51	20.59	1.08	25.49	1.66

FIGURE B.15: Enhanced lift design spreadsheet for HPKA case study (Part 2).

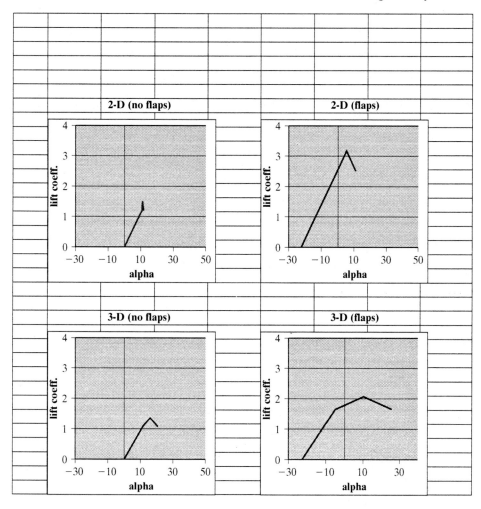

FIGURE B.16: Enhanced lift design spreadsheet for HPKA case study (Part 3).

B.9 REFINED WEIGHT ANALYSIS

The refined weight analysis for the HPKA was performed with the spreadsheet **refwt.xls**. The results are shown in Figures B.17 to B.19.

For all of the component weights, general aviation was considered the appropriate category for the weight estimates. Starting with the main wing, a design load factor of 2.53 was used. This was based on load factors that were determined for climb and instantaneous and sustained turn rates for the specified wing loading analyzed in the **wingld.xls** spreadsheet. The design gross weight was the estimated take-off weight, which along with the fuel weight came from the **itertow.xls** spreadsheet. The fuel was assumed to be stored in the wing. Based on these, the wing weight was estimated to be 70.38 lbs.

Wing Weight						
Fighter	*Transport*	*Gen. Av.*		**A**	10	
0.0103	0.0051	0.0090		**K_dw**	1	
1.0000	1.0000	1.0000		**K_vs**	1	
1.0000	1.0000	1.0000		**n**	2.53	
0.5000	0.5570	0.4900		**q**	101	lb/f^2
0.5000	0.5770	0.4900		**S_w**	57	f^2
0.6220	0.6490	0.7580		**(Sf/Sw)**	0.48	
0.7850	0.5000	0.6000		**S_f**	27.36	f^2
−0.4000	−0.4000	−0.3000		**t/c**	0.15	
1.0000	1.0000	0.0000		**W_dg**	1142	lbs
0.0500	0.1000	0.0040		**W_fw**	121	lbs
−1.0000	−1.0000	−0.9000		Λ	0	deg
0.0400	0.1000	0.0000		λ	1	
0.0000	0.0000	0.0060				
0.0000	0.0000	0.0035				

Fighter	105.31	lbs
Transport	61.13	lbs
Gen. Av.	70.38	lbs

Horizontal Tail Weight						
Fighter	*Transport*	*Gen. Av.*				
0.5503	0.0379	0.0092		**A_ht**	2	
−2.0000	−0.2500	0.0000		**b_ht**	4.04	f
0.2600	0.6390	0.4140		**F_w**	0.72	f
0.2600	0.1000	0.4140		**K_y**	5.01	f
0.8060	0.7500	0.8960		**L_ht**	16.7	f
0.0000	−1.0000	0.0000		n	2.53	
0.0000	0.7040	0.0000		q	101	lbs/f^2
0.0000	−1.0000	0.0340		**S_ht**	8.16	f^2
0.0000	0.1160	0.0430		**(t/c)ht**	0.06	
0.0000	0.0000	−0.1200		W_dg	1142	lbs
0.0000	0.0000	−0.0200		(Λ)ht	15	deg
0.0000	0.0000	0.1680		(λ)ht	0.77	

Fighter	17.12	lbs
Transport	3.62	lbs
Gen. Av.	5.15	lbs

Vertical Tail Weight						
Fighter	*Transport*	*Gen. Av.*				
0.4520	0.0026	0.0076		**A_vt**	1.2	
1.0000	1.0000	0.2000		**H_ht**	3.19	f
0.5000	0.2250	1.0000		**H_vt**	3.19	f
0.4880	0.5560	0.3760		**K_z**	14.4	f
0.4880	0.5360	0.3760		**K_rht**	1	
0.7180	0.5000	0.8730		**L_vt**	14.4	f
0.3410	0.0000	0.0000		**M**	0.35	
−1.0000	−0.5000	0.0000		n	2.53	

FIGURE B.17: Refined weight estimate spreadsheet for HPKA case study (Part 1).

The horizontal tail is part of a T-tail design. Its weight was estimated as only 5.15 lbs. The vertical tail supports the horizontal tail. Its weight estimate is approximately twice that of the horizontal tail, or 9.45 lbs.

The fuselage will have a single canopy for an entrance. The structural diameter, D, was taken as the maximum height of the fuselage, 4.9 feet. The fuselage cabin is

0.3480	0.0000	0.0000		q	101	lbs/f^2		
0.2230	0.3500	0.3570		S_r/S_vt	0.03			
1.0000	0.0000	0.0000		S_r	0.26	f^2		
0.2500	0.0000	0.0390		S_vt	9	f^2		
−0.3230	−1.0000	−0.2240		(t/c)_vt	0.06			
0.0000	−0.5000	−0.4900		W_dg	1142	lbs		
0.0000	0.8750	0.0000		(Λ)vt	30	deg		
0.0000	0.0000	0.1220		(λ)vt	0.78			

Fighter	9.34	lbs
Transport	10.27	lbs
Gen. Av	9.45	lbs

Fuselage Weight

Fighter	Transport	Gen. Av.						
0.4990	0.3280	0.0520		b_w	23.88	f		
1.0000	1.0600	1.0000		D	4.9	f		
1.0000	1.1200	1.0000		K_ws	0.000			
0.3500	0.5000	0.1770		K_dwf	1			
0.2500	0.5000	0.1770		K_door	1.06			
0.5000	0.3500	−0.0720		K_lg	1.12			
0.0000	0.0000	−0.0510		L_f	24	f		
0.8490	−0.1000	0.0720		L_t	16.7	f		
0.0000	0.3020	1.0860		n	2.53			
0.6850	0.0000	0.0000		q	101			
0.0000	0.0400	0.0000		S_f	225.5	f^2		
0.0000	0.0000	0.2410		S_vt	9	f^2		
0.0000	0.0000	11.9000		W_f	24	f		
				V_pr	0	f^3		
				W_p	11.9			
				W_dg	1142	lbs		
				Λ	0	deg		
				λ	1			

Fighter	1,231.94	lbs
Transport	278.93	lbs
Gen. Av	191.82	lbs

Main Landing Gear Weight

Fighter	Transport	Gen. Av.						
1.0000	0.0106	0.0344		K_cb	1			
1.0000	1.0000	1.0000		K_mp	1			
1.0000	1.0000	1.0000		K_tpg	1			
0.2500	0.8880	0.7680		L_m	88	in		
0.2500	0.2500	0.7680		n	2.53			
0.9730	0.4000	0.4090		N_mw	2	wheels		
0.0000	0.3210	0.0000		N_mss	2	struts		
0.0000	−0.5000	0.0000		V_s	91.53	f/s		
0.0000	0.1000	0.0000		W_l	1,033	lbs		

FIGURE B.18: Refined weight estimate spreadsheet for HPKA case study (Part 2).

unpressurized so that the factor, W_p, was set to zero. Based on these input values, the weight of the fuselage was estimated to be 191 lbs.

The design will use a conventional tricycle landing gear. The main gear will have two struts, each with one wheel. Its weight was estimated to be 90.4 lbs. The nose gear would consist of a single strut and wheel. It was estimated to weigh 32.2 lbs.

Fighter	557.55								
Transport	52.8								
Gen. Av.	90.42								
	Nose Landing Gear Weight								
Fighter	*Transport*	*Gen. Av.*							
1.0000	0.0320	1.0000		**K_np**	1				
1.0000	1.0000	0.0153		**L_n**	44	in			
0.2900	0.6460	0.5660		n	2.53				
0.2900	0.2000	0.5660		**N_nw**	1	wheels			
0.5000	0.5000	0.8450		**W_l**	1033	lbs			
0.5250	0.4500	0.0000							
Fighter	64.97	lbs							
Transport	22.63	lbs							
Gen. Av.	32.18	lbs							
	Summary								
	Fighter	Transport	Gen. Av.						
Wing	105.31	61.13	70.38						
Horiz. Tail	17.12	3.62	5.15						
Vert. Tail	9.34	10.27	9.45						
Fuselage	1,231.94	278.93	191.82						
Main Gear	557.55	52.8	90.42						
Nose Gear	64.97	22.63	32.18						
Total	1,986.24	429.37	399.40						

	Case Study			
Wstr	571	lbs		
Parts	**Weights**	**W/Wstr**		
Wing	70.38	12.33	%	
Horiz. Tail	5.15	0.9	%	
Vert. Tail	9.45	1.66	%	
Fuselage	191.82	33.59	%	
Main Gear	90.42	15.84	%	
Nose Gear	32.18	5.64	%	
Ins. Eng.	150.00	26.27	%	
Total	549.40	**96.22**	%	
Other	21.60	3.78	%	
Target	159.88	28	%	

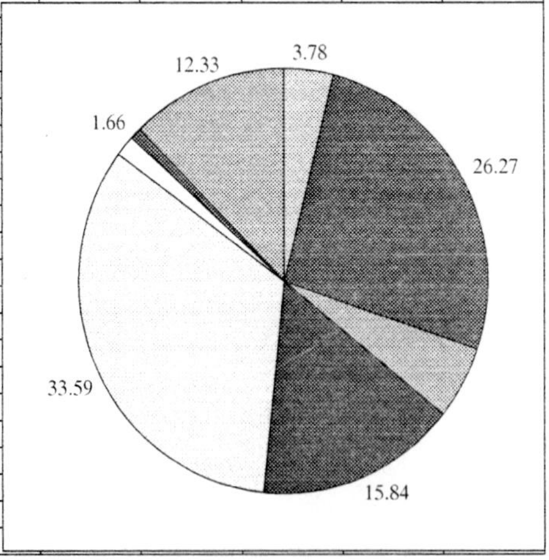

(Pie chart labels: 12.33, 3.78, 1.66, 26.27, 33.59, 15.84)

FIGURE B.19: Refined weight estimate spreadsheet for HPKA case study (Part 3).

The weight summary is given in the bottom table of Figure B.19. For this, the appropriate numbers had been copied down from the individual estimates. The pie chart gives a graphical representation of the weight breakdown as a percentage of the total structure weight. The structure weight was taken from the **itertow.xls** spreadsheet. The installed engine weight was estimated to be 150 lbs.

The sum of the weights of the major components plus the target weight is approximately 4 percent less than the initial structure weight estimate. The "target" weight represents the added weight due to numerous minor components. Based on historical data, this corresponds to 14 percent of the take-off weight, which is 159.88 lbs. The target weight exceeds the "other" weight by approximately 140 lbs. We would like the difference to be within 5 percent of the initial structure weight estimate ($\pm$29 lbs). In addition, it would be preferable if the other weights were <u>more</u> than the target since that would indicate that the component weights were overestimated.

As it happens in this case, the component weights need to be reduced. The largest percentage components are the fuselage, wing, and landing gear. The wing and fuselage weights could be reduced by approximately 30 percent by making extensive use of composites, for example, by constructing a large portion out of molded fiberglass-reinforced resins. That savings would correspond to approximately 80 lbs.

Further weight reductions could come from the landing gear by using higher strength-to-weight materials, such as a Magnesium alloy for the struts and wheel hubs. This could give a weight savings of approximately 60 lbs, which would put the component weights on target with the initial estimate. Overall however, this will increase the cost of the aircraft, which would be one of the design trade-offs.

If the weights of these components could not be reduced by these or other means, the structure factor used in the initial weight estimate will have to be increased. If the take-off weight is not increased, this would result in a decrease in the range or payload.

B.10   STATIC STABILITY ANALYSIS

The static stability analysis is performed using the **stab.xls** spreadsheet. The result is shown in Figures B.20 to B.22.

The refined weights from the previous section were used to calculate the location of the center of gravity of the aircraft. These were input into the top portion of the spreadsheet in the table labeled "Load Summary." The fuel is assumed to be distributed in the wing. Payload is primarily the pilot and one passenger. Their weight is distributed over and slightly forward of the main wing. The engine is mounted to the front of the fuselage. The other weight component is left to be zero, since it is assumed to be uniformly distributed and, therefore, will not affect the longitudinal stability.

Based on these input values, the location of the center of gravity was determined to be at $x_{cg}/L = 0.33$ or 7.83 feet from the nose of the fuselage. This is slightly forward of the center of lift making it statically stable. The static margin in this case was 0.07.

The analysis of the longitudinal stability found the effect of the wing to be stabilizing and the horizontal tail to be slightly destabilizing. There was no inlet effect on the longitudinal stability from the propeller. The overall longitudinal stability coefficient, $C_{M_\alpha} = -0.89$, makes this aircraft stable. This is a good feature for a recreational aircraft and the coefficient magnitude would make it quite responsive to longitudinal maneuvers with a low trim drag to improve sport flying.

	Longitudinal Stability					
Fuselage Length						
L (f)	24					
Wing Center of Lift						
L_ctr (x/L)	0.35					
m.a.c. (ft)	8.4					
Load Summary (fuselage)						
Load Type	Magnitude	x/L_start	x/L_end	resultant	M @ C_lift	dw
	(lbs)			x/L	f-lb (+ cw)	
Fuel	121	0.32	0.42	0.37	58.08	40.33
Payload	450	0.2	0.4	0.3	−540	90
Fus. Struct.	191	0	1	0.5	687.6	9.1
Engine(s)	150	0	0.12	0.06	−1044	44.12
Wing Struct.	70.38	0.32	0.42	0.37	33.78	23.46
Horiz. Tail	5.2	0.93	1.03	0.98	78.62	1.73
Vert. Tail	9.1	0.86	1.3	1.08	159.43	0.93
Other	0	0	1	0.5	0	0
Σ L	996.68			Σ M	−566.48	
Tail Lift (req)	−37.47	0.93	1.03	0.98	−566.48	−12.49
Center of Gravity						
X_cg / L	0.33					
X_cg (ft)	7.83	f				
Static Margin						
S.M.	0.07	stable				
Longitudinal Stability Coefficient:						
Wing Parameters:						
S_w	57	f^2				
(C_L_α)_w	0.095	(deg)^−1				
x_w	−0.57	f				
cbar	8.4	f				
Horiz. Tail Parameters:						
(C_L_α)_ht	0.0460	(deg)^−1				
de/dα	0.2	Fig. 11.3				
η_ht	0.933					
l_ht	15.69	f				
S_ht	8.16	f^2				
Engine Parameters						
mdot	0	lbm/s				
l_i	6.87	f				

FIGURE B.20: Static stability analysis spreadsheet for HPKA case study (Part 1).

rho	0.0475	lbm/f^3				
V	370.3	f/s				
$d\beta/d\alpha$	1					
Calculations						
V_bar_hs	0.27					
inlet effect	0	unstable				
wing effect	−0.37	stable	check: C_M_α=−S.M.*C_L_α			−0.37
h. tail effect	0.53	unstable				
C_M_α	**−0.89**	stable				
Directional Stability Coefficient:						
Wing Parameters:						
A_w	10					
Λ	0	deg				
λ	1					
S_w	57	f^2				
b	23.9	f				
z_w	−2.6	f				
C_L (cruise)	0.1978					
Fuselage Parameters:						
h	4.51	f				
w	4.9	f				
Vol_f.	143.4	f^3				
Vertical Tail Parameters:						
(C_L_α)_vs	0.0302	(deg)^−1				
l_vs	14.4	f				
S_vs	9	f^2				
Λ_vs	30	deg				
Calculations						
V_bar_vs	0.0951					
$(1+d\sigma/d\beta)q/q$	0.84	Eq[11.42]				
v. tail effect	0.14	Eq[11.40]	stable			
fuse. effect	−0.13	Eq[11.44]	unstable			
wing effect	0	Eq[11.43]	stable			
C_n_β	**0.0130**	stable				
C_L_β	**−0.0130**	stable				
	Rudder Sizing					
Input Parameters						
δ_r	20	deg				
β	11.5	deg				
Asym. T	0	lbs				

FIGURE B.21: Static stability analysis spreadsheet for HPKA case study (Part 2).

S_w	57	f^2				
b	23.9	f				
C_n_ β	0.0130					
diam_e	0	f				
V_T-O	109.8	f/s				
rho_T-O	0.0754	lbm/f^3				
Calculations						
1.2V_T-O	131.76	f/s				
0.2V_T-O	21.96	f/s				
q	20.33	lbs/f^2				
D_e	0	lbs				
C_n_ δR:				Abs( )		
Asy. Power	0.0000	[rad]^−1	Eq[11.47]	0.0000		
Cross Wind	−0.0075	[rad]^−1	Eq[11.50]	0.0075		
dα_0L/dδ_r	0.0505		Eq[11.51]			
C_R/C_VS	3.0000	%	Fig. 11.9			

FIGURE B.22: Static stability analysis spreadsheet for HPKA case study (Part 3).

The results of the directional stability analysis found the wing and vertical tail to be stabilizing and the fuselage to be destabilizing. The vertical tail, however, slightly offsets the effect of the fuselage so that overall, it is directionally stable. The estimated values of the directional stability coefficients should make the design quite responsive to roll and turning maneuvers.

Finally, with regards to sizing the rudder, since there is only one engine, the only criterion is based on the cross-wing condition. In this case, the *minimum* rudder area was estimated to be approximately 3 percent of the vertical stabilizer. Based on comparison aircraft, the rudder area is more typically 5 to 15 percent.

B.11   COST ANALYSIS

The cost estimate was made using the CERs given in Chapter 11, which have been incorporated into the **cost** spreadsheet. The cost breakdown from the spreadsheet is shown in Figures B.23 to B.26.

The parameters used in determining the cost are the structure weight, maximum velocity, number of development and production aircraft, and the production rate. The costs are based on year 2000 dollars.

The only change that was made to the spreadsheet was for the cost of the engine. The spreadsheet is standardly set to estimate this cost for turbo-jet and turbofan engines. The HPKA is designed for a single reciprocating engine with a propeller. Therefore, a fixed cost estimate was inserted in place of the calculated value.

In order to reduce the unit price, a total production of 500 planes was planned, and the RDT&E costs were amortized over the first 400 planes. However, the unit cost estimate is still excessively large; approximately four times larger than comparable aircraft.

		Cost Analysis			
Input Parameters:				**Constants:**	
W_s	571	lbs		CPI (70−98)	3.95
V_max	370.3	f/s		CPI (86−98)	1.49
N_D	1	aircraft		CPI (98−00)	1.06
N_P	500	aircraft			
R_D	0.5	aircraft/mo.			
R_P	1	aircraft/mo			
Year	2000				
V_max	219.25	knots			
CR_E	94	$/hr			
CR_T	100	$/hr			
CR_ML	80	$/hr			
CR_QC	88	$/hr			
Research, Development, Test and Evaluation					
		Based on 1970	**Based on 1986**		
Airframe Engineering:					
	H_E (hrs)	33,999.76	133,299.81		
	C_E ($)	$3,195,978	$12,530,183		
Development Support:					
	C_D ($)	$417,776	$8,509,379		
Manufacturing Labor:					
	H_ML (hrs)	55,355.03	23,501.15		
	C_ML ($)	$4,428,403	$1,880,092		
Manufacturing Materials:					
	C_MM	$244,809	$235,601		
Tooling:					
	H_T (hrs)	69,332.36	50,938.54		
	C_T ($)	$6,933,236	$5,093,854		
Quality Control:					
	(C or O)		O	Cargo or Other Aircraft	
	H_QC	7,196.15	3,055.15		
	C_QC	$633,262	$268,853		
Flight Test:					
	C_FT	$15,601	$1,988,793		

FIGURE B.23: Cost analysis spreadsheet for HPKA case study (Part 1).

It is necessary to point out that the CERs on which the cost estimate was based are the result of correlations to existing production aircraft. A small aircraft like the proposed HPKA is not likely to be represented well in this group. Therefore, either a new set of CERs are needed, or a correction factor is needed to bring the cost estimate in line with those of comparable small aircraft.

Engine:					
	Cost/eng	3,000			
	N_eng	1			
	C_EN ($)	$3,000	$3,000		
Subtotal (RDT&E):		$15,872,064	$30,509,755		
Profit:					
	Percent	10	10		
	C_P	$1,587,206	$3,050,975		
Total (RDT&E):		$17,459,270	$33,560,730		
Acquisition:					
		Based on 1970	Based on 1986		
Airframe Engineering:					
	H_E (hrs)	106,019.90	367,080.79		
	C_E ($)	$9,965,871	$34,505,594		
Manufacturing Labor:					
	H_ML (hrs)	1,436,869.73	1,262,202.01		
	C_ML ($)	$114,949,579	$100,976,160		
Manufacturing Materials:					
	C_MM	$33,605,623	$33,779,531		
Tooling:					
	H_T (hrs)	209,581.36	261,141.31		
	C_T ($)	$20,958,136	$26,114,131		
Quality Control:					
	(C or O)		O	Cargo or Other Aircraft	
	H_QC	186,793.07	164,086.26		
	C_QC	$16,437,790	$14,439,591		
Engine:					
	Cost/eng	3,000			
				turbofan or turbojet	
	N_eng	1			
	C_EN ($)	$1,500,000	$1,500,000		
Subtotal (A):		$197,416,998	$211,315,007		
Profit:					
	Percent	10	10		
	C_P	$19,741,700	$21,131,501		
Total (A):		$217,158,698	$232,446,508		

FIGURE B.24: Cost analysis spreadsheet for HPKA case study (Part 2).

Total Profit:		$21,328,906	$24,182,476		
Aircraft Unit Cost:					
RDT&E amortized					
unit no.:		400	400		
Cost/Prod. Aircraft		$43,648	$83,902		
Unit cost:					
1	400	**$477,966**	**$548,795**		
401	500	**$434,317**	**$464,893**		
Summary Totals:		**Based on 1970**	**Based on 1986**		
	C_E ($)	$13,161,849	$47,035,777		
	C_D ($)	$417,776	$8,509,379		
	C_ML ($)	$119,377,981	$102,856,252		
	C_MM ($)	$33,850,432	$34,015,132		
	C_T ($)	$27,891,372	$31,207,984		
	C_QC ($)	$17,071,051	$14,708,444		
	C_EN ($)	$1,503,000	$1,503,000		
	C_P ($)	$21,328,906	$24,182,476		

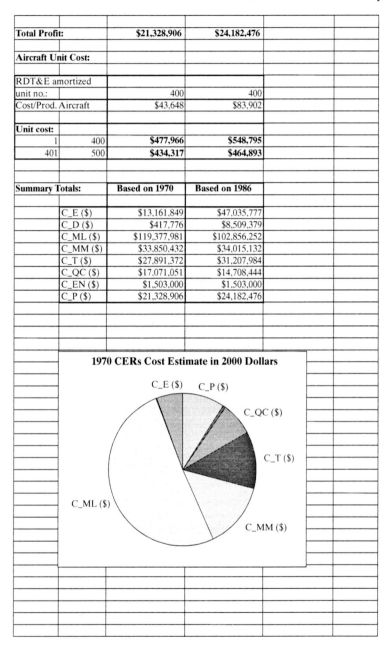

1970 CERs Cost Estimate in 2000 Dollars

C_E ($)   C_P ($)

C_QC ($)

C_T ($)

C_ML ($)

C_MM ($)

FIGURE B.25: Cost analysis spreadsheet for HPKA case study (Part 3).

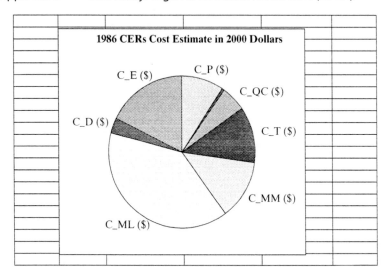

FIGURE B.26: Cost analysis spreadsheet for HPKA case study (Part 4).

TABLE B.2: HPKA proposed and final characteristics.

	Proposed	Final	Diamond Katana	Pulsar Sport 150	Dyn'Aero MCR01 VLA
Max. Speed (kts)	150	150	80	118	100
Cruise Speed (kts)	250	250	157	135	197
Max. TO Weight (lbs)	1800	1142	1610	1200	992
Empty Weight (lbs)	1000	571	1085	680	485
Fuel Capacity (gal)	30	29	20.1	20	23.2
Range (nm)	1000	1000	500	521	894
TO Distance (f)	1500	1340	1560	—	1395
Landing Distance (f)	1500	1580	1490	—	—
Aspect Ratio	7	10	10	6	8.45

B.12 SUMMARY

The overall design was relatively successful. As indicated in Table B.2, most of the characteristics that were initially proposed were closely met in the design. This design should then be competitive in the market of comparable aircraft.

APPENDIX C

Case Study: KC-42 Tanker/Transport Aircraft

DESIGN PROPOSAL: KC-42 TANKER/TRANSPORT

We propose to design a tanker aircraft for mid-air refueling of military aircraft. It will have a higher subsonic cruise Mach number and longer range and be able to operate at higher altitudes compared to existing tanker aircraft. This design is a response to the new joint-strike-fighter design that will have a limited range and, therefore, will need frequent air refueling to extend its operating radius. The KC-42 will also be convertible to a cargo or troop transport to add to its versatility.

The proposed characteristics of the KC-42, along with those of the KC-10 tanker, are given in the following table. The principle design drivers will be longer range, higher cruise speed, and comparable cost.

Proposed characteristics of the KC-42 versus those of the KC-10

	KC-42	KC-10
Cruise Mach No.	0.85	0.825
Cruise Altitude	35,000	35,000
Maximum Ceiling	55,000	—
Max. T-O Weight (lbs)	640,000	555,000
Empty Weight (lbs)	230,000	241,000
Range (nm)	4400	3800
T-O Distance (ft)	11,000	10,257
Landing Distance (ft)	6000	—
Wing Loading	167	149

C.1 TAKE-OFF WEIGHT ESTIMATE

The take-off weight is estimated using the **itertow.xls** spreadsheet. The spreadsheet is shown in Figure C.1.

The design drivers that heavily affect the take-off weight are operating range, cruise speed and altitude, thrust-specific fuel consumption, aspect ratio, and payload weight. The TSFC was taken from engines now used on tanker aircraft. The aspect ratio was increased slightly over present tankers to improve the range. The structure factor of 0.36 was based on military cargo aircraft given in Table 2.1 in the text, using a nominal value of the take-off weight.

The payload is indicated as a combination of expendable and nonexpendable. The expendable payload is the fuel use for mid-air refueling of aircraft. This is expected to be used to refuel aircraft after the KC-42 has flown to its design operating radius. The nonexpendable payload represents payload fuel that is trapped in lines and cavities. This corresponds to approximately 3 percent of the total fuel by weight.

With the input values used in the spreadsheet, the take-off weight was estimated to be 638,950 lbs. This is close to the original estimate. The structure weight is estimated to be 230,022 lbs, which is slightly less than the KC-10.

C.2 WING LOADING SELECTION

The effect of the wing loading on different performance characteristics was evaluated using the **wingld.xls** spreadsheet. The spreadsheet output is shown in Figures C.2 and C.3.

In the design of the KC-42, an emphasis was placed on efficient cruise. Therefore, the optimum wing loading for maximum cruise efficiency was chosen. This gave a wing loading of 167 lbs/f^2. Based on the weight of the aircraft at the start of cruise, this wing loading gave a wing area of 3610 f^2. This wing loading is slightly larger than that of the KC-10.

At the end of cruise, following its complete mission, the wing loading is reduced to 70.3 lbs/f^2. If the aircraft were allowed to drift up in altitude so that lift balanced weight, it would rise to an altitude of approximately 36,800 feet.

Based on the wing area and take-off weight, the take-off wing loading is 177 lbs/f^2. Using the indicated thrust, which is based on engines now commonly used on tankers, and a modest lift coefficient of 1.6, the take-off distance is estimated to be 10,557 feet.

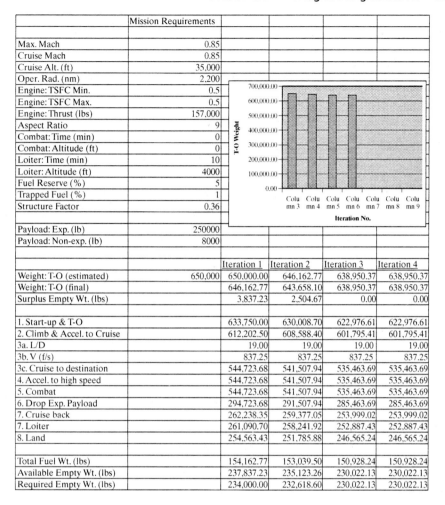

	Mission Requirements				
Max. Mach	0.85				
Cruise Mach	0.85				
Cruise Alt. (ft)	35,000				
Oper. Rad. (nm)	2,200				
Engine: TSFC Min.	0.5				
Engine: TSFC Max.	0.5				
Engine: Thrust (lbs)	157,000				
Aspect Ratio	9				
Combat: Time (min)	0				
Combat: Altitude (ft)	0				
Loiter: Time (min)	10				
Loiter: Altitude (ft)	4000				
Fuel Reserve (%)	5				
Trapped Fuel (%)	1				
Structure Factor	0.36				
Payload: Exp. (lb)	250000				
Payload: Non-exp. (lb)	8000				

		Iteration 1	Iteration 2	Iteration 3	Iteration 4
Weight: T-O (estimated)	650,000	650,000.00	646,162.77	638,950.37	638,950.37
Weight: T-O (final)		646,162.77	643,658.10	638,950.37	638,950.37
Surplus Empty Wt. (lbs)		3,837.23	2,504.67	0.00	0.00
1. Start-up & T-O		633,750.00	630,008.70	622,976.61	622,976.61
2. Climb & Accel. to Cruise		612,202.50	608,588.40	601,795.41	601,795.41
3a. L/D		19.00	19.00	19.00	19.00
3b. V (f/s)		837.25	837.25	837.25	837.25
3c. Cruise to destination		544,723.68	541,507.94	535,463.69	535,463.69
4. Accel. to high speed		544,723.68	541,507.94	535,463.69	535,463.69
5. Combat		544,723.68	541,507.94	535,463.69	535,463.69
6. Drop Exp. Payload		294,723.68	291,507.94	285,463.69	285,463.69
7. Cruise back		262,238.35	259,377.05	253,999.02	253,999.02
7. Loiter		261,090.70	258,241.92	252,887.43	252,887.43
8. Land		254,563.43	251,785.88	246,565.24	246,565.24
Total Fuel Wt. (lbs)		154,162.77	153,039.50	150,928.24	150,928.24
Available Empty Wt. (lbs)		237,837.23	235,123.26	230,022.13	230,022.13
Required Empty Wt. (lbs)		234,000.00	232,618.60	230,022.13	230,022.13

FIGURE C.1: Take-off weight spreadsheet for KC-42 case study.

This is about 300 feet longer than that of the KC-10. Similarly, based on the wing loading at landing, the landing distance is estimated to be 5572 feet.

Climb was taken to be just following take-off. The climb Mach number was taken to be 0.3, and the climb rate was set at 500 f/min. Analysis for these conditions resulted in a required wing loading of 101 lbs/f^2. Since this is less than the specified wing loading, these conditions could be met in the design.

The acceleration was also considered for conditions following take-off where it relates to acceleration to cruise Mach number. A positive load factor of 3.5 was specified. This value is typical of an aircraft of this type. With these as input, the necessary wing loading was found to be 158 lbs/f^2, which again is within the value of the wing loading that will occur at this part of the mission.

The conditions for analyzing the instantaneous and sustained turn rates were based in those at the start of cruise. For the instantaneous turn rate, the coefficient of lift was

Take-off			Landing	
H (f)	1000		H (f)	1000
CL(max)	1.6		CL(max)	1.6
T(max) (lb)	157500		T(max) (lb)	157500
W_TO (lb)	638950		W_L (lb)	246565
S (f^2)	3610.62		W/S (lb/f^2)	68.29
W/S (lb/f^2)	176.96		S (f^2)	3610.62
SIGMA	0.97		SIGMA	0.97
T/W	0.25		T/W	0.64
TOP	460.74		LP	43.83
S_TO (f)	10556.71		S_TO(f)	5571.54
Cruise Start			**Cruise End**	
CD_0	0.0350		CD_0	0.0350
A	9		A	9
H (f)	35,000		H (f)	36,800
Cruise Mach	0.85		Cruise Mach	0.85
W (lb)	601795		W (lb)	253,999
k	0.0442		k	0.0442
V (f/s)	837.25		V (f/s)	832.05
rho (lbm/f^3)	0.0243		rho (lbm/f^3)	0.0220
q (lbf/f^2)	264.92		q (lbf/f^2)	236.76
W/S (lb/f^2)	166.67		W/S_optimum	148.96
S (f^2)	3610.62		W/S_actual	70.35
Climb			**Acceleration**	
H (f)	4,000		H (f)	15,000
Climb Mach	0.300		Cruise Mach	0.85
dH/dt (f/min)	500		n	3.5
V (f/s)	327.12			
G (rad)	0.03		V (f/s)	895.05
Gamma (deg)	1.46		rho (lbm/f^3)	0.05
rho (lbm/f^3)	0.07		q (lbf/f^2)	622.59
q (lbf/f^2)	113.78		W/S (lb/f^2)	158.27
T/W min	0.104			
T/W min − G	0.08			
W/S_ + (lb/f^2)	101.24			
W/S_ − (lb/f^2)	101.24			

FIGURE C.2: Wing loading spreadsheet for KC-42 case study (Part 1).

varied from 0.7 to 1.6. This gave values of turn rates from 1.1 to 5.2 deg/s. For this range, the load factor varied from 1.1 to 2.6, which was less than the maximum design load factor of 3.5.

Based on the load factor of 2.6, the sustained turn rate was found to be 4.53 deg/s. For $n = 3.5$, the sustained turn rate was 5.8 deg/s.

Turn-Inst.			Turn-Sustained	
H (f)	35,000		H (f)	35,000
Cruise Mach	0.85		Cruise Mach	0.85
CL_max	1.6		n	2.55
W/S (lb/f^2)	166		T/W_max	0.2
			W/S (lb/f^2)	166
V (f/s)	837.25			
rho (lbm/f^3)	0.0243		V (f/s)	837.25
q (lbf/f^2)	264.92		rho (lbm/f^3)	0.0243
psi_dot (rad/s)	0.09		q (lbf/f^2)	264.92
psi_dot (deg/s)	5.18		psi_dot (rad/s)	0.08
n	2.55		psi_dot (deg/s)	4.53
Ceiling			Glide	
W (lb)	601795		W (lb)	601795
S (f^2)	3610		S (f^2)	3610
H (f)	40,000		H (f)	35,000
Cruise Mach	0.85		Cruise Mach	0.85
W/S (lb/f^2)	166.7		W/S (lb/f^2)	166.7
V (f/s)	822.8		V (f/s)	837.25
rho (lbm/f^3)	0.02		rho (lbm/f^3)	0.02
q (lbf/f^2)	188.28		q (lbf/f^2)	264.92
CL_required	0.89		V_min_D (f/s)	704.1
			M_min_D	0.71

FIGURE C.3: Wing loading spreadsheet for KC-42 case study (Part 2).

The maximum flight ceiling was based on conditions at the start of cruise. For this, the input altitude was varied until the required lift coefficient needed to fly at that altitude was equal to the lift coefficient at minimum drag. This lift coefficient was found to be 0.885 and gave a ceiling of 40,000 feet. At the end of cruise, as a result of the lower wing loading, the flight ceiling increased to 55,000 feet.

Finally, for the conditions at the start of cruise, the Mach number for the maximum glide rate was determined to be 0.71.

C.3 WING DESIGN

The spreadsheet for the wing design is shown in Figures C.4 and C.5. This uses the wing area, cruise speed and altitude, and aspect ratio from the previous spreadsheets. Based on these, the range of lift coefficients from the start of cruise to the end of cruise was found to be from 0.629 to 0.266. For efficient cruise, an airfoil was selected that had a drag bucket width that encompassed this range of design C_l values.

Other factors in the selection of the wing section were having a maximum t/c of from 12 to 15 percent. This range was consistent with historical aircraft and would provide a large internal volume for storing fuel, flaps, and landing gear in the wings.

Three types of airfoil sections were considered. These are summarized in Table C.1. All of them have a drag bucket that encompassed the range of the design C_l values. The thinner airfoil has the higher L/D ratio of the three, but offers the smallest amount of internal volume. Since the difference was small, the thicker airfoil sections were selected.

Design Parameters				Airfoil Data					
M	0.85			Name	NACA 4415				
S	3611	ft^2		Cl_{max}	1.6				
A	9.0			$C_{l\alpha}$	0.113	1/deg			
Λ_{LE}	30	deg		a.c.	0.25	c			
t/c	0.15			α_{0L}	−4	deg			
λ	0.15			Cd0	0.0065				
W c-start	601.795	lbf/f^2		r_{lc}	0.0024	c			
W c-end	253.999	lbf/f^2		Cl_{minD}	0 − 0.7				
q c-start	264.92	lbf/f^2		(t/c)max	0.35	c			
q c-end	236.76	lbf/f^2							
Cl c-start	0.629								
Cl c-end	0.266								
Calculations				**Sweep Angles**					
b	180.3	ft			x/c	$\Lambda_{x/c}$ (deg)			
M_{eff}	0.74			LE	0.00	30.0			
c_r	34.8	ft		1/4C	0.25	26.3			
c_t	5.2	ft		a.c	0.25	26.4			
m.a.c.	23.7	ft		(t/c)max	0.35	24.8			
				TE	1.00	14.0			
β	0.68								
$C_{L\alpha}$	0.102	1/deg		**Viscous Drag**					
C_{Lo}	0.41			V_eff	725.08	f/s			
α_{trim}	2.1	deg		q_eff	198.69	lbf/f^2			
C_{Ltrim}	0.629			Re_mac	3.91E+07				
k	0.04			sqrt(Re)	6249.08				
C_D	0.025			Cf	2.32E−03				
L/D	24.77			S_wet	7420.61	ft^2			
				F	1.66				
				Q	1				
Total Drag	24291.47	lbf		C_{D0}	0.00790				
Plotting:									
Spanwise View									
x	y								
0	0								
34.8	0								
57.27	90.14								
52.04	90.14								
0	0								
Lift Curves									
α	Cl	α	CL						
−4	0	−4	0						
10.16	1.6	10.16	1.45						

FIGURE C.4: Wing design spreadsheet for KC-42 case study (Part 1).

Air Properties					
Cruise Alt.	35,000	ft			
V	837.25	f/s			
ρ	0.0243	lbm/f^3			
q	264.92	lbf/f^2			
μ	1.07E−05	lbm/(f-s)			
ν (cruise)	4.40E−04	f^2/s			

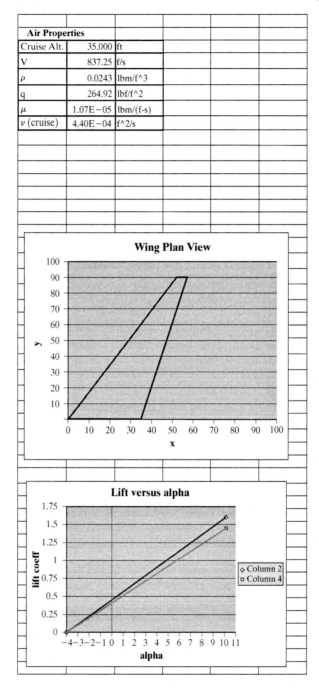

FIGURE C.5: Wing design spreadsheet for KC-42 case study (Part 2).

TABLE C.1: Comparison of candidate wing section types for the KC-42.

Characteristic	NACA 4412	**NACA 4415**	NACA $64_2 - 415$
$(t/c)_{max}$	12%	**15%**	15%
$C_{l_{max}}$	1.5	**1.6**	1.5
C_{d_0}	0.006	**0.0065**	0.005
$C_{l_{min D}}$	0.0–0.8	**0.0–0.7**	0.1–0.7
α_{trim}	2.2°	**2.1°**	3.1°
C_{D_0}	0.007543	**0.007898**	0.007898
L/D	25.12	**24.77**	24.77
Total Drag (lbs)	23,953	**24,293**	24,293

Comparing the two thicker airfoils, the L/D ratios are identical. However, the NACA 4415 has a slightly higher maximum lift coefficient. Therefore, it was selected for the KC-42 design.

Both the sweep angle, Λ_{LE}, and the taper ratio, λ, were chosen based on historical trends. For the design Mach number of 0.85, a sweep angle of 30 degrees was typical. For this sweep angle, a taper ratio of 0.15 was fairly common.

The lift coefficient versus angle of attack for this wing is fairly close to the 2-D wing, as shown in the plot in the spreadsheet in Figure C.5. This is mainly a result of the large aspect ratio and relatively small sweep angle.

For the viscous drag calculations, an interference factor of one was chosen since the wings are expected to be "clean" without any external stores. The overall drag on the wing was estimated to be 24,293 lbs. This gave a drag coefficient of $C_D = 0.025$. The L/D ratio was 24.77, which is considerably larger than the value of 19 that was estimated in the wing loading spreadsheet.

The overall dimensions of the wing are a 180.3 ft span, 34.8 ft root chord, and a 5.2 ft tip chord. The planform shape of the wing can be seen in the plot in the spreadsheet in Figure C.5.

C.4 FUSELAGE DESIGN

The main design driver for the fuselage is to have sufficient volume to hold the payload of fuel. Using a nominal value of 0.02 f^3/lb (Table 5.9) for the density of the fuel and a maximum fuel payload of 400,000 lbs, the required volume is approximately 8000 cubic feet.

The best location for the fuel is in the wings and the carry-through structure. Bladder tanks were chosen to carry the fuel for safety considerations. These utilize approximately 80 percent of the available volume. Based on this, each wing half-span will accommodate approximately 1840 f^3. The wing carry-through will accommodate approximately 2100 f^3. This leaves a volume of 2200 f^3 that is required in the fuselage. This will be designed to fit under the cabin floor. It will be distributed so that two-thirds will be forward of the main wing carry-through location.

In deciding the shape of the fuselage, comparison aircraft were considered. This is shown in Table C.2. Based on these, the fuselage of the KC-42 is predominately circular

TABLE C.2: Comparison of different fuselage shapes.

	KC-42	KC-135	KC-10
Cross-Section Shape	circle	ellipse	circle
Length (f)	153	145.5	170.5
Max. Diameter (f)	17	12.3	19.75
Wing Span (f)	180.3	130.8	165.38
Max. Fuel Payload (lbs)	400,000	200,000	356,000

in cross-section, with the cockpit portion being elliptic. A fuselage length-to-diameter ratio of nine was chosen, which is the same as the KC-10. The maximum diameter will be 17 feet, which is smaller than the KC-10. Therefore the overall fuselage length of the KC-42 will be 153 feet, which is slightly shorter than that of the KC-10.

The tail and nose of the fuselage are shaped like right cones. The lower portion of the fuselage tail is angled upward so that it will not contact the ground during the roll maneuver at take-off. The design of the fuselage nose gives an over-nose viewing angle of 29 degrees.

The location of the wing attachment on the fuselage is 20 percent of the fuselage length from the nose and 10 percent of the fuselage diameter below the longitudinal centerline.

The chosen landing gear configuration consists of two four-wheel bogeys under the wings and a two-wheel bogey under the fuselage centerline. The centerline wheel arrangement is similar to that on the MD-11. (See Figure 5.9.) A nose gear with a two-wheel bogey completes the landing gear arrangement. Based on Table 5.1, the diameter of the main gear wheels are 51.5 inches, and they are 20 inches wide. The main gear are stored in the wing. The wheel base is 80 feet, based on comparison aircraft. The nose wheels are 31 inches in diameter and 12 inches wide.

The fuselage data were input into the spreadsheet **fuse.xls** to determine the drag force. This is shown in Figure C.6.

In order to accommodate the elliptic cross-section portions of the fuselage in determining the perimeter, $P(x)$, and the wetted surface area, $S(x)$, the forms given in Eqs. [5.7 & 5.11], respectively, were used. The plot of the perimeter in the spreadsheet shows it to be smooth, without jumps.

Based on the conditions at cruise, the total drag on the fuselage is 3897 lbs. This corresponds to a drag coefficient, using the wing as the reference area, of 0.00407. This corresponds to only 16 percent of the drag on the wing.

C.5   TAIL DESIGN

A T-tail design was selected for the KC-42. This type of tail gives the aircraft good spin recovery by moving the rudder out of the wake of the horizontal tail during a spin.

The design of the tail was done using the **tail.xls** spreadsheet. These are shown in Figures C.7 and C.8.

The spreadsheet for the vertical tail design is shown in Figure C.7. For the placement of the vertical tail, the distance, l_{VT}, was selected to be 75 feet. This is approximately

			Fuselage Design					
Flight Regime Data:								
Cruise Mach	0.85							
Cruise Alt. (ft)	35,000							
V(f/s)	837.25							
ρ(lbm/f^3)	0.0243							
q (lbf/f^2)	264.92							
μ (lbm/(f-s))	1.1E−05							
ν (cruise) (f^2/s)	4.4E−04							
Dimension Data:			**Form Factors:**					
D-max (ft)	17		F	1.1048				
L/D	9		Q	1.0000				
L (ft)	153		$F*Q$	1.1048				
S (f^2)	3611							
Viscous Drag Calculations:			*Von-Karman Ogive Fuselage Shape*					
x/L	x (ft)	H (ft)	W (ft)	P (ft)	Sw(ft^2)	Re$_x$	C$_F$	Drag (lbf)
0.00	0.00	0.00	0	0.0				
0.10	15.30	15.00	13.9	45.4	694.6	2.9E+07	2.39E−03	485
0.20	30.60	17.00	17	53.4	817.1	5.8E+07	2.15E−03	515
0.30	45.90	17.00	17	53.4	817.1	8.7E+07	2.03E−03	486
0.40	61.20	17.00	17	53.4	817.1	1.2E+08	1.95E−03	467
0.50	76.50	17.00	17	53.4	817.1	1.5E+08	1.89E−03	453
0.60	91.80	17.00	17	53.4	817.1	1.7E+08	1.85E−03	442
0.70	107.10	17.00	17	53.4	817.1	2.0E+08	1.81E−03	433
0.80	122.40	16.00	16	50.3	769.1	2.3E+08	1.78E−03	400
0.90	137.70	8.75	8.75	27.5	420.6	2.6E+08	1.75E−03	215
1	153.00	0.00	0	0.0	0.0	2.9E+08	1.73E−03	0
Totals:					**6787.0**			**3897**

Wave Drag Calculations:	
A_max	226.98
CDW	0
Drag (lbf)	**0**
Total Drag:	**3897**
(lbf)	
Equiv. CD	0.00407

FIGURE C.6: Fuselage design spreadsheet for KC-42 case study.

45 percent of the fuselage length, which is common for aircraft of this type. Also based on historical trends, the vertical tail volume coefficient, C_{VT}, was 0.08. Based on the spreadsheet, these gave a vertical tail area of 755 f^2.

The section shape of the vertical tail was chosen to be a NACA 0012. This has a reasonably low base drag and wide drag bucket. The 12 percent maximum t/c is less than that of the main wing so that it is less likely to stall.

A leading-edge sweep angle of 35 degrees was chosen based on historical trends. These also led to the choice of the taper ratio of 0.9 and the aspect ratio of 1.1. The

Main Wing Reference			Air Properties				
b	180.3	ft	Cruise Alt.	35.000	ft		
m.a.c.	23.7	ft	V	837.25	f/s		
S	3611	ft^2	ρ	0.0243	lbm/f^3		
M	0.85		q	264.92	lbf/f^2		
Λ_{LE}	30	deg	μ	107.0E−7	lbm/(f-s)		
t/c	0.15		ν (cruise)	439.6E−6	f^2/s		
λ	0.15						

Vertical Tail							
Design Parameters			**Airfoil Data**				
Cvt	0.08		Name	NACA 0012			
Lvt	70.0	ft	Cl$_{max}$	1.5			
Λ_{LE}	35	deg	Cl$_\alpha$	0.104	1/deg		
t/c	0.12		a.c.	0.26	c		
λ	0.90		α_{0L}	0	deg		
Avt	1.10		Cd	0.006			
Calculations			**Sweep Angles**			**Viscous Drag**	
Svt	744	ft^2		x/c	$\Lambda_{x/c}$ (deg)	V_eff	685.84	f/s
b	28.6	ft	LE	0.00	35.0	q_eff	177.76	lbf/f^2
c$_r$	27.4	ft	1/4 chord	0.25	33.1	M_eff	0.7	
c$_t$	24.6	ft	(t/c)max	0.35	32.3	Re_mac	406.1E+5	
m.a.c.	26.0	ft	TE	1.00	27.0	sqrt(Re)	6372.58	
β	0.53					Cf	2.32E−03	
C$_{L\alpha}$	**0.029**	1/deg				S_wet	1517.46	ft^2
						F	1.67	
						Q	1.05	
Total Drag	**1638.880**	**lbf**				C$_{D0}$	**0.0083**	

Spanwise View		
x	y	
0	0	
27.4	0	
41.93	28.6	
20.03	28.6	
0	0	

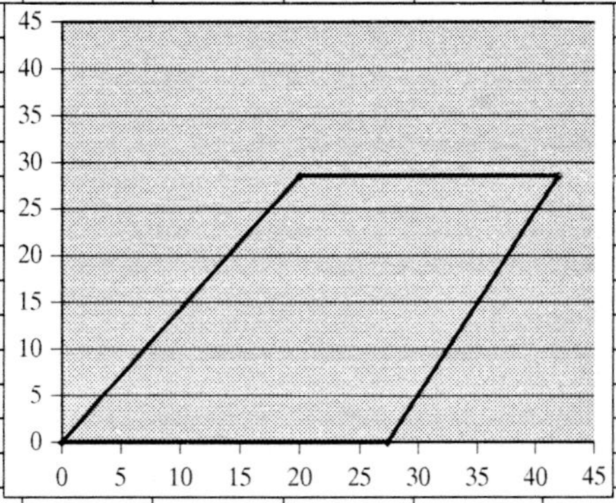

FIGURE C.7: Tail design spreadsheet for KC-42 case study (Part 1).

Horizontal Tail								
Design Parameters			**Airfoil Data**					
Cht	1.00		Name	*NACA 0012*				
Lht	80.0	ft	Cl_{max}	*1.5*				
Λ_{LE}	31	deg	Cl_{α}	*0.1*	1/deg			
t/c	0.12		a.c.	*0.26*	c			
λ	0.50		α_{0L}	*0*	deg			
Aht	3.20		Cd	*0.006*				
Calculations			**Sweep Angles**			**Viscous Drag**		
Sht	1069.76	ft^2		x/c	$\Lambda_{x/c}$ (deg)	V_eff	717.66	f/s
b	58.5	ft	LE	0.00	31.0	q_eff	194.64	lbf/f^2
c_r	24.4	ft	1/4 chord	0.25	26.4	M_eff	0.73	
c_t	12.2	ft	(t/c)max	0.35	24.5	Re_mac	309.5E+5	
m.a.c.	19.0	ft	TE	1.00	10.4	sqrt(Re)	5563.41	
β	0.53					Cf	2.40E−03	
$C_{L\alpha}$	**0.070**	1/deg				S_wet	2181.67	ft^2
						F	1.71	
						Q	1.05	
Total Drag	**2494.813**	lbf				C_{D0}	**0.0088**	

Spanwise View

x	y
0	0
24.4	0
29.77	29.3
17.58	29.3
0	0

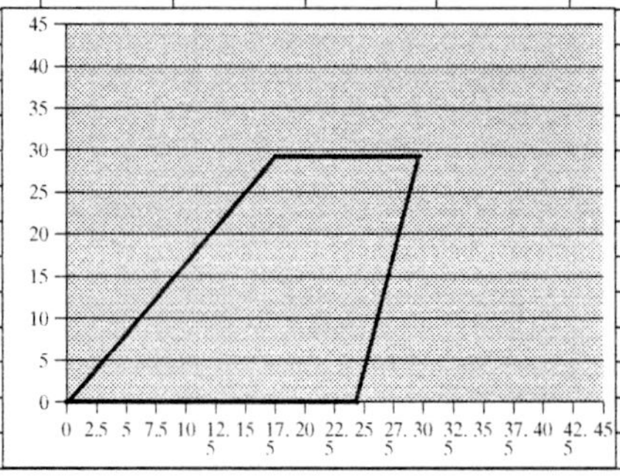

FIGURE C.8: Tail design spreadsheet for KC-42 case study (Part 2).

result gave a span of 28.6 feet and a mean chord of 26 feet. The planform view of the vertical tail is shown in the plot in Figure C.7. Based on the surface area, the drag on the vertical tail was estimated to be 1640 lbs.

The spreadsheet for the horizontal tail design is shown in Figure C.8. For the placement of the horizontal tail, the distance, l_{HT}, was selected to be 80 feet. Based on historical trends, the horizontal tail volume coefficient, C_{HT}, was 1.0. Based on the spreadsheet, these gave a horizontal tail area of 1017 f^2.

As with the vertical tail, a NACA 0012 section shape was chosen. The leading-edge sweep angle was set at 31 degrees, which is slightly more than that of the main wing. Since this is part of a T-tail design, the root chord of the horizontal tail corresponds to the tip chord of the vertical tail. Manipulating the taper and aspect ratios of the horizontal tail accomplished this and gave horizontal tail span of 58.5 feet and a mean chord of 19 feet. The planform view of the horizontal tail is shown in the plot in Figure C.8. Based on the surface area, the drag on the horizontal tail was estimated to be 2495 lbs.

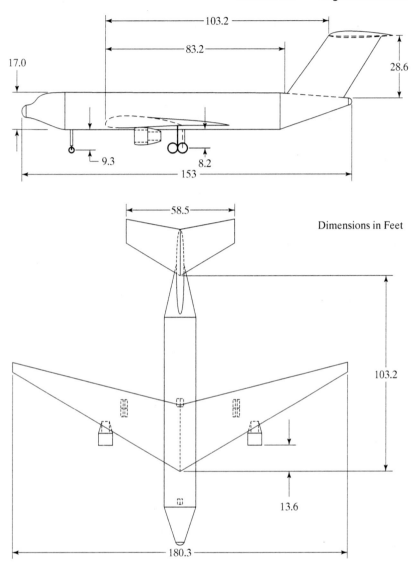

FIGURE C.9: Scale three-view drawing of the KC-42 case study.

As a check on the stall recovery of the design, at a main wing stall angle of 15 degrees, the horizontal tail was completely out of the wake of the main wing.

With the wing, fuselage and tail designed, the basic external shape of the aircraft is known. The exterior view of the KC-42 is shown in a scale drawing in Figure C.9.

C.6   ENGINE SELECTION

The KC-42 is designed to be powered by two high bypass turbo-jet engines. These will be sized based on providing the thrust that equals the drag at cruise conditions.

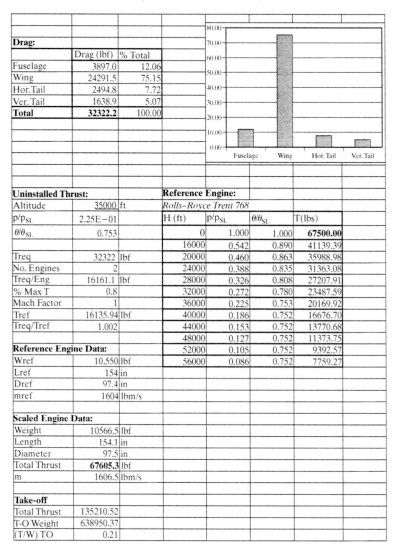

FIGURE C.10: Engine selection spreadsheet for KC-42 case study.

The drag of each of the main components (wing, fuselage, and tail) is plotted in the **engine.xls** spreadsheet shown in Figure C.10. As usual, the majority of the drag comes from the wing. The total drag is 32,322 lbs. Dividing the total drag by the two engines specifies the thrust per engine to be 16,161 lbs. When the change in static pressure and temperature at altitude are factored in, as well as assuming a continuous operation thrust of 80 percent of the maximum value, the required sea-level thrust per engine should be approximately 65,000 lbs.

There are several engines that can provide the necessary sea-level thrust. These are listed in Table C.3. Any of the listed engines could be used in this design. The

TABLE C.3: Comparison of engines considered for the KC-42.

	Pratt & Whitney 4168	Rolls-Royce Trent 768	General Electric CF6-80E1A2	General Electric CF6-80E1A3
T_{ref} (lbs)	68,999	67,500	67,500	72,000
W_{ref} (lbs)	14,350	10,550	10,627	10,627
L_{ref} (in)	163.1	154	173.5	173.5
D_{ref} (in)	107	97.4	96	96

Rolls-Royce engine provides the necessary thrust and has the lowest weight of the four. Therefore, this engine was chosen.

The reference data for this engine were put into the spreadsheet. The scaling factor, T_{req}/T_{ref}, was 1.002, which for all intents makes this engine an exact fit.

The total thrust provided at sea level is 135,210 lbs. Based on the take-off weight, this gives a thrust-to-weight ratio of 0.21. This value is on the low side for comparable aircraft such as the KC-10 with $T/W = 0.267$ and the KC-135 with $T/W = 0.268$. However, this should be sufficient to reach the take-off distance requirements set out in the design proposal.

The placement of the two engines can be seen in the drawing in Figure C.9.

C.7   TAKE-OFF AND LANDING ANALYSIS

The take-off and landing analysis used the **to-l.xls** spreadsheet. This used several parameters from previous design steps including the wing area and aspect ratio, take-off weight, and maximum trust. The spreadsheet is shown in Figures C.11 and C.12.

The KC-42 is intended to take off from built-up air fields. Therefore, the friction coefficient is representative of those conditions. Values of the landing gear parameters were based on the comparison aircraft. The flaps are intended to be slotted flaps. When deployed, they are expected to increase the wing area by approximately 20 percent. This gives a flapped wing area of 4332.7 f^2 that is used in the spreadsheet.

The take-off climb angle is set to 3 degrees, and the obstacle height was specified as 35 feet. With these, the estimated total take-off distance was 13,777 feet. In this a majority of the distance is in the ground roll portion.

The take-off distance is somewhat larger than the value given in the design proposal, as well as first estimated based on the wing loading. This is largely due to the lower T/W at take-off. As an example, if the General Electric CF6-80E1A3 engine listed in Table C.3 were used, its larger thrust would reduce the distance to 12,104 feet.

The landing analysis is done in the bottom half of the spreadsheet. The landing weight assumed a worst case scenario in which all of the payload and one-half of the fuel is on board. Thrust reversers corresponding to 45 percent of the thrust are used in the analysis.

The resulting landing distance is 5765 feet. This is under the 6000 ft value specified in the design proposal. If the 60 percent factor is used to account for pilot variability, the landing distance is 9224 feet.

Take-Off								
CD_0	0.0468		mu_TO	0.03				
A	9		T_max (lb)	135000				
H (f)	0		f_LG	81.65				
CL_G	2.23		A_LG (f^2)	5				
W_TO (lb)	638.950		deltCD_0_flap	0.024				
S (f^2)	4332.7		gamma_CL (deg)	3				
			H_obstacle (f)	35				
k	0.04		T/W	0.21				
rho (lbm/f^3)	0.08		f1 (f/s^2)	5.84				
W/S (lb/f^2)	147.47		deltCD_0_LG	0.09				
S (f^2)	4332.7		f2 (f^−1)	−0.000051				
V_T-O (f/s)	283.18		R_TR (f)	16603.16				
q_T_O (lb/f^2)	95.23		H_TR (f)	22.75				
			S_G (f)	11825.2				
			S_R (f)	849.55				
			S_TR (f)	868.94				
			S_CL (f)	233.67				
			S_T-O (f)	13777.36				
Landing								
W_L (lb)	563486		D_50 (lb)	164354.19				
W/S (lb/f^2)	130.05		gamma_A (deg)	−16.96				
V_50 (f/s)	288.1		gamma_A _act	−3				
V_TD (f/s)	254.86		R_TR (f)	12144.86				
q_50 (lb/f^2)	98.56		H_TR (f)	16.64				
q_TD (lb/f^2)	77.13		f1 (f/s^2)	−22.67				
mu _L	0.6		f2 (f^−1)	0.0003159				
T_L (lb)	−58562							
			S_A (f)	636.49				
			S_TR (f)	635.61				
			S_FR (f)	764.57				
			S_B (f)	3728.57				
			S_L (f)	5765.23				
			1.6(S_L) (f)	9224.37				

FIGURE C.11: Take-off and landing spreadsheet for KC-42 case study (Part 1).

C.8   ENHANCED LIFT DESIGN

Passive enhanced lift devices consisting of slotted trailing-edge flaps and leading-edge slats were chosen for the KC-42. These were designed to achieve the maximum lift coefficient of 2.2 that was used in the take-off analysis.

The flaps correspond to 60 percent of the wing area and 20 percent of the wing chord. The maximum flap deflection was chosen to be 40 degrees. The estimate of this flap arrangement was done using the **flap.xls** spreadsheet. This is shown in Figures C.13 to C.15.

The data for the basic wing were taken from the **wing.xls** spreadsheet. Values of $C_{L_{max}}$ and α_s for the basic wing were found to be 1.28 and 11 degrees, respectively. The

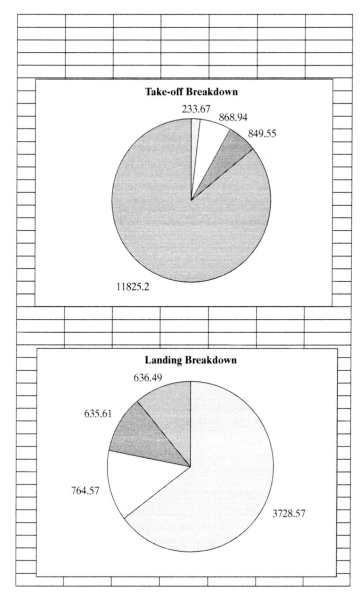

FIGURE C.12: Take-off and landing spreadsheet for KC-42 case study (Part 2).

aspect-ratio criterion determined the wing to be a high aspect ratio. Therefore, only that portion of the spreadsheet is relevant.

For the 3-D wing, the trailing-edge flaps gave $C_{L_{\max}} = 2.2$ and $\alpha_s = 8°$. The effect of the leading-edge flap increased $C_{L_{\max}}$ to 2.43. This is higher than the value used in the take-off analysis and, therefore, makes those estimates conservative.

The added drag due to the flaps corresponded to $\Delta C_{D_0} = 0.024$, which was the value used in the take-off and landing analysis.

Wing Data:								
Airfoil	NACA	4415						
ΛLE	30	deg						
λ	0.15							
t/c	0.15							
T-O Mach No.	0.1							
β	0.99							
A	9							
$\Lambda_{t/c}$	24.8	deg						
$C_{l\alpha}$ (no flap)	0.113	1/deg						
$C_{L\alpha}$ (no flap)	0.102	1/deg						
α_{0L}	−4	deg						
Clmax	1.6							
α_s	15	deg						
Trailing-edge Flap Design:								
Flap type	slot	slot,plane or split						
S_f/S_w	0.60							
δ_f	40	deg						
c_f/c	0.2							
Delta $\alpha 0L$:								
	Plane Flap							
K'	0.6	Fig. 9.3						
dCl/dδ_f	0.38	Fig. 9.4						
$\Delta\alpha_{0l}$	−14.14	deg						
	Single Slotted & Fowler Flap							
dα/dδ_f	−0.33	Fig. 9.5						
$\Delta\alpha_{0l}$	−15	deg						
	Split Flaps							
k	1	Fig. 9.6						
ΔCl	1.13	Fig. 9.7						
$\Delta\alpha_{0l}$	−10.000	deg						
Aspect Ratio Criterion:								
C_1	0.38	Fig. 9.8						
High A criteria	3.36	High						
Basic Wing-High Aspect Ratio:								
Δy	3.7	%	Fig. 9.10					
$C_{L.max}/C_{lmax}$	0.8	Fig. 9.9						
$C_{L.max}$	1.28							
$\Delta\alpha_{CL.max}$	2.5	deg	Fig. 9.11					
α_s	11.05	deg						

FIGURE C.13: Enhanced lift design spreadsheet for KC-42 case study (Part 1).

C.9 REFINED WEIGHT ESTIMATE

The refined weight analysis for the KC-42 was performed with the spreadsheet **refwt.xls**. The results are shown in Figures C.16 to C.18.

For all of the component weights, the transport was considered the appropriate category for the weight estimates. Starting with the main wing, a design load factor of 3.5 was used. This was based on load factors that were determined for climb, and instantaneous

Basic Wing-Low Aspect Ratio:							
$(C1 + 1)$?	10.77						
$(C_{Lmax})_{base}$	1.2		Fig. 9.12				
C_2	0		Fig. 9.14				
$(C2 + 1)$?	5.2						
ΔC_{Lmax}	−0.02		Fig. 9.13				
C_{Lmax}	1.19						
$(\alpha_{CLmax})_{base}$	34	deg	Fig. 9.15				
A cos(?	8.14						
$\Delta\alpha_{CLmax}$	1	deg	Fig. 9.16				
α_s	35	deg					

Effect of Trailing-edge Flap:							
Flap type	slot		slot, plane or split				
α_{0l}	−19						
Basic 3-D α_s	11.05	deg					
Basic 3-D C_L	1.28						
2-D $\Delta\alpha_s$	−2.9	deg	Fig. 9.18				
2-D $\alpha_{s\ flapped}$	12.1	deg					
$(C_{lmax})_{flapped}$	3.51						
ΔC_{lmax}	1.91						
$K\Delta$	0.87						
ΔC_{Lmax}	1.00						
C_{Lmax}	2.28						
3-D $\alpha_{s\ flapped}$	8.15	deg					

Leading-edge flap CL Max:							
ΔC_{lmax}	0.3		Table 9.1				
ΔC_{Lmax}	0.16						
C_{Lmax}	2.43						

Trailing-edge flap Added drag:							
k_1	1		Fig. 9.20				
k_2	0.04		Fig. 9.21				
ΔC_{D0}	0.0240						

Lift Curve Plotting:							
	2-D (no flaps)		2-D (flaps)		3-D (no flaps)		3-D (flaps)	
	α	C_l	α	C_l	α	C_L	α	C_L
	−4	0	−19	0	−4	0	−19	0
	7.33	1.28	5.88	2.81	6.04	1.02	−1.14	1.82
	15	1.6	12.1	3.51	11.05	1.28	8.15	2.28
	22.67	1.28	18.32	2.81	16.06	1.02	17.43	1.82

FIGURE C.14: Enhanced lift design spreadsheet for KC-42 case study (Part 2).

and sustained turn rates for the specified wing loading were analyzed in the **wingld.xls** spreadsheet. The design gross weight was the estimated take-off weight, which along with the fuel weight came from the **itertow.xls** spreadsheet. The mission fuel was assumed to be stored in the wing. Based on these, the wing weight was estimated to be 59,262 lbs.

The horizontal tail is part of a T-tail design. Its weight was estimated as only 6283 lbs. The vertical tail supports the horizontal tail. Its weight was estimated to be 4912 lbs.

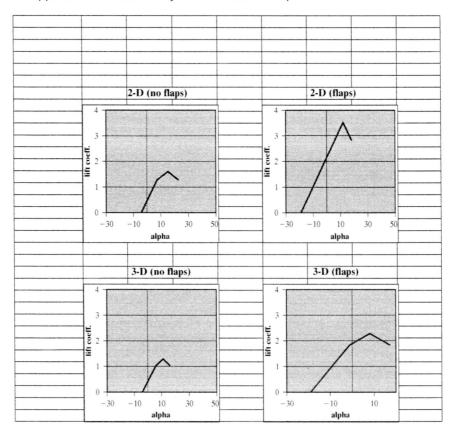

FIGURE C.15: Enhanced lift design spreadsheet for KC-42 case study (Part 3).

The fuselage will have a single cargo door and approximately 30 percent of the fuselage volume pressurized. The structural diameter, D, was taken as the maximum height of the fuselage, 17 feet. Based on these input values, the weight of the fuselage was estimated to be 30,742 lbs.

The design will use a conventional tricycle landing gear. The main gear will have two struts, each with four wheels. The weight at landing was assumed for a worst case in which the aircraft retained all of its payload and half of its mission fuel. Based on this, the total weight of the main landing gear was estimated to be 23,677 lbs. The nose gear would consist of a single strut and two wheels. It was estimated to weigh 3078 lbs.

The weight summary is given in the bottom table of Figure C.18. For this, the appropriate numbers had been copied down from the individual estimates for the transport category. The pie chart gives a graphical representation of the weight breakdown as a percentage of the total structure weight. The structure weight was taken from the **itertow.xls** spreadsheet. The installed engine weight was estimated to be 13,221 lbs.

The sum of the weights of the major components plus the target weight is within 1 percent of the initial structure weight estimate. Therefore, the initial estimate, which was based on the structure factor, was quite good.

Wing Weight					
Fighter	Transport	Gen. Av.			
0.0103	0.0051	0.0090	A	9	
1.0000	1.0000	1.0000	K_dw	1	
1.0000	1.0000	1.0000	K_vs	1	
0.5000	0.5570	0.4900	n	3.5	
0.5000	0.5770	0.4900	q	265	lb/f^2
0.6220	0.6490	0.7580	S_w	3611	f^2
0.7850	0.5000	0.6000	(Sf/Sw)	0.6	
-0.4000	-0.4000	-0.3000	S_f	2166.6	f^2
1.0000	1.0000	0.0000	t/c	0.15	
0.0500	0.1000	0.0040	W_dg	638950	lbs
-1.0000	-1.0000	-0.9000	W_fw	184000	lbs
0.0400	0.1000	0.0000	Λ	30	deg
0.0000	0.0000	0.0060	λ	0.15	
0.0000	0.0000	0.0035			

Fighter	47,638.90	lbs
Transport	59,262.23	lbs
Gen. Av.	46,531.56	lbs

Horizontal Tail Weight					
Fighter	Transport	Gen. Av.			
0.5503	0.0379	0.0092	A_ht	3.2	
-2.0000	-0.2500	0.0000	b_ht	58.5	f
0.2600	0.6390	0.4140	F_w	6	f
0.2600	0.1000	0.4140	K_y	24	f
0.8060	0.7500	0.8960	L_ht	80	f
0.0000	-1.0000	0.0000	n	3.5	
0.0000	0.7040	0.0000	q	265	lbs/f^2
0.0000	-1.0000	0.0340	S_ht	1070	f^2
0.0000	0.1160	0.0430	(t/c)ht	0.12	
0.0000	0.0000	-0.1200	W_dg	638950	lbs
0.0000	0.0000	-0.0200	(Λ)ht	31	deg
0.0000	0.0000	0.1680	(λ)ht	0.5	

Fighter	5,602.14	lbs
Transport	6,283.70	lbs
Gen. Av.	7,077.75	lbs

Vertical Tail Weight					
Fighter	Transport	Gen. Av.			
0.4520	0.0026	0.0076	A_vt	1.1	
1.0000	1.0000	0.2000	H_ht	28.6	f
0.5000	0.2250	1.0000	H_vt	28.6	f
0.4880	0.5560	0.3760	K_z	70	f
0.4880	0.5360	0.3760	K_rht	1	
0.7180	0.5000	0.8730	L_vt	70	f
0.3410	0.0000	0.0000	M	0.85	
-1.0000	-0.5000	0.0000	n	3.5	

FIGURE C.16: Refined weight spreadsheet for KC-42 case study (Part 1).

C.10   STATIC STABILITY AND CONTROL

The static stability analysis is done using the **stab.xls** spreadsheet. The result is shown in Figures C.19 to C.21.

The refined weights from the previous section were used to calculate the location of the center of gravity of the aircraft. These were input into the top portion of the spreadsheet in the table labeled "Load Summary." The majority of the fuel (as payload and assumed by the mission) is assumed to be distributed in the wing. The weight of the "other" category in the refined weight analysis is assumed to be distributed evenly over the whole length of the fuselage.

0.3480	0.0000	0.0000	q	265	lbs/f^2
0.2230	0.3500	0.3570	S_r/S_vt	0.3	
1.0000	0.0000	0.0000	S_r	223.2	f^2
0.2500	0.0000	0.0390	S_vt	744	f^2
−0.3230	−1.0000	−0.2240	(t/c)_vt	0.12	
0.0000	−0.5000	−0.4900	W_dg	638950	lbs
0.0000	0.8750	0.0000	(Λ)vt	35	deg
0.0000	0.0000	0.1220	(λ)vt	0.9	
Fighter	1,751.35	lbs			
Transport	4,912.49	lbs			
Gen. Av	4,300.49	lbs			

Fuselage Weight					
Fighter	*Transport*	*Gen. Av.*			
0.4990	0.3280	0.0520	b_w	180.3	f
1.0000	1.0600	1.0000	D	17	f
1.0000	1.0000	1.0000	K_ws	0.58	
0.3500	0.5000	0.1770	K_dwf	1	
0.2500	0.5000	0.1770	K_door	1.06	
0.5000	0.3500	−0.0720	K_lg	1	
0.0000	0.0000	−0.0510	L_f	153	f
0.8490	−0.1000	0.0720	L_t	80	f
0.0000	0.3020	1.0860	n	3.5	
0.6850	0.0000	0.0000	q	265	
0.0000	0.0400	0.0000	S_f	5202	f^2
0.0000	0.0000	0.2410	S_vt	744	f^2
0.0000	0.0000	31.9666	W_f	17	f
			V_pr	8000	f^3
			W_p	31.97	
			W_dg	638950	lbs
			Λ	30	deg
			λ	0.15	
Fighter	70,127.01	lbs			
Transport	30,742.50	lbs			
Gen. Av	19,705.00	lbs			

Main Landing Gear Weight					
Fighter	*Transport*	*Gen. Av.*			
1.0000	0.0106	0.0344	K_cb	1	
1.0000	1.0000	1.0000	K_mp	1	
1.0000	1.0000	1.0000	K_tpg	1	
0.2500	0.8880	0.7680	L_m	119	in
0.2500	0.2500	0.7680	n	3.5	
0.9730	0.4000	0.4090	N_mw	4	wheels
0.0000	0.3210	0.0000	N_mss	2	struts
0.0000	−0.5000	0.0000	V_s	234	f/s
0.0000	0.1000	0.0000	W_l	563,486	lbs

FIGURE C.17: Refined weight spreadsheet for KC-42 case study (Part 2).

Based on all of the input values, the location of the center of gravity was determined to be at $x_{cg}/L = 0.39$ or 59.08 feet from the nose of the fuselage. This is forward of the center of lift making it statically stable. The static margin in this case was 0.22.

The analysis of the longitudinal stability found the effect of the wing to be stabilizing and the horizontal tail to be slightly destabilizing. The inlet effect on the longitudinal stability was destabilizing. The overall longitudinal stability coefficient, $C_{M_\alpha} = -2.67$, makes this aircraft nicely stable. This is a good feature for a long-range transport aircraft.

Fighter	3919.63							
Transport	23877.77							
Gen. Av.	16595.09							

Nose Landing Gear Weight

Fighter	*Transport*	*Gen. Av.*					
1.0000	0.0320	1.0000	**K_np**	1			
1.0000	1.0000	0.0153	**L_n**	111.6	in		
0.2900	0.6460	0.5660	n	3.5			
0.2900	0.2000	0.5660	**N_nw**	2	wheels		
0.5000	0.5000	0.8450	**W_l**	563486	lbs		
0.5250	0.4500	0.0000					

Fighter	1,017.21	lbs
Transport	3,078.43	lbs
Gen. Av.	3,005.45	lbs

Summary

	Fighter	Transport	Gen. Av.
Wing	47,638.90	59,262.23	46,531.56
Horiz. Tail	5,602.14	6,283.70	7,077.75
Vert.Tail	1,751.35	4,912.49	4,300.49
Fuselage	70,127.01	30,742.50	19,705.00
Main Gear	3919.63	23877.77	16595.09
Nose Gear	1,017.21	3,078.43	3,005.45
Total	130,056.24	128,157.11	97,215.35

Case Study

Wstr	230,022	lbs

Parts	**Weights**	**W/Wstr**	
Wing	59,262.23	25.76	%
Horiz. Tail	6,283.70	2.73	%
Vert.Tail	4,912.49	2.14	%
Fuselage	30,742.50	13.37	%
Main Gear	23,877.77	10.38	%
Nose Gear	3,078.43	1.34	%
Ins.Eng.	13,221.00	5.75	%
Total	**141,378.11**	**61.46**	%
Other	88,643.89	38.54	%
Target	89,453.00	38.89	%

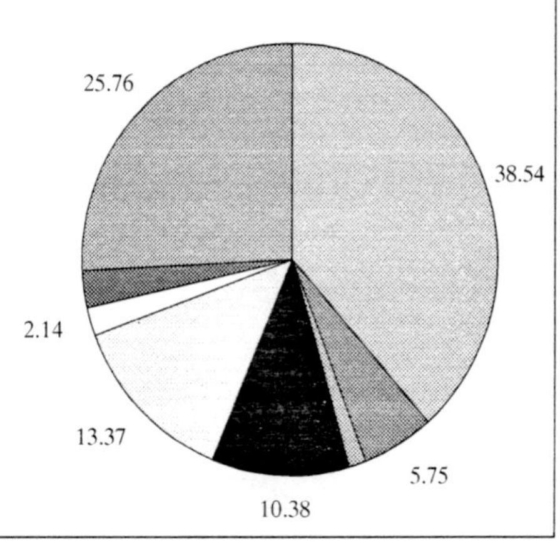

FIGURE C.18: Refined weight spreadsheet for KC-42 case study (Part 3).

The results of the directional stability analysis found the wing and vertical tail to be stabilizing and the fuselage to be destabilizing. The vertical tail, however, slightly offsets the effect of the fuselage so that overall, it is directionally stable, with $C_{n_\beta} = -0.1464$.

Finally, with regards to sizing the rudder, the one engine out and cross-wind conditions were examined. The cross-wind condition was the worse case of the two. Based on this, the *minimum* rudder area was estimated to be approximately 3.5 percent of the vertical stabilizer.

	Longitudinal Stability					
Fuselage Length						
L (f)	153					
Wing Center of Lift						
L_ctr (x/L)	0.42					
m.a.c. (ft)	23.7					
Load Summary (fuselage)						
Load Type	Magnitude	x/L_start	x/L_end	resultant	M @C_lift	dw
	(lbs)			x/L	f-lb (+ cw)	
Fuel	400000	0.13	0.55	0.34	−4896000	42553.19
Payload	8000	0.1	0.6	0.35	−85680	727.27
Fus.Struct.	30742	0	1	0.5	376282.08	1463.9
Engine(s)	13221	0.35	0.35	0.35	−141596.91	13221
Wing Struct.	59262	0.26	0.48	0.37	−453354.3	10974.44
Horiz. Tail	6283	0.93	1.13	1.03	586392.39	1256.6
Vert. Tail	4912	0.8	1	0.9	360737.28	982.4
Other	88643	0	1	0.5	1084990.32	4221.1
ΣL	611063			ΣM	−3168229.14	
Tail Lift (req)	−33946.52	0.93	1.13	1.03	−3168229.14	−6789.3
Center of Gravity						
X_cg / L	0.39					
X_cg (ft)	59.08	f				
Static Margin						
S.M.	0.22	stable				
Longitudinal Stability Coefficient:						
Wing Parameters:						
S_w	3611	f^2				
(C_L_α)_w	0.102	(deg)^−1				
x_w	5.18	f				
cbar	23.7	f				
Horiz. Tail Parameters:						
(C_L_α)_ht	0.0700	(deg)^−1				
de/dα	0.2	Fig. 11.3				
η_ht	1.000					
l_ht	98.51	f				
S_ht	1070	f^2				
Engine Parameters						
mdot	1546.2	lbm/s				
l_i	7.1	f				

FIGURE C.19: Stability analysis spreadsheet for KC-42 case study (Part 1).

rho	0.0243	lbm/f^3				
V	827	f/s				
dβ/dα	2.75					
Calculations						
V_bar_hs	1.23					
inlet effect	0.0003	unstable				
wing effect	1.28	unstable	check: C_M_ α = − S.M.*C_L_ α			−1.28
h. tail effect	3.95	unstable				
C_M_ α	−2.67	stable				
Directional Stability Coefficient:						
Wing Parameters:						
A_w	9					
Λ	30	deg				
λ	0.15					
S_w	3611	f^2				
b	180.3	f				
z_w	5.55	f				
C_L (cruise)	0.6000					
Fuselage Parameters:						
h	17	f				
w	17	f				
Vol_f	26800	f^3				
Vertical Tail Parameters:						
(C_L_α)_vs	0.0290	(deg)^−1				
l_vs	77.13	f				
S_vs	744	f^2				
Λ_vs	35	deg				
Calculations						
V_bar_vs	0.0881					
(1+dα/dβ)q/q	1.28	Eq[11.42]				
v. tail effect	0.19	Eq[11.40]	stable			
fuse. effect	−0.05	Eq[11.44]	unstable			
wing effect	0.01	Eq[11.43]	stable			
C_n_ β	0.1464	stable				
C_L_ β	−0.1464	stable				
	Rudder Sizing					
Input Parameters						
δ_r	20	deg				
β	11.5	deg				
Asym. T	65069	lbs				

FIGURE C.20: Stability analysis spreadsheet for KC-42 case study (Part 2).

S_w	3611	f^2				
b	180.3	f				
C_n_ β	0.1464					
diam_e	7.97	f				
V_T-O	288.9	f/s				
rho_T-O	0.0754	lbm/f^3				
Calculations						
1.2V_T-O	346.68	f/s				
0.2V_T-O	57.78	f/s				
q	140.72	lbs/f^2				
D_e	8424.24	lbs				
C_n_ δR:				Abs( )		
Asy. Power	0.0023	[rad]^−1	Eq[11.47]	0.0023		
Cross Wind	−0.0842	[rad]^−1	Eq[11.50]	0.0842		
d α_0L/d δ r	0.6388		Eq[11.51]			
C_R/C_VS	3.5000	%	Fig. 11.9			

FIGURE C.21: Stability analysis spreadsheet for KC-42 case study (Part 3).

C.11 COST ESTIMATE

The cost estimate was made using the CERs given in Chapter 11, which have been incorporated into the **cost** spreadsheet. The cost breakdown from the spreadsheet is shown in Figures C.22 to C.25.

The parameters used in determining the cost are the structure weight, maximum velocity, number of development and production aircraft, and the production rate. The costs are based on year 2000 dollars.

It was assumed that only one RDT&E aircraft would be produced. The number of acquisition aircraft would total 100. The profit was set at 10 percent. For quality control, the aircraft was designated as cargo (C). The engine was designated as Turbofan (TF). Finally, the cost of RDT&E was set to be amortized over the first 70 aircraft.

TABLE C.4: KC-42 proposed and final characteristics.

	Proposed	Final
Max. TO Weight (lbs)	640,000	638,950
Empty Weight (lbs)	230,000	230,022
Cruise Mach	0.85	0.85
Range (nm)	4400	4400
TO Distance (f)	11,000	13,777
Landing Distance (f)	6000	5765
Unit Cost (Y2000 $)	$80M	$60–$129

		Cost Analysis			
Input Parameters:				**Constants:**	
W_s	230,000	lbs		CPI (70–98)	3.95
V_max	234	f/s		CPI (86–98)	1.49
N_D	1	aircraft		CPI (98–00)	1.06
N_P	100	aircraft			
R_D	0.06	aircraft/mo.			
R_P	1	aircraft/mo			
Year	2000				
V_max	138.55	knots			
CR_E	94	$/hr			
CR_T	100	$/hr			
CR_ML	80	$/hr			
CR_QC	88	$/hr			
Research, Development, Test and Evaluation					
		Based on 1970	**Based on 1986**		
Airframe Engineering:					
	H_E (hrs)	1,940,501.04	9,349,271.56		
	C_E ($)	$182,407,098	$878,831,526		
Development Support:					
	C_D ($)	$32,994,362	$205,100,381		
Manufacturing Labor:					
	H_ML (hrs)	3,653,293.28	2,575,044.80		
	C_ML ($)	$292,263,463	$206,003,584		
Manufacturing Materials:					
	C_MM	$11,464,341	$44,430,096		
Tooling:					
	H_T (hrs)	3,912,167.19	3,912,583.22		
	C_T ($)	$391,216,719	$391,258,322		
Quality Control:					
	(C or O)		C	Cargo or Other Aircraft	
	H_QC	474,928.13	195,703.40		
	C_QC	$41,793,675	$17,221,900		
Flight Test:					
	C_FT	$8,744,973	$9,580,513		

FIGURE C.22: Cost estimate spreadsheet for KC-42 case study (Part 1).

Based on these input values, the unit cost of the first 70 aircraft in year 2000 dollars was found to be between $75 to $129 million based on the 1970 and 1986 CERs, respectively. After those, the unit price would drop to between approximately $60 to $101 million. This is quite competitive with the other comparison aircraft. For example, in year 2000 dollars the KC-10 costs $94.6 million, and the KC-135 costs $56.9 million.

Engine:					
	T_max (lbs)	65,070			
	(TJ or TF)	TF		turbofan or turbojet	
	N_eng	2			
	C_EN ($)	$11,460,953	$11,460,953		
Subtotal (RDT&E):		$972,345,583	$1,763,887,275		
Profit:					
	Percent	10	10		
	C_P	$97,234,558	$176,388,727		
Total (RDT&E):		**$1,069,580,141**	**$1,940,276,002**		
Acquisition:					
		Based on 1970	**Based on 1986**		
Airframe Engineering:					
	H_E (hrs)	4,507,273.18	19,805,133.51		
	C_E ($)	$423,683,679	$1,861,682,550		
Manufacturing Labor:					
	H_ML (hrs)	40,802,289.99	49,292,947.56		
	C_ML ($)	$3,264,183,200	$3,943,435,805		
Manufacturing Materials:					
	C_MM	$439,895,058	$1,760,667,129		
Tooling:					
	H_T (hrs)	8,880,090.81	13,136,013.55		
	C_T ($)	$888,009,081	$1,313,601,355		
Quality Control:					
	(C or O)		C	Cargo or Other Aircraft	
	H_QC	5,304,297.70	3,746,264.01		
	C_QC	$466,778,198	$329,671,233		
Engine:					
	T_max (lbs)	65,070			
	(TJ or TF)	TF		turbofan or turbojet	
	N_eng	2			
	C_EN ($)	$11,460,953	$11,460,953		
Subtotal (A):		$5,494,010,168	$9,220,519,025		
Profit:					
	Percent	10	10		
	C_P	$549,401,017	$922,051,902		
Total (A):		**$6,043,411,185**	**$10,142,570,927**		

FIGURE C.23: Cost estimate spreadsheet for KC-42 case study (Part 2).

Total Profit:		$646,635,575	$1,098,440,630	
Aircraft Unit Cost:				
RDT&E amortized				
unit no.:		70	70	
Cost/Prod. Aircraft		$15,279,716	$27,718,229	
Unit cost:				
1	70	**$75,713,828**	**$129,143,938**	
71	100	**$60,434,112**	**$101,425,709**	
Summary Totals:		**Based on 1970**	**Based on 1986**	
	C_E ($)	$606,090,777	$2,740,514,076	
	C_D ($)	$32,994,362	$205,100,381	
	C_ML ($)	$3,556,446,662	$4,149,439,389	
	C_MM ($)	$451,359,399	$1,805,097,225	
	C_T ($)	$1,279,225,800	$1,704,859,677	
	C_QC ($)	$508,571,873	$346,893,133	
	C_EN ($)	$22,921,905	$22,921,905	
	C_P ($)	$646,635,575	$1,098,440,630	

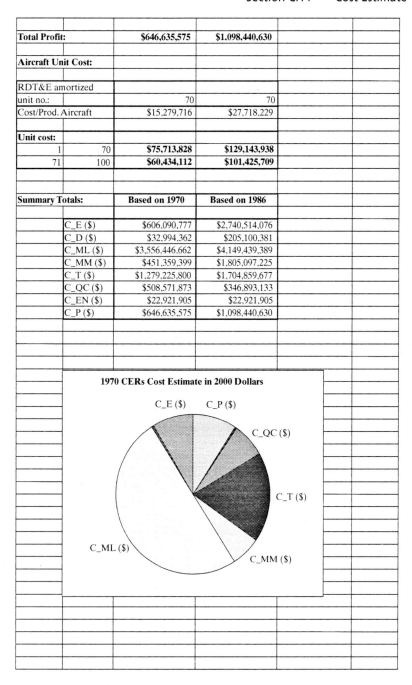

1970 CERs Cost Estimate in 2000 Dollars

FIGURE C.24: Cost estimate spreadsheet for KC-42 case study (Part 3).

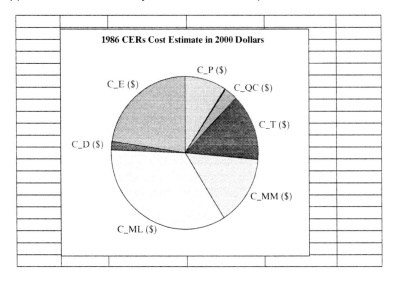

FIGURE C.25: Cost estimate spreadsheet for KC-42 case study (Part 4).

C.12   DESIGN SUMMARY

The KC-42 was able to meet or exceed almost all of its original design drivers made in the design proposal. These are summarized in Table C.4. The design was able to carry the proposed payload at the desired range and have a slightly lower take-off weight. It would use two engines that have characteristics that perfectly match existing engines without any scaling. Its flight characteristics are statically stable, which is a desired characteristic for a long-range transport. Finally, the estimated unit price is very competitive with that of comparable aircraft.

Bibliography

Abbott, I., and A. Von Doenhoff. "Characteristics of Airfoil Sections" NACA Report 824, 1945.

Abbott, I., and A. Von Doenhoff. "Theory of Wing Sections" Dover Publications, Inc., New York, 1949.

Bertin, J., and M. Smith. "Aerodynamics for Engineers" Prentice Hall, 1998.

Ellison, D. "USAF Stability and Control Handbook (DATCOM)" AF Flight Dynamics Lab, AFFDL/FDCC, Wright-Patterson AFB, Ohio, 1969.

"Generalized Method of Propeller Performance" Rept. PDB 6101A, United Aircraft Corp., 1963.

Harmon, S., and I. Jeffreys. "Theoretical Lift and Damping in Roll of Thin Wings with Arbitrary Sweep and Taper at Supersonic Speeds" NACA TN-2114, 1950.

Harris, R. "An Analysis and Correlation of Aircraft Wave Drag" NASA TM-X-947, 1964.

Hess, R., and H. Romanoff. "Aircraft Airframe Cost Estimation Relationships" Rand Corp., Rept. R-3255-AF, 1987.

Jackson, A. "Preliminary Design Weight Estimation Program" AeroCommander Division, Rept. 511–009, 1971.

"Jane's All The World Aircraft" Ed. M. Lambert, Jane's Information Group, Ltd., Surry, UK.

Jones, R. "Theory of Wing-Body Drag at Supersonic Speeds" NACA Report 1284, 1956.

Jumper, E. "Wave Drag Prediction Using Simplified Supersonic Area Rule" *J. Aircraft*, Vol. 20, pp. 893–895, 1983.

Kuchemann, D. "The Aerodynamic Design of Aircraft" Pergamon Press, Oxford, 1978.

Large, J. "Estimating Aircraft Turbine Engine Costs" Rand Report RM-6384/1-PR, 1970.

Levenson, G., H. Boren, D. Tihansky, and F. Timson. "Cost-estimating Relationships for Aircraft Airframes" Rand Report R-761-PR, 1971.

Loffin, L. "Subsonic Aircraft: Evolution and the Matching of Size to Performance" NACA Ref. 1060, 1980.

Lowry, J., and E. Polhamus. "A Method for Predicting Lift Increments Due to Flap Deflection at Low Angles of Attack in Incompressible Flow" NASA TN-3911, 1957.

Mattingly, J. "Elements of Gas Turbine Propulsion" McGraw-Hill, 1996.

Nelson, Robert C. "Flight Stability and Automatic Control" 2d Edition McGraw-Hill, New York, 1998.

Nicolai, L. "Fundamentals of Aircraft Design" Univ. of Dayton, Dayton, OH, 1975.

Perkins, C., and R. Hage. "Airplane Performance Stability and Control" John Wiley and Sons, New York, 1949.

Perry, D., and J. Azar. "Aircraft Structures" McGraw-Hill, 1982.

Philips, W. "Appreciation and Prediction of Flying Qualities" NACA 927, 1949.

Purser, P., and J. Campbell. "Experimental Verification of Simplified V-tail Theory and Analysis of Available Data on Complete Models of V-tails" NACA 823, 1945.

Raymer, D. "Aircraft Design: A Conceptual Approach" AIAA Educational Series, Washington, D.C., 1992.

Schrenk, O. "A Simple Approximation Method for Obtaining the Spanwise Lift Distribution" NACA TM-948, 1940.

Sears, W. "On Projectiles of Minimum Drag" Quarterly Math Series, Vol. 4, No. 4, 1947.

Silverstein and Katzoff, "Design Chart for Predicting Downwash Angles and Wake Characteristics Behind Plane and Flapped Wings" NACA TR 648.

Staton, R. "Cargo/Transport Statistical Weight Estimation Equations" Vought Aircraft Report 2-59320/9R-50549, 1969.

Staton, R. "Statistical Weight Estimation Methods for Fighter/Attack Aircraft" Vought Aircraft Report 2-59320/8R-50475, 1968.

Young, A. "The Aerodynamic Characteristics of Flaps" ARC R&M 2622, 1953.

Index